LINEAR MOTION ELECTROMAGNETIC SYSTEMS

LINEAR MOTION ELECTROMAGNETIC SYSTEMS

I. BOLDEA
S. A. NASAR

A Wiley-Interscience Publication
JOHN WILEY & SONS
New York • Chichester • Brisbane • Toronto • Singapore

Library of Congress Cataloging in Publication Data

Boldea, I.
Linear motion electromagnetic systems.

"A Wiley-Interscience publication."

Includes index.
1. Electric motors, Linear. 2. Magnetic levitation vehicles. I. Nasar, S. A. II. Title.

TK2537.N36 1985 621.31′042 84-25772
ISBN 0-471-87451-5

Printed in the United States of America

10 9 8 7 6 5 4 3 2 1

PREFACE

This book is a sequel to the author's earlier book Linear Motion Electric Machines (Wiley-Interscience, 1976). Much progress has been made in the area of linear electric machines since the publication of the earlier work. The present book not only updates the material on linear motion electromagnetic devices and systems, but has a much broader scope in that this book includes topics not treated in the first book. Thus, the present book deals with the principles, construction guidelines, static and dynamic performance analyses, theory and certain design criteria pertaining to linear motion electromagnetic energy conversion systems. Topics presented here include: linear motion induction, synchronous and dc electric machines; linear magnetohydrodynamic (mhd) energy converters; linear inductor machines; magnetic levitators; and linear oscillators. Among the applications of these devices and systems are: material handling systems; sliding doors operated by linear induction motors; short-stroke linear oscillators; lifting electromagnets; electromagnetic launchers; propulsion and magnetic levitation of high-speed ground vehicles; electromagnetic pumps; and mhd generators.

Compared to their rotary counterparts, linear motion machines and their applications are relatively new. Nevertheless, the literature on the latter subject is rich. Thousands of technical papers and several books have appeared on the subject over a short span of thirty years. Our book attempts to present a practical and unified approach to the entire field of linear motion energy converters. While maintaining a level of mathematical rigor, we have presented the results in forms which should be useful to designers, practicing engineers, researchers, and students in the field. Thus, in each chapter, construction guidelines are followed by mathematical analysis, steady state and dynamic performance calculations and design criteria. Much of this material is illustrated by worked-out numerical examples of practical interest.

The book has twelve chapters. The first three chapters present the basic governing equations and their solutions, classification and potential applications of a large class of linear motion electromagnetic devices. The next chapter is devoted to low-speed linear induction motors. The potential application of these low-speed linear induction motors is enormous.

Linear-motion mhd machines--induction and dc homopolar--are treated in detail in Chapter 5. Electromagnetic field equations and fluid-flow equations are given, from which pump (or motor) and generator performance characteristics are evaluated. Numerical examples are included to facilitate the use of the results derived in this chapter.

Linear induction motors, exclusively for transportation applications, are presented in Chapter 6, and linear inductor motors are discussed in Chapter 7. Chapters 8 and 9 relate to linear synchronous motors with conventional field excitations and with superconducting field windings, respectively.

Various types of linear motion levitators are studied in Chapter 10 and 11. The book ends with a presentation of the electrodynamic wheel in Chapter 12. As in the preceding chapters, construction, analysis of static and dynamic performance, design aspects control schemes are presented. Most of analyses and design procedures throughout the book are illustrated by numerical examples.

The work reported here was, in part, supported by the various grants from the National Science Foundation (NSF) afforded to one of the authors (S.A.N.) over a twenty year period. He wishes to acknowledge the support of his researches in linear electric machines by NSF. We wish to express our gratitude to Persis Ann Elwood for her diligent assistance in the manuscript preparation. The capable typing of the manuscript by Vickie Lynn Brann is gratefully acknowledged.

Ion Boldea
Syed A. Nasar

Lexington, Kentucky
May 1985

CONTENTS

LINEAR MOTION ELECTROMAGNETIC SYSTEMS

CHAPTER 1

INTRODUCTION TO LINEAR MOTION ELECTROMAGNETIC SYSTEMS

For virtually, every rotating electric machine there exists a linear motion counterpart. However, there are no rotary counterparts for certain linear motion machines. To be general, therefore, we have labeled all types of linear motion electric machines linear motion electromagnetic systems (LMESs). As we see later, such a designation includes linear motion electromagnetic devices that should not be termed electric machines in the conventional sense. This book pertains to LMESs wherein the motion of one member with respect to some other member of the device is horizontal and/or vertical. Traditionally, a force tending to produce motion in a horizontal direction is termed thrust, propulsion force, or traction force. On the other hand, a force in a plane perpendicular to the direction of thrust is called normal force. Because we have explicitly mentioned the existence of thrust and normal forces, we wish to emphasize at the outset that the LMESs have found applications only in the motoring, rather than the generating (except for regenerative braking) mode. The presence of a normal force in an LMES is not considered abnormal, and is one of the features distinguishing an LMES from a rotating machine (in which the presence of a normal force is certainly undesirable).

It is almost a general principle that when the topological features of an electric machine are modified--such as developing a linear motion machine is developed from its rotary counterpart--the operating characteristics and the design criteria of the new machine change considerably. For instance, as a result of the topological changes in the magnetic circuit of a machine, new electromagnetic phenomena come into play that cannot be fully explained by conventional rotating machine theory. Consequently, older methods of analysis have to be modified, and sometimes new theories have to be developed. In this book, we present the methods of analysis and certain design procedures applicable to LMESs. Before proceeding with analytical techniques, it is worthwhile to consider some of the common types of LMESs to appreciate the major differences introduced in transforming the rotating machine to its linear motion counterpart.

1.1. Certain Examples of LMESs

Three of the most common types of rotating machines are:

1. Induction machines;

2. Synchronous machines; and

3. DC commutator machines.

The linear motion counterpart of an induction machine--known as the linear induction machine (LIM)--is shown in Fig. 1.1. In Fig. 1.2 we see the linear counterpart of a salient-pole synchronous machine, known as the linear synchronous machine (LSM).

Referring now to Figs. 1.1 and 1.2, notice that the linear machine has been obtained by an imaginary process of "cutting" and "unrolling" the rotary machine. We designate the "stator" of the linear machine as its primary and the "rotor" as the secondary. The primary,

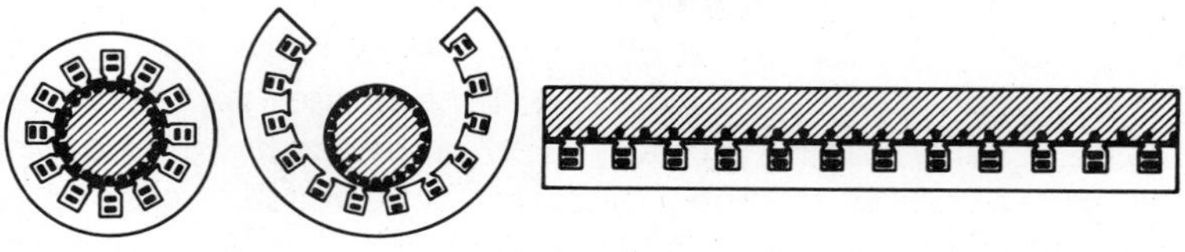

Fig. 1.1. Imaginary process of unrolling a conventional motor to obtain a linear induction motor.

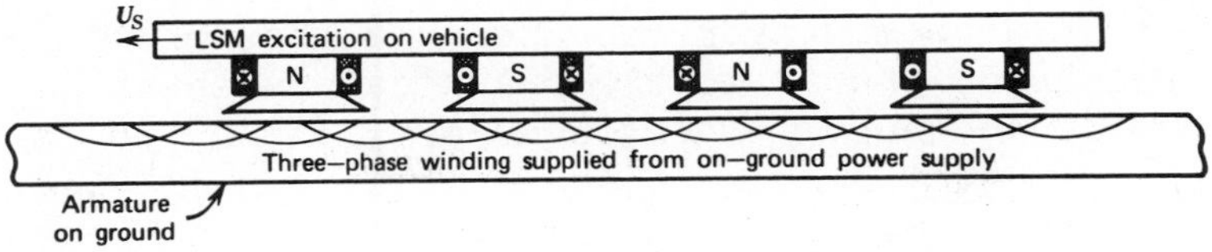

Fig. 1.2. Salient-pole LSM.

which includes the core and the windings, now has a finite length. It has a beginning and an end. The presence of these two ends leads to the phenomenon of end effects, which does not exist in a rotating machine. End effects are a consequence of the topological changes shown in Figs. 1.1 and 1.2. Up to this point we have identified end effects and the presence of normal forces in certain LMESs as two of several pertinent, unique features. Whereas in Figs. 1.1 and 1.2 we have shown the LIM and the LSM respectively, to evolve from a rotary induction machine and a rotary salient-pole synchronous machine, numerous topological variations are practicable in linear machines.

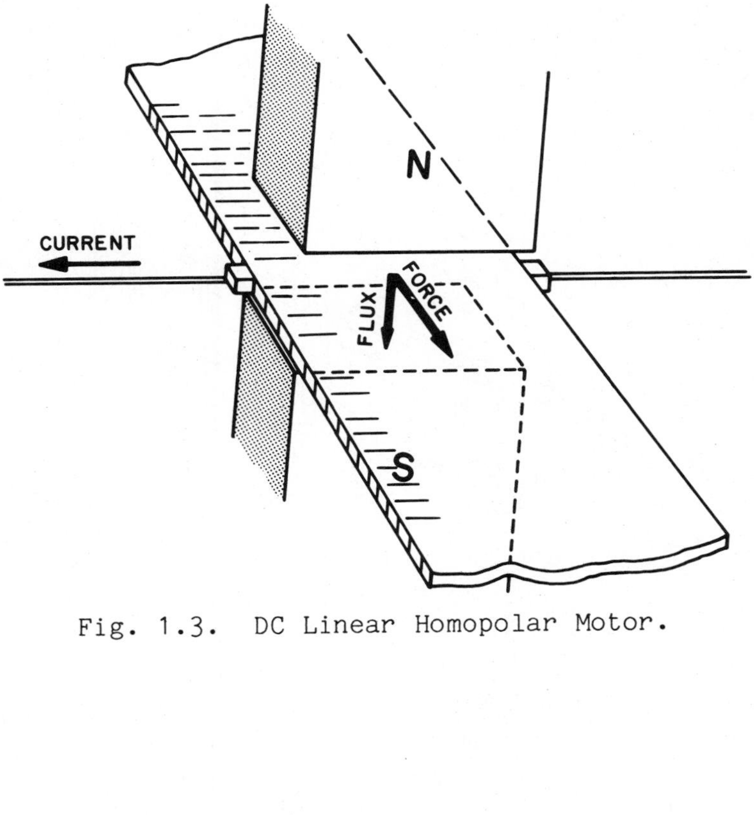

Fig. 1.3. DC Linear Homopolar Motor.

Linear dc commutator machines, although theoretically feasible, are of little practical interest and are not considered here. Other forms of linear dc motors are the linear dc homopolar motor, shown in Fig. 1.3, and the linear brushless dc motor, which is schematically represented in Fig. 1.4. The linear brushless dc motor is essentially a thyristor-commutated dc motor. It provides propulsion and levitation simultaneously. As shown in Fig. 1.4, the armature coil system is stationary. The armature coils are split into a number of groups and the coil groups are staggered, as shown in the diagram. These coils are arranged upright with respect to a moving magnet system, causing the magnetic flux to cross the lower and vertical parts of the coil. The vertical displacement between magnets and coils gives rise to levitation forces. The armature current is switched by the thyristors from one coil group to another in succession as the magnet moves along the armature coils.

Two other types of linear motors of importance are the linear synchronous homopolar motor (LSHM) and the linear reluctance motor (LRM). These, respectively shown in Figs. 1.5 and 1.6, belong to the class of LSMs.

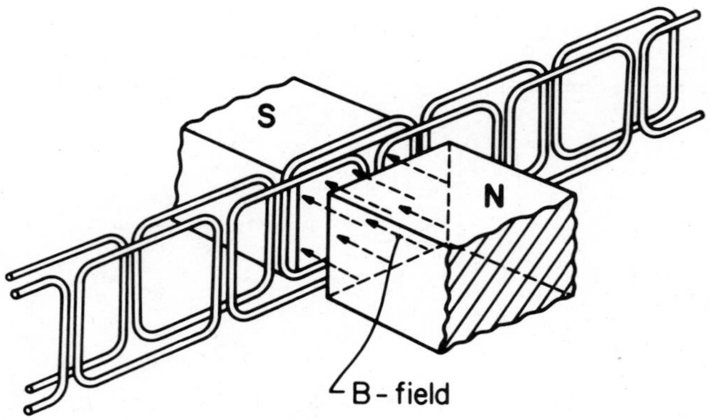

Fig. 1.4. Field Magnet and Armature Coils.

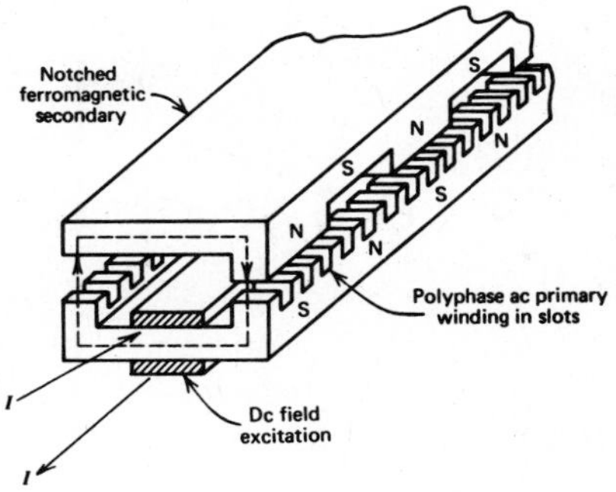

Fig. 1.5. An LSHM Layout.

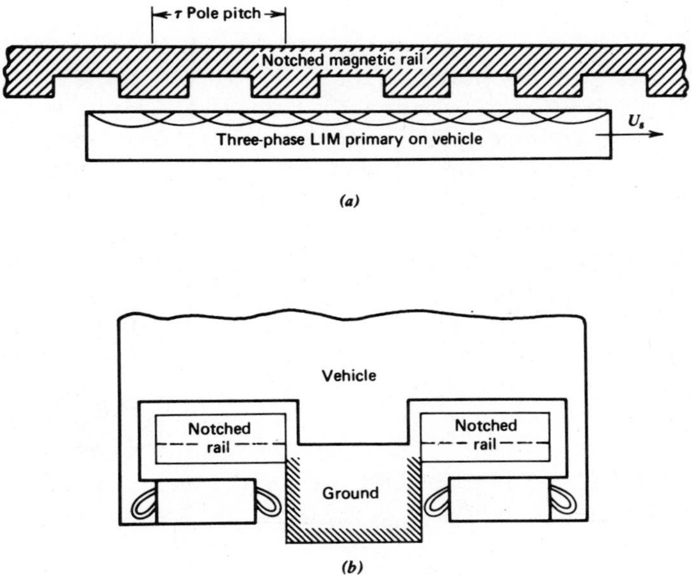

Fig. 1.6. Conventional linear reluctance motor (LRM). (a) Longitudinal view, (b) transverse view.

In an LSHM, a dc field winding and a three-phase armature winding are placed on a common U-shaped core constituting the primary (Fig. 1.5). The secondary is made of a solid ferromagnetic structure notched along the direction of motion. The primary is laminated to reduce the eddy currents owing to the ac excitation, but the dc flux, being transverse, is not affected by the laminations. Notice that the notched secondary structure has a period 2τ, where τ is the pole pitch of the three-phase primary winding. An LSHM with a segmented secondary is also feasible (e.g., the segmented-secondary LRM shown in Fig. 1.7), but this configuration is not considered here. The LSHM is not self-starting. It has to be started in the same manner as the usual LSM and brought up to the synchronous speed. Once synchronized, the LSHM operates like a salient-pole LSM. The LSHM is usually electronically controlled.

The LRM is the linear counterpart of the rotary reluctance motor, and it does not have any field winding. The stator, or primary, of the LRM is similar to that of a synchronous motor. In other words, the primary may have a polyphase or a single-phase winding placed on a ferromagnetic core. But there are two common types of rotor, or secondary, configurations: (1) the conventional LRM (Fig. 1.5) and (2) the LRM with segmented secondary (Fig. 1.7) in which the secondary is made of rectangular blocks of a ferromagnetic material, embedded in a nonmagnetic material (e.g., concrete). The thrust produced by the LRM is similar to the torque due to saliency in a rotary salient-pole synchronous machine. The LRM is not a self-starting motor, except when the induced eddy currents in the secondary at standstill are sufficient to provide a starting force. The acceleration of the LRM in such cases will not be rapid.

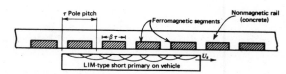

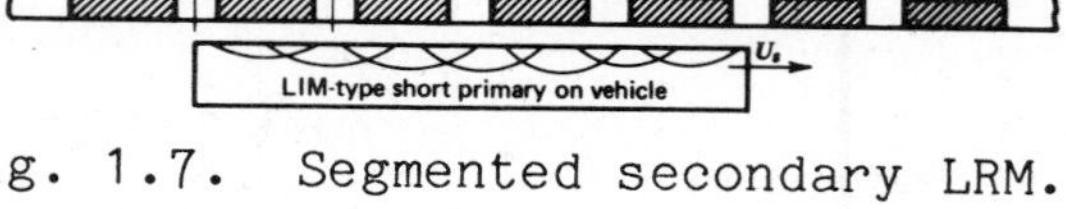

Fig. 1.7. Segmented secondary LRM.

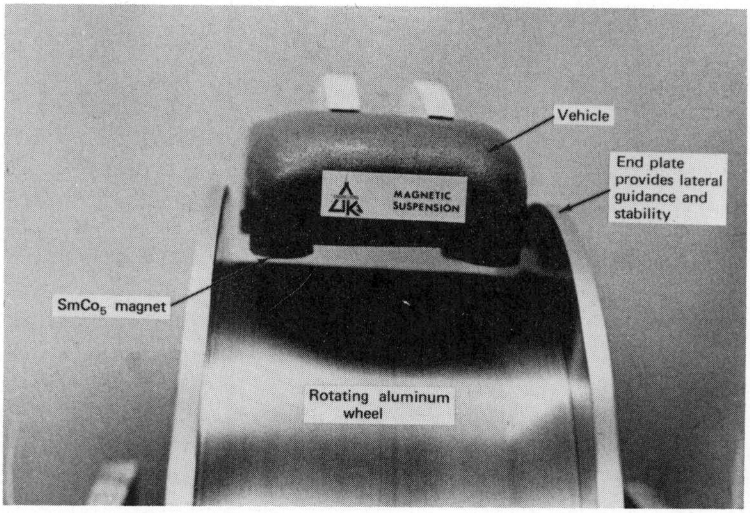

Fig. 1.8. Magnetic suspension demonstration model.

The examples shown in Figs. 1.1 to 1.7 are linear machines belonging to the category of machines mainly producing propulsion forces. The presence of any normal forces in these machines is incidental. Linear machines intended primarily for producing normal forces belong to the class of linear levitation machines (LLMs). An induction-type LLM is shown in Fig. 1.8. The operation is based on the principle that, if a magnet moves over a conducting sheet, the eddy currents induced in the sheet will cause a normal force of repulsion between the magnet and the sheet. In the model shown in Fig. 1.8, which demonstrates the existence of a normal force a "vehicle" having four samarium-cobalt ($SmCo_5$) permanent magnets is levitated by a rotating aluminum drum. Interactions between the sides of the magnets and the rims at the two ends of the drum produce forces normal to the rims and thereby provide lateral stability to the system. There exist numerous other types of LLMs--linear machines exclusively designed to produce normal forces. These machines are discussed in later chapters.

We close this section by observing that the machines shown in Figs. 1.1 to 1.8 are a few examples of LMESs.

1.2. Classification of LMESs

In Section 1.1 we pointed out that, in principle, for every rotary machine there is a corresponding linear electric machine (LEM). The LIMs and LSMs are the ones of practical interest, and these form the primary subject matter of this book.

As we see later, because the topology of the magnetic circuit of a LEM can be varied easily, there are many more diverse types of LIMs and LSMs than rotating induction and synchronous motors. Consequently, it might become unmanageable to encompass all different types of LEMs in our study unless a systematic classification of such machines is made. We attempt to present such a classification here to aid our subsequent study of LEMs.

We recall from Section 1.1 that some LEMs develop two mutually perpendicular forces--one in the direction of motion and the other normal to the direction of motion. The normal force may be an attraction or a repulsion force between the primary and the secondary. A machine in which the net normal force is such that the secondary tends to be suspended over the primary may be used mainly for suspension and is called a <u>linear levitation machine</u> (LLM). On the other hand, a machine used primarily for producing thrust is called a <u>linear motor</u> (rather than a thrust machine!).

Both LIMs and LSMs may be used either as levitation machines or as linear motors, and the categories of linear motor and levitation machine are very broad. However, we classify the LIMs and LSMs from topological considerations. Because dc linear machines are not very common, we make only a brief reference to such devices. We treat levitation machines as a separate entity, whether such machines are induction type or synchronous type.

1.2.1. Linear Induction Motors

The most common of the LIMs is the polyphase LIM. As in a rotary polyphase induction motor, the airgap magnetic field in a polyphase LIM is a traveling magnetic field. The airgap field of a LIM, unlike that of a polyphase rotary induction motor, often has a forward component, a backward component, and a pulsating component owing to the discontinuities in the magnetic circuit. However, the forward component is predominant and produces the useful force by interacting with the currents induced in the secondary. Again, in contrast with its rotary counterpart, a LIM may have a moving primary (with a fixed secondary) or a moving secondary (the primary being stationary).

Depending on the relative lengths of the primary and the secondary, a LIM may be a short-primary LIM (Fig. 1.9) or a short-secondary LIM (Fig. 1.10). Also the LIM may have two primaries face to face to obtain a double-sided LIM (DSLIM), as in Fig. 1.9. If the LIM has only one primary (Fig. 1.10), it is called a single-sided LIM (SLIM).

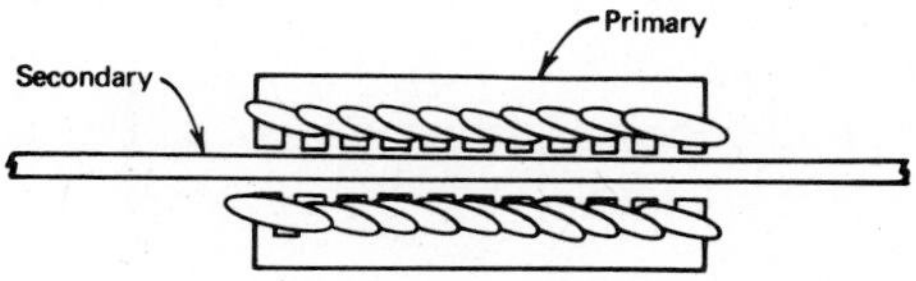

Fig. 1.9. Short-primary DSLIM.

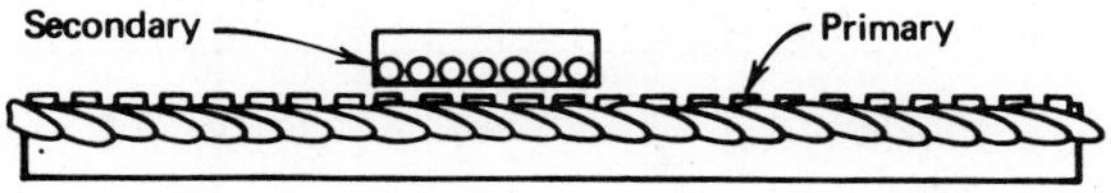

Fig. 1.10. Short-secondary SLIM.

The secondary of a LIM may consist of a secondary plate or sheet, sometimes backed by a ferromagnetic material (known as the *back iron*), or the secondary may be of the usual cage type. A wound-type secondary is rather uncommon. Finally, a liquid metal may constitute the secondary, such as in a liquid-metal pump.

As in a rotary motor, a LIM may have three phases or two or one. The primaries of all these LIMs are essentially similar to the stator winding of the rotary machine, except that often the primary of a LIM has half-filled end slots (Fig. 1.11). The starting mechanism in a single-phase LIM is similar to that in a single-phase conventional induction motor. Thus a single-phase LIM may be a shaded-pole type, or it may have an auxiliary starting winding incorporating a capa-

In the discussions so far, we have considered LIMs as derived from rotary induction motors by cutting them and splitting. Such motors are obviously *flat LIMs*. If the flat primary is rerolled about an axis parallel to the direction of the field motion, as in Fig. 1.12, an entirely different form of cylindrical structure is produced, and the field now travels along the bore of the primary. Such motors are *tubular motors* or *axial flux motors*. One obvious advantage of a tubular motor is that it does not have any end connections. There are a number of ways in which a tubular motor may be wound, and a three-phase tubular motor is illustrated in Fig. 1.13.
citor.

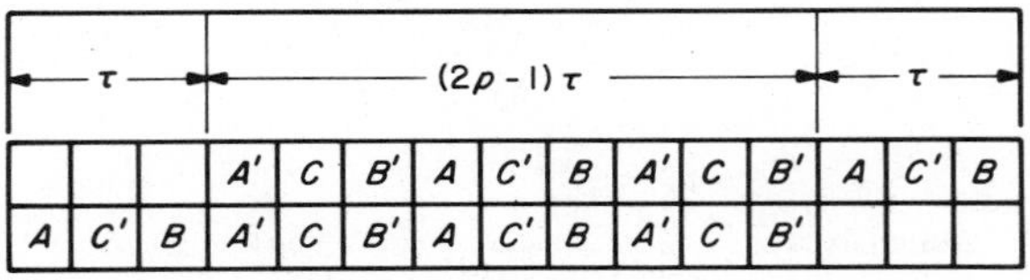

Fig. 1.11. Primary of an LIM with half-filled end slots.

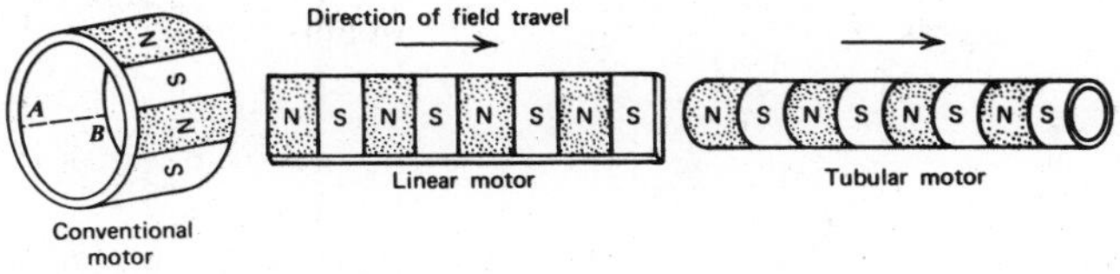

Fig. 1.12. Development of a tubular motor.

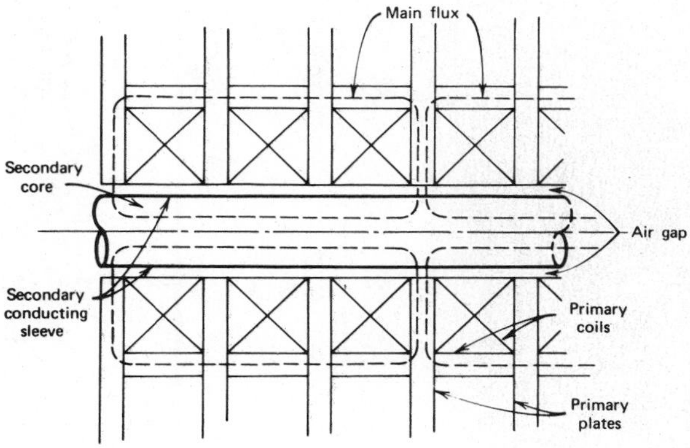

Fig. 1.13. A three-phase tubular motor.

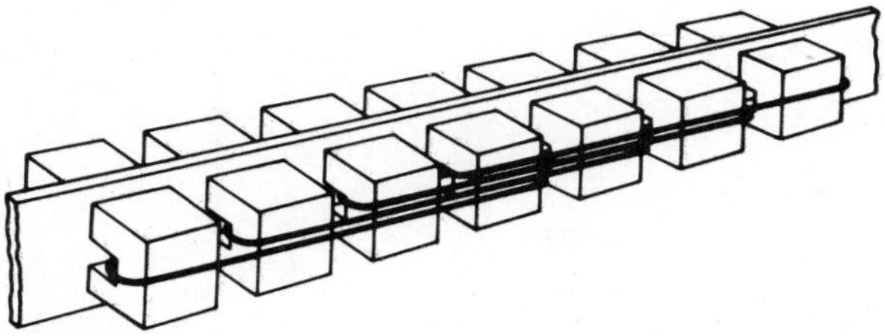

Fig. 1.14. Double-sided C-core TFLIM with distributed winding. Courtesy IEE (London).

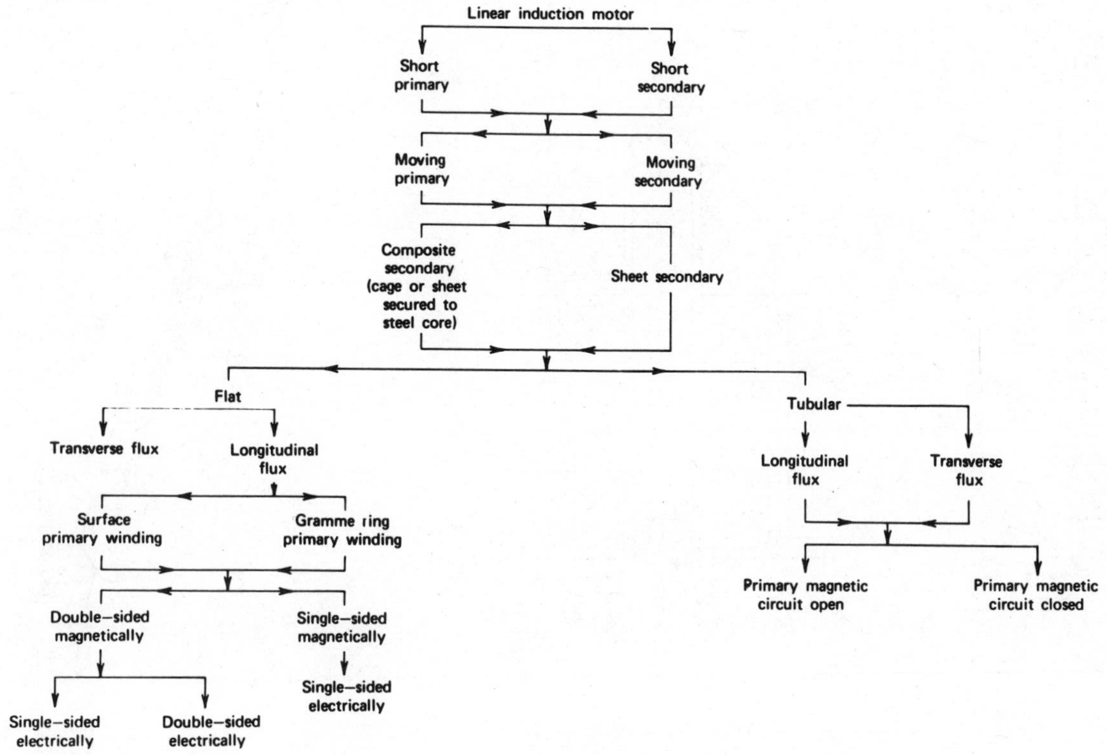

Fig. 1.15. Classification of LIMs.

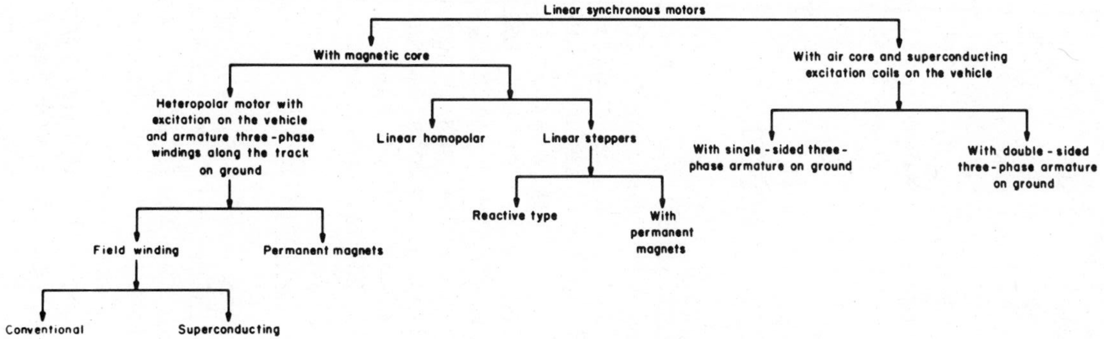

Fig. 1.16. Classification of LSM's.

The flat LIM and the tubular LIM belong to the class of motors in which the magnetic flux in the iron is in the direction of motion (Fig. 1.9). Such motors may also be called longitudinal flux motors. Electromagnetic circuits may, however, be modified such that the flux path in the iron is in a plane perpendicular to the direction of motion, resulting in a transverse flux motor. One possible arrangement of the transverse flux motor is shown in Fig. 1.14. From topological considerations, a classification of LIMs is outlined in Fig. 1.15. Although the foregoing list is by no means complete, for the present, most LIMs of practical significance are included in it.

1.2.2. Linear Synchronous Motors

The LSM is similar in principle to its rotary counterpart. The LSM came into prominence in the last decade as a propulsion device for high-speed ground transportation (HSGT). As in a conventional synchronous motor, the LSM has polyphase armature excitation and dc field excitation. The field excitation may be of a conventional type (Fig. 1.2), or the LSM may have a superconducting field winding. Although in principle it is irrelevant whether the LSM has a moving armature or a moving field winding, it appears that the latter type is more practical. A summary of classifications of LSMs, including LRMs and LSHMs, appears in Fig. 1.16.

1.2.3. DC Linear Motors

We have alluded to the dc linear motor (DCLM) in Section 1.1. This motor is not among the most commonly used linear motors. A linear motor proposed for short-stroke applications and in essence similar to rotary acyclic (or homopolar) machines is shown in Fig. 1.17. A mild steel core of circular cross section is wound with a single-layer surface winding to form the armature. The direction of the winding is reversed at the midpoint and the winding is encapsulated, resulting in an arrangement having no commutator or brushes. The armature is encircled by the field unit consisting of a mild steel outer shell and two end plates that enclose the cylindrical field winding. When current flows in this winding, main field flux is established in the core of the armature, which completes its path by way of the airgap pole shoes, end plates, and outer shell as in Fig. 1.17. The radial component of airgap flux interacts with the armature current to produce an axial unidirectional force at each pole.

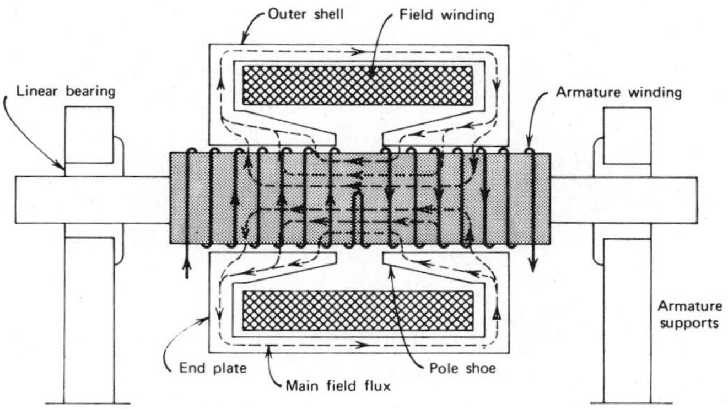

Fig. 1.17. A dc linear actuator. Courtesy IEE (London).

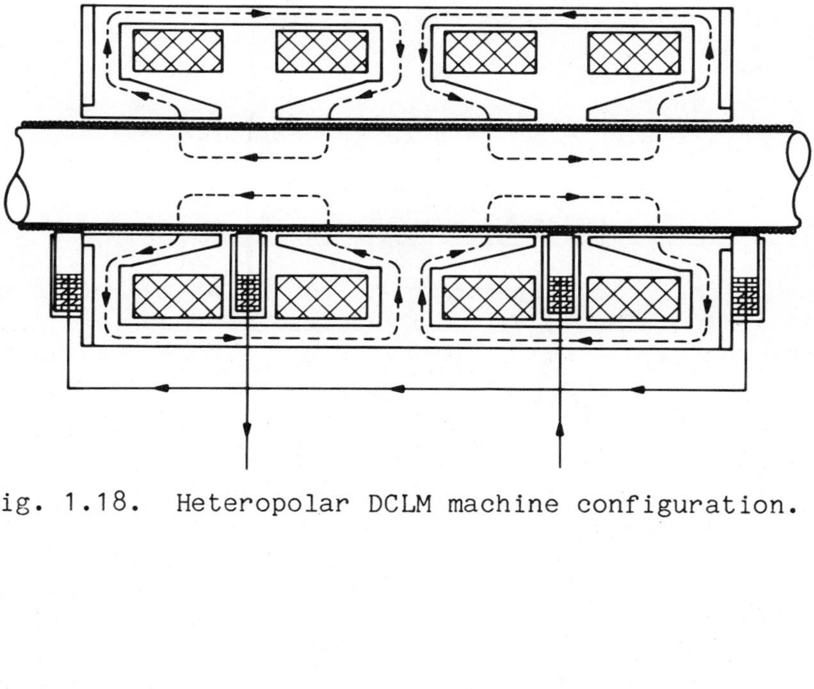

Fig. 1.18. Heteropolar DCLM machine configuration.

The DCLM just described is the homopolar DCLM. A heteropolar DCLM is also feasible. The heteropolar DCLM machine configuration is such that the output thrust and displacement are produced by the armature current-carrying conductors placed in a dc magnetic field. In the example of Fig. 1.18, the armature is the stationary member and the field unit is movable. A single-layer winding of enameled copper wire is bonded to the surface of a mild steel core. The outer face of the winding is machined to expose the copper to provide an extended linear commutator. Current is fed into the armature by brushes attached to the field unit. The brushes are connected across a low-voltage source by means of slip rails beneath the armature. The excited main field winding establishes the flux in the armature core. Circumferential current in the armature winding cuts the airgap radial flux at each pole orthogonally, thereby producing an axial force.

The cylindrical symmetry of the DCLM offers a number of advantages, such as (1) absence of end turns and complete utilization of the armature winding between the brushes, and (2) uniform airgap, resulting in the neutralization of the attraction force between the armature and the field unit.

Another class of DCLMs is the brushless DCLM, which is shown in Fig. 1.4 and has been discussed earlier. We return to these DCLMs in a later chapter.

1.2.4. Linear Levitation Machines

Reference has been made to the existence of normal forces in LEMs. Fig. 1.19 classifies machines having predominantly normal forces as linear levitation

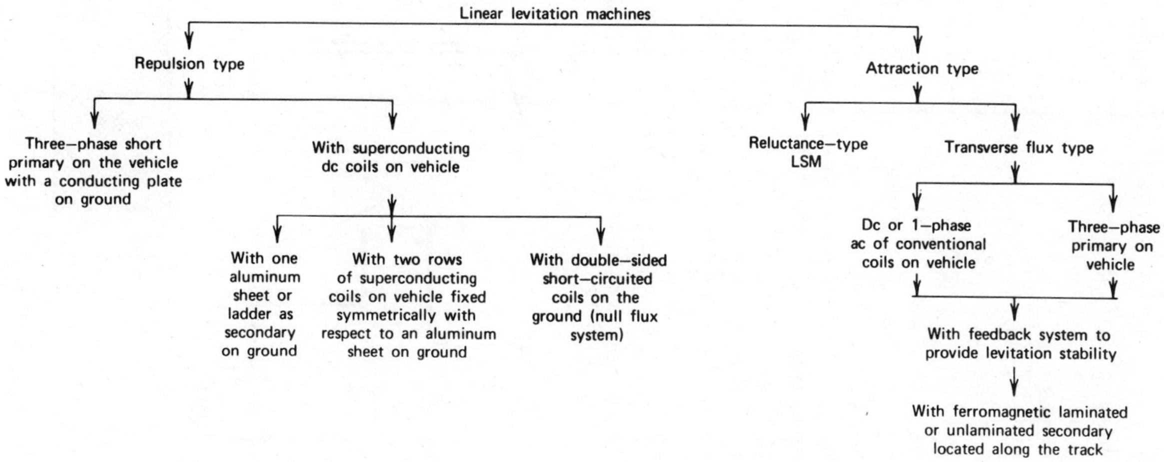

Fig. 1.19. Classification of LLM's.

machines (LLMs). For HSGT it is important that the vehicle not make any contact with track, and for such applications a number of suspension schemes have been proposed using levitation machines. We discuss some of these machines very briefly in this section.

There are attraction and repulsion levitation machines. Generally, the attraction-type linear machine is inherently unstable. Under certain conditions, however, such devices may be stabilized. An ac attraction-type levitation machine is depicted in Fig. 1.20. The value of the capacitor is such that the circuit is slightly overtuned. Thus for a small range of the airgap, the net force between the fixed secondary and the moving primary with the ac magnet serves to restore equilibrium. Two obvious disadvantages of this system are the tight constraint on the airgap tolerance and on the amount of perturbation permissible and the existence of a drag (or braking) force because of induced eddy currents in the secondary. The eddy currents may be reduced by using a laminated core.

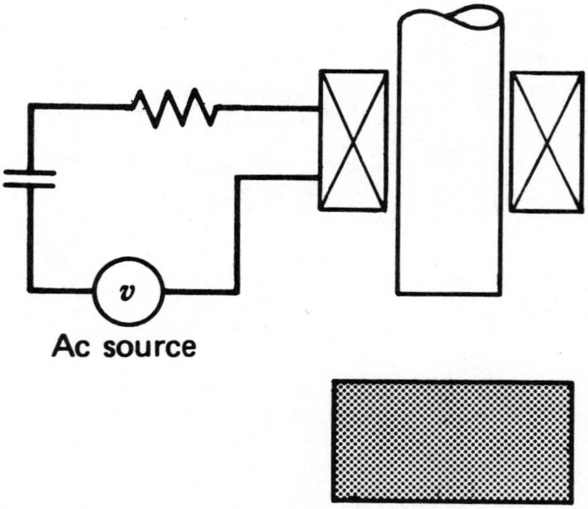

Fig. 1.20. Ac attraction-type levitation machine.

The small perturbations on the airgap of the machine may be overcome by using a dc attraction-type levitation machine having a feedback dependent on the airgap, as in Fig. 1.21. The drag force, however, is present in this machine also.

A third attraction-type levitation machine is the three-phase transverse flux machine (Fig. 1.22). This machine has no or very little drag force, but it is unstable. Feedback or other means of stabilization may be used.

The repulsion-type levitation machine is generally an induction device. The induction-type machine is based on the principle that, if a magnet is moved over a conducting sheet, the eddy currents induced in the sheet will cause a force of repulsion between the magnet and the sheet. In the model of Fig. 1.8, demonstrating such a force, as well as stability, a "vehicle" having four samarium-cobalt ($SmCo_5$) permanent magnets is levitated by a rotating aluminum drum. The rims at the two ends

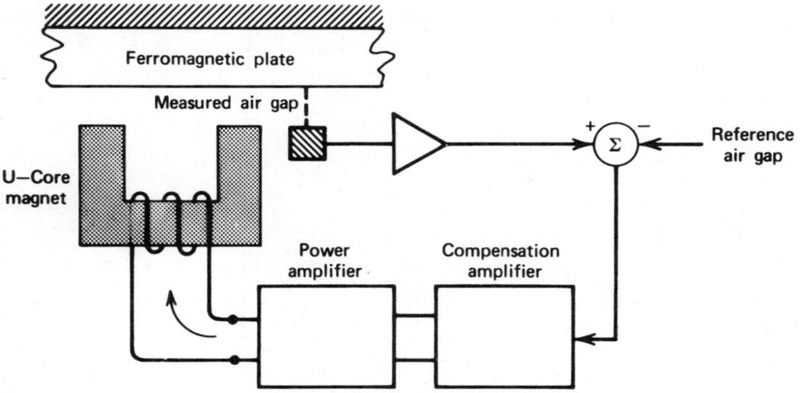

Fig. 1.21. Dc attraction-type levitation machine (with feedback stabilization).

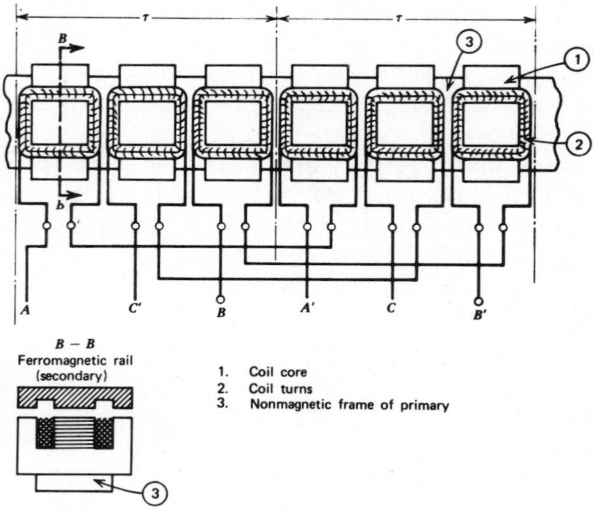

Fig. 1.22. Three-phase transverse flux linear levitation machine.

of the drum provide lateral stability. In practice the "magnet" may be a superconducting electromagnet and the "sheet" may be replaced by a ladder network or by short-circuited individual coils. As in the attraction-type machine, the induction-type repulsion levitation machine suffers from the presence of a drag force.

A special repulsion levitation system has also been proposed as a transverse flux LIM having an aluminum sheet secondary that may develop thrust and lateral guidance. Also, normal forces developed in LIMs and LSMs are often utlized for levitation, espectially in HSGT applications.

References

1. S. A. Nasar, Electromagnetic Energy Conversion Devices and Systems, Prentice-Hall, Englewood Cliffs, N.J., 1970, chapters 2, 4, and 5.

2. S. A. Nasar and I. Boldea, Linear Motion Electric Machines, Wiley-Interscience, New York, 1976, chapters 1 and 4.

3. E. R. Laithwaite, Induction Machines for Special Purposes, Chemical Publishing Company, New York, 1966, chapter 1.

4. I. Boldea and S. A. Nasar, "Optimum goodness criterion for linear induction motor design," Proc. IEE, Vol. 122, 1975, pp. 922-924.

CHAPTER 2

CERTAIN APPLICATIONS OF LINEAR MOTION ELECTROMAGNETIC SYSTEMS

Before the advent of linear motors, rotary motors with rotary-to-linear converters of some kind were used to produce linear motion. The most obvious advantage of a linear motor is that it has no gears and requires no mechanical rotary-to-linear converters. Thus compared with rotary motors having mechanical gears and similar devices, the linear motor is robust and more reliable. Some of the other advantages of linear motors over rotary motors are as follows:

1. High acceleration and deceleration and less wear of the wheels and the track where acceleration and deceleration take place.

2. Mechanical and electrical protection and the ability to withstand a hostile environment.

3. Ease in maintenance, repair, and replacement.

4. Ability to exert thrust on the secondary without mechanical contacts, as well as convenient control of thrust and speed.

5. Existence of the normal force, which can be advantageous in levitation machines.

In this chapter we review some of the possible applications of linear motors. It is convenient to divide linear motion electromagnetic systems (LMESs) into the following categories from an applications standpoint: (1) force machines, (2) power machines, and (3) energy machines. As may be expected, LMESs find applications in the production of thrust, propulsion force, and levitation force.

Based on this classification, force machines are short-duty machines operating essentially at standstill or at very low speeds, and efficiency is not a major consideration with regard to overall performance. These machines are most useful in small sizes. Power machines often operate at medium or high speeds and are continuous-duty machines; they must have high efficiencies. Energy machines are short-duty machines and have found applications as accelerators and impact extruders.

Fig. 2.1a. Primary of a double-sided LIM (DSLIM). Courtesy U.S. Department of Transportation.

Fig. 2.1b. A test vehicle propelled by DSLIM (LIMRV). Courtesy U.S. Department of Transportation.

2.1. Ground Transportation Applications

Medium- and high-speed applications of linear electric machines (LEMs) are mainly as propulsion and levitation systems for ground transportation. Rotary motors are a poor choice for use at speeds above 250 km/hr because of adhesion and other mechanical considerations. On the other hand a linear motor offers a means of propulsion that is ideally suited for speeds exceeding 250 km/hr. For speeds at which mechanical contact is undesirable, it has been proposed to use a levitation machine for a magnetic suspension system. For both propulsion and suspension of HSGT vehicles, test vehicles having linear motors have been designed and built in the United States, Great Britain, West Germany, France, and Japan.

Fig. 2.2. A low speed levitated short-haul vehicle.

An HSGT vehicle may use either a linear induction motor (LIM) or a linear synchronous motor (LSM). A system using a short-primary LIM on the vehicle (with on-board power source) with an aluminum reaction rail (or track) as the secondary has been tested and seems to be one of the promising solutions. Figures 2.1a and 2.1b show the primary and general views, respectively, of a 1800-kW test vehicle built in the United States. A fully automated short-haul people mover used at an airport is shown in Figure 2.2.

Air core LSMs, with superconducting field magnets, are also feasible for HSGT. Figure 2.3 shows the arrangements of magnets and windings used in a Canadian HSGT project. For propulsion, a number of rectangular superconducting magnets are arranged along the underside of the vehicle. These magnets interact with current-carrying motor windings embedded in the surface of the guideway. Levitation force is a repulsion-type force, and is produced by the interaction between the levitation magnets and the eddy currents induced in the levitation strip by the motion of those magnets. A major advantage of the LSM is that it can operate efficiently with a vehicle clearance of 15 to 20 cm, compared to the 3-cm allowable airgap of a LIM of similar power ratings. Another advantage of the LSM is that all the propulsion power is supplied to the stationary embedded primary windings and the on-board power is required only for the excitation of the magnets.

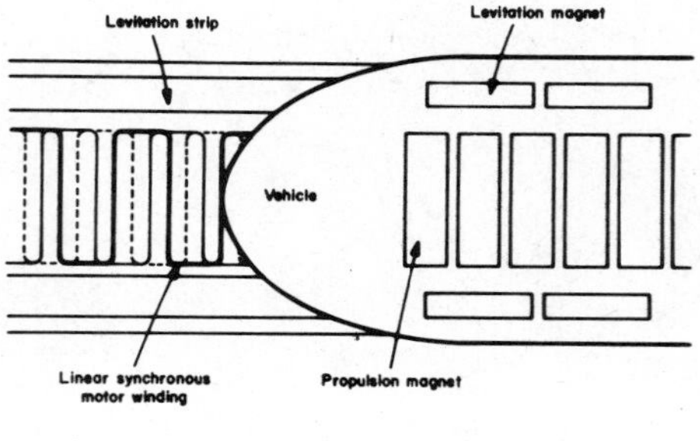

Fig. 2.3. LSM.

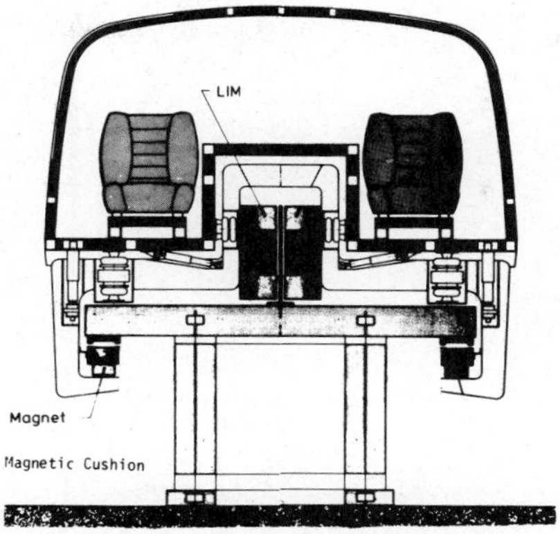

Fig. 2.4. Cross section of Krauss-Maffei experimental vehicle and guideway. Courtesy U.S. Department of Transportation.

Because of high speeds, wheel contacts are undersirable. A levitation machine is used to provide the suspension of a high-speed vehicle. As mentioned in Chapter 1, a levitation system may be either (1) repulsion type or (2) attraction type. An attraction-type suspension system developed in West Germany appears in Fig. 2.4 and Fig. 2.5 shows a more recent system implemented in Japan in JNR. The force of attraction between the on-board electromagnet and the steel guide rails provides lift and guidance. A gap of 15 mm can be maintained between magnet and rail with a power expenditure of about 1 kW/ton of suspended weight. A feedback system monitors the gap and adjusts the current in the electromagnet to overcome the inherent instability of an attraction system. The rating of the test vehicle is about 350 kW.

The repulsion-type suspension system is also considered promising. Small-scale laboratory models have been tested. Preliminary test results on a full-scale HSGT vehicle are also available.

Fig. 2.5. A vehicle developed by JNR.

The preceding discussions indicate that in high-speed applications LMESs provide the thrust and/or levitation force to suspend the vehicle. In some instances the linear motor is used for propulsion only, and the vehicle may be supported either by rail or air cushion. To ensure propulsion, the linear motor must provide sufficient thrust to (1) balance the drag of the vehicle at constant speed, (2) accelerate the vehicle from zero to cruising speed, and (3) help brake the vehicle.

The estimated drags of a 100-passenger vehicle, for the various vehicle suspensions, are given in Fig. 2.6. The propulsion power requirements of single-vehicle and multivehicle trains (including a small margin for acceleration) are listed in Table 2.1.

The large power requirements make it obvious that high efficiency, power factor, and voltage--as well as

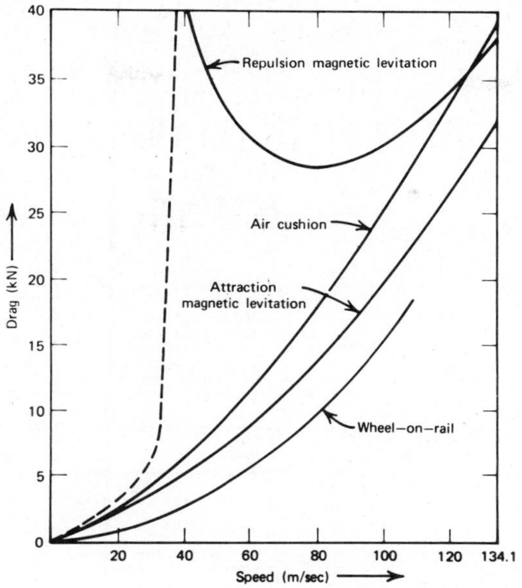

Fig. 2.6. Drag of a 100-passenger vehicle. Courtesy Mr. Matthew Guarino, Jr.

Table 2.1. Propulsion Power Requirements

System	Megawatts required 100 passengers	400 passengers	800 passengers
Rail (400 km/hr)	3.0	8.0	13
Air cushion (480 km/hr)	6.0	16.5	30
Attraction-type machine (480 km/hr)	4.7	12	21
Repulsion-type machine (480 km/hr)	5.7	16	29

low specific weight and specific volume--are important considerations in the choice and design of linear motors for high-speed applications. The power factor has a major impact on the power conditioning subsystem.

From the viewpoint of the linear motor's interaction with the guideway, it is clear that a single-sided motor is superior to a double-sided one because (1) it greatly simplifies vehicle switching, and (2) it may allow simpler and more economical guideway construction: compatibility of the single-sided configuration with the guideway is generally easier to achieve.

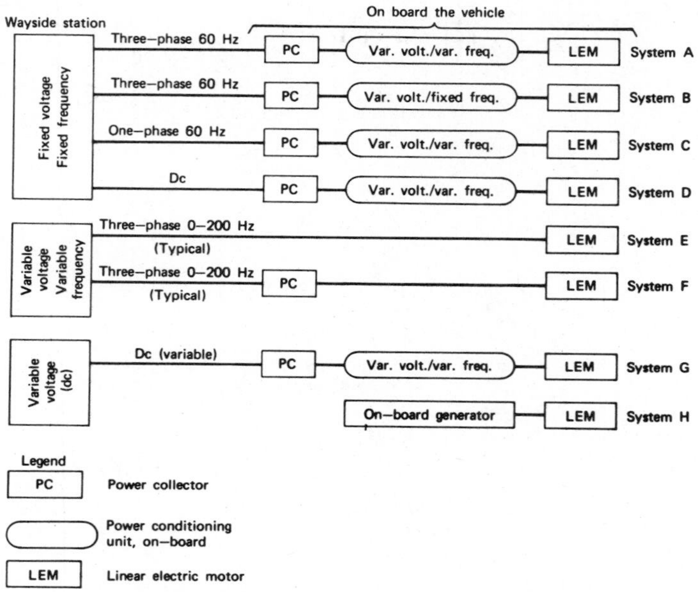

Fig. 2.7. Possible linear motor propulsion systems. Courtesy Mr. Matthew Guarino, Jr.

Fig. 2.8. TLRV (Grumman/AiResearch). Courtesy U.S. Department of Transportation.

Fig. 2.9. Aerotrain, PTACV (France). Courtesy U.S. Department of Transportation.

The various possible linear motor applications are depicted in Fig. 2.7, which also shows the power collection-distribution and power conditioning provisions (as black boxes). Hardware examples of applications of some of these systems are as follows:

1. System A: TLRV (Fig. 2.8).

2. System B: PTACV, Aerotrain (Fig. 2.9).

3. System E: Japanese "ML-100" vehicle (Fig. 2.10a) and the German Transrapid (Fig. 2.10b).

4. System H: LIMRV (Fig. 2.1b).

Systems A, B, and E require a minimum of three power rails. System E (active guideway) is technically the most desirable and also the most expensive. System H is extremely heavy, making it the least desirable for levitated vehicles from a technical viewpoint. However, the important cost reduction due to the elimination of the power collection-distribution may justify its use in specific cases. The most desirable characteristics of linear motor propulsion and suspension systems for ground transportation are summarized in Table 2.2.

Fig. 2.10a. ML-100 vehicle (Japan National Railroad). Courtesy U.S. Department of Transportation.

Fig. 2.10b. German transrapid vehicle and guideway.

2.2. Low-Speed and Standstill Applications

The single-sided linear induction motor (SLIM) is by far the most widely used linear motor. The secondary of the SLIM is either sandwich type or of steel only.

Table 2.2. Recommended Goals for Linear Electric Motors

Suspension	Electrical systems (ranked)	Linear motor characteristics[a]: Power factor of unity to 0.95 lag	Efficiency over 90%	High voltage	Ruggedness of reaction rail	Low motor weight	High, steady attraction force	High, steady guidance force	Decoupling of forces and thrust	Large air gap
Wheel on rail	C		b	c	a	c	b	c	c	c
	D	b								
	H	c								
	F	a	b	a	a	b	d	c	b	b
	G	b								
	D									
	A									
	B									
	H	c		c						
Repulsion-type machine	E	a	b	a	—	b	d	c	b	a
	F				a					b
	G	b								
	D									
	A									
	H	c		c						
Attraction-type machine[b]	F	a	b	a	a	b	a	a	a	a
	G	b								
	D									
	A									
	E									
	H	c		c						

[a]Key: a = extremely desirable, b = very desirable, c = desirable, d = not desirable, can be tolerated. See Fig. 2.7 for definition of electrical systems. [b]Integrated propulsion-levitation system.

The sandwich secondary consists of aluminum strips 3 mm thick, attached to mild steel plates that are 6 mm thick. Empirically, the best thrust-to-size ratio is obtained when the width of the secondary w_s and the core width w_c are related to each other by

$$w_s = w_c + \frac{\lambda}{\pi} \qquad (1)$$

where λ is twice the pole pitch. As we see later, such a SLIM with back iron has been proposed for high-speed applications also. But at present, in practice, its use is limited to low-speed work.

SLIMs with steel secondaries have found applications in the speed range of 0.7 to 1.5 m/sec, in the field of mechanical handling equipment for example. Such a SLIM has pronounced (normal) attraction force between the primary and the secondary. The normal force may be as high as 10 times the thrust. A typical variation of the normal force with the airgap is shown in Fig. 2.11.

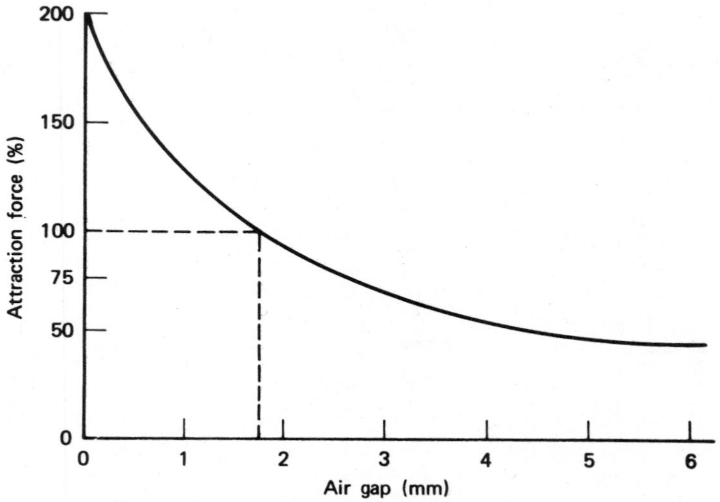

Fig. 2.11. Attraction/airgap characteristic. Intersection indicates position of nominal working airgap. Courtesy IEE (London).

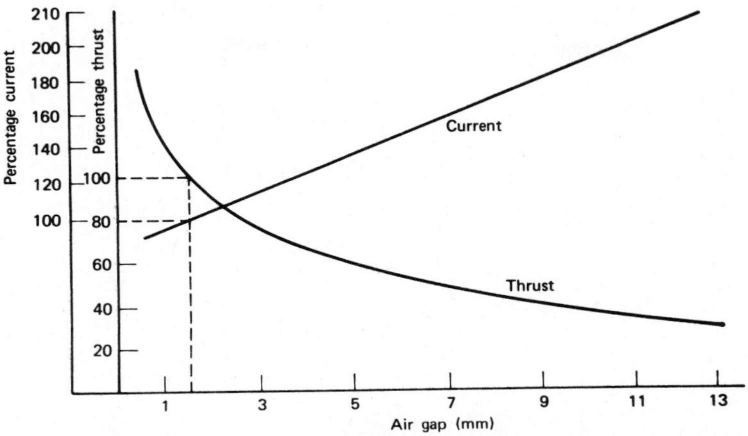

Fig. 2.12. Effect of airgap on thrust and line current. Courtesy IEE (London).

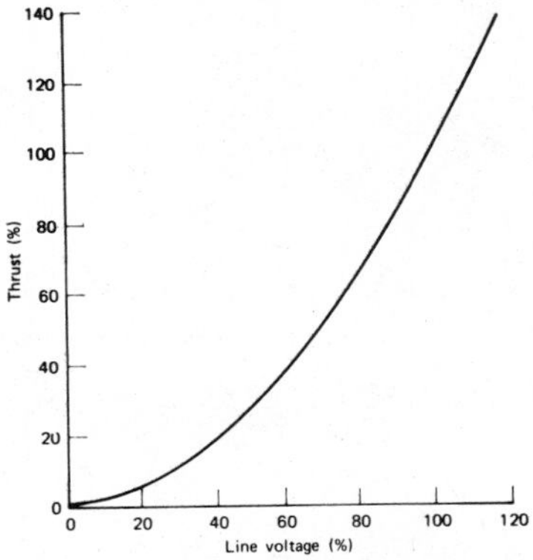

Fig. 2.13. Thrust/line-voltage characteristic. Courtesy IEE (London).

Certain other typical characteristics of LIMs for low-speed applications are illustrated for typical industrial flat LIMs in Figs. 2.12 to 2.14. Tubular LIMs are generally used for short-stroke (0.5 to 2 m) applications. Typical thrust characteristics are similar to that given in Fig. 2.13 and the input power to thrust ratio is of the order of 20 VA/N at 60 Hz.

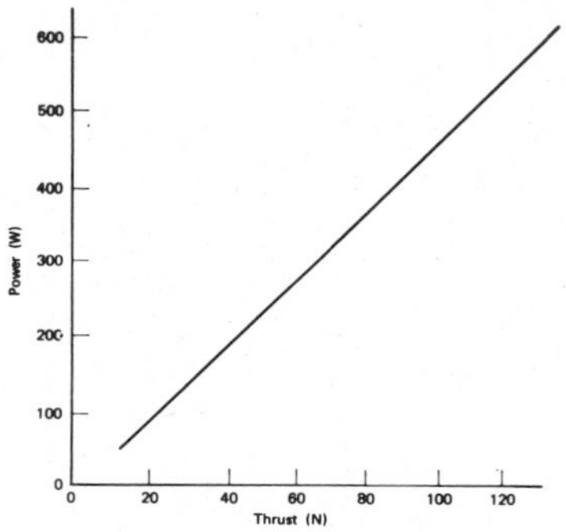

Fig. 2.14. Thrust/input-power characteristic. Courtesy IEE (London).

Fig. 2.15. Aluminum strip tension for winding on roll. Courtesy Linear Motors Ltd.

Because low-speed and standstill applications of linear motors are very numerous, it is neither practical nor desirable to catalog them here. Nevertheless, a few typical applications are presented in Figs. 2.15 to 2.20. In Fig. 2.15 a LIM is used in the tensioning of aluminum strip for coiling. The LIM thrust against the

Fig. 2.16. Linear motor disk drive unit installed on a crane. Courtesy Linear Motors Ltd.

motion of the strip enables tension to be applied without contact, thus avoiding scratching of the surface. Typically, nine DSLIMs are used, and the aluminum foil (or strip) constitutes the secondary. For the application shown in Fig. 2.15, the aluminum foil thickness is between 0.5 and 1.4 mm, and each LIM has a continuous thrust rating of 60 N.

Fig. 2.17. Linear motor-driven roller conveyor handling steel sections. Courtesy Linear Motors Ltd.

Another application of a LIM appears in Fig. 2.16 as a linear disk motor, useful in achieving the traction motion of an overhead crane. The low shaft speed of 150 to 200 rpm enables a single gear reduction of 4 : 1 between the motor and the crane wheels to be fitted. The linear disk motor also can be used in marine ship propulsion.

Figure 2.17 shows an application of LIMs in conveyors. For such a horizontal application carrying steel stock sections, SLIM primaries placed at intervals, depending on the length of material handled, give a simple drive thrust, with variable speed if desired. For a typical coal-carrying conveyor, a number of LIM primaries may be spaced about 3 m apart; the airgap is

Fig. 2.18. Linear motor sliding door operator. Courtesy Skinner Polynoid.

about 0.5 cm, and the primary is fed at 10 Hz for a speed range of 2 to 5 m/sec. The LIM has a starting thrust of 8.4 kN and a continuous thrust of 4 kN.

A flat linear motor about 21 cm long and 9 cm wide, used as an automatic door operator, is illustrated in Fig. 2.18. A flat LIM can serve for pipe-lifting beam in steel works, as in Fig. 2.19, and tubular LIMs can be employed as actuators in mechanical handling, as in Fig. 2.20.

Other applications of linear motors range from turntables to aerodynamic disk-reading heads in a computer, and from induction stirrers for molten metals to shuttle propulsion and package winding in the textile industry.

Fig. 2.19. Linear motor-powered grabs on pipe-lifting beam in steel works. Courtesy Linear Motors Ltd.

Fig. 2.20. Application of tubular motor in bottle transfer. Courtesy Skinner Polynoid.

2.3. Energy LMESs

One of the earliest LIM applications was as an energy machine to launch aircraft. The machine was called Electropult. The primary winding was mounted on a carriage, and the secondary consisted of a winding in slots in a ferromagnetic structure. The motor was very similar to the primitive SLIM with extended secondary. Two full-scale tracks were built, one 5/8 mile long, the other just over a mile. Figure 2.21 shows the primary unit on the runway. Current collection was by means of brushes running in the slots alongside the secondary members. An aircraft attached to the primary unit by a sling appears in Fig. 2.22. The motor developed 10,000 hp and attained speeds exceeding 225 mph. A 10,000-lb jet aircraft was accelerated to 117 mph in a 540-ft run in 4.2 sec from rest. The system was finally abandoned on the grounds of high initial cost.

Other applications of LIMs as energy machines include accelerators for very-high-velocity projectiles and actuators in high-voltage circuit breakers. LIMs have also been used to simulate the conditions of a car crash. A LIM capable of accelerating vehicles weighing as much as 10,000 lb to any speed up to 40 mph has been used as the prime mover (to drive the vehicle). A typical impact barrier for a car-crash indoor facility (Fig. 2.23), and a view of the LIM track (Fig. 2.24) are illustrative. The LIM and a spare unit are shown in Fig. 2.25. The aluminum alloy reaction rail or secondary stands vertically in a pit. The reaction rail is approximately 25 cm thick and 75 cm high with steel flanges top and bottom. The primary has an inverted U-frame containing the two winding blocks, one on either side of the secondary. Current is fed to the primary by means of carbon brushes that run along collector rails mounted on insulators on the walls of the pit. At the floor level, on top of the moving component, is a coupling to which the test vehicle is attached. Cams at the ends of the track (or secondary) trigger the coupling to release the vehicle. Another interesting application of the LIM as an energy machine has been proposed for impact extrusion.

Fig. 2.21. The "Electropult." Courtesy Westinghouse Engineer.

The few examples mentioned in this chapter are indicative of the future of linear motors. Because they are simple in construction, inexpensive, and very reliable, they will find more and more applications in situations in which mechanical gears and rotary-to-linear converters are undesirable.

Fig. 2.22. Primary unit on its track. Courtesy Westinghouse Engineer.

Fig. 2.23. Typical 30-mph barrier impact test, using LIM. Courtesy Motor Industry Research Association.

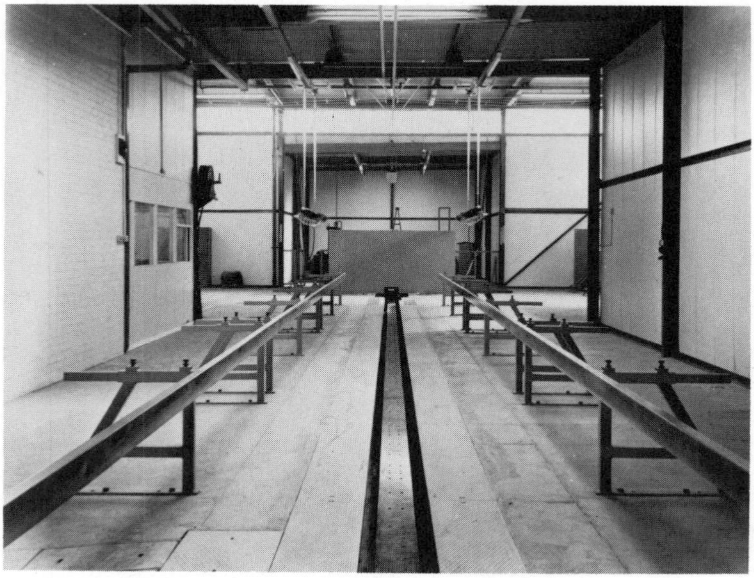

Fig. 2.24. View of LIM track for car-crash test facility, looking toward barrier. Courtesy Motor Industry Research Association.

2.4. A Comparative Quantitative Evaluation

In the design of linear motors, there are a large number of parameters that can be varied to meet a given set of specifications. For example, in a conveyor application the major performance criteria are (1) the speed, (2) the thrust, (3) the thrust per unit surface area, and (4) the figure of merit as measured by thrust per square root of input power (as large as possible). For comparing various motors, we assume reasonably well-designed motors having an airgap flux density of

Fig. 2.25. Moving component of LIM, spare unit off track (courtesy Motor Industry Research Association).

0.2 T (or 2000 gauss) and an electrical loading of 2.25 x 10^5 A/m (or 5715 A/in.). On this basis, the data available in the literature are summarized in Table 2.3.

Table 2.4 summarizes the performances of various motors considered in Chapter 2.

Table 2.3. A Quantitative Comparison

Motor type	Thrust/area, $lb/in.^2$	Thrust/$\sqrt{\text{input power}}$, lb/$\sqrt{\text{watt}}$
DC homopolar	0.15[a]	4.2
DC brushless	1.65[b]	0.75
LSM field-excited or PM	5.5[c]	4.8
Linear synchronous homopolar motor (LSHM)	2.4[d]	3.4
LSRM	2.5[e]	11.5[f]
DSLIM	6.8[g]	0.32
SLIM	3.4[h]	0.30

[a] A current density of 10^6 A/m^2 is assumed. The magnet pole-face area is 9.45 in. x 4.72 in. The force is independent of speed.
[b] This result is for an airgap flux density of 0.87 T. The force is at zero speed.
[c] This value is for a very large LSM operating at the maximum power condition. The airgap flux density is close to 1 T.
[d] This result is for a 0.9 MW machine, and is the maximum force density.
[e] This is the maximum thrust density.
[f] This value has not been substantiated experimentally.
[g] This value is for a very large (2000 hp) DSLIM.
[h] The thrust for a SLIM may be taken as half of that for the DSLIM.

Table 2.4. Summary of Overall Performance Ratings

Motor type	Thrust density	Controllability	Efficiency	Power/weight	Reliability
DC homopolar	Low	Good	Low	High	Low
DC brushless	Low	Cumbersome	High	Low	High
LSM	High	Good	Medium	Medium	Medium
LSHM	Medium	Good	High	High	Medium
LSRM	Medium	Good	Medium	Medium	High
DSLIM	High	Good	Medium	Medium	High
SLIM	Medium	Good	Medium	Medium	High

References

1. E. R. Laithwaite and S. A. Nasar, "Linear-motion electrical machines," Proc. IEEE, Vol. 58, 1970, pp. 531-542.

2. W. Johnson, E. R. Laithwaite, and R. A. C. Slater, "Experimental impact-extrusion machine driven by linear induction motor," Proc. IME, Vol. 179, 1964-1965, pp. 15-35.

3. "A wound rotor motor 1400 ft. long," Westinghouse Eng., Vol. 6, 1946, p. 160.

4. G. F. Nix and E. R. Laithwaite, "Linear induction motors for low speed and standstill applications," Proc. IEE, Vol. 113, 1966, pp. 1044-1056.

5. Y. Sundberg, "Magntic traveling fields for metallurgical processes," IEEE Spectrum, Vol. 6, May 1969, pp. 79-88.

6. M. Guarino, Jr., "Integrated linear electric motor propulsion system's for high speed transportation," International Symposium on Linear Electric Motors, Lyon, France, May 1974.

7. G. V. Sadler and A. W. Davey, "Applications of linear induction motors in industry," Proc. IEE, Vol. 118, 1971, pp. 765-775.

8. H. Alscher, I. Boldea, A. R. Eastham and M. Iguchi, "Propelling passengers faster than a speeding bullet," IEEE Spectrum, Vol. 21, No. 8, August 1984, pp. 57-64.

9. International Conference on "Maglev transport--Now and for the future," I. Mech. E. Conference Publications, 1984-12, I. Mech. E., London, 1984.

CHAPTER 3

METHODS OF ANALYSIS AND DESIGN

3.1. Introduction

In principle, the methods of analysis for rotating electric machines are also applicable to linear motion electromagnetic systems (LMESs). These methods include: (1) lumped-parameter circuit analysis; (2) distributed-parameter field analysis; and (3) a combination of (1) and (2). It must be pointed out that the three approaches are not independent of each other. Rather, the classification of the methods is merely for the sake of convenience. In fact, we often start with the field analysis to obtain the equivalent circuit of the device at hand. Otherwise, the field analysis directly yields some of the device characteristics as demonstrated in the following sections.

3.2. Governing Equations for Idealized Linear Induction Motors (LIMs)

As discussed in earlier chapters, the principle of operation of a LIM is the same as that of a rotary induction motor. We have already shown how a LIM is developed from a conventional rotary induction motor. Recall from Chapter 1 how the secondary of a LIM, corresponding to the rotor, was shown to have the bars of the cage of the original rotary motor. In practice, however, a cage-type secondary is rarely used in LIM. Rather, almost invariably, the secondary of a LIM is made of a solid conducting plate or sheet, often backed by iron. With such a configuration of the secondary, having a continuous material medium, it is convenient to analyze the machine by using the electromagnetic field equations. Such an approach is straightforward in principle, but often intricate in details, as we see later. The steps involved in the field analysis are as follows:

(1) choose a physical model that is amenable to mathematical analysis, (2) formulate the appropriate field equations and the boundary conditions, (3) solve the field equations subject to the specified boundary conditions, and (4) interpret the results. We now turn to the details in carrying through these steps.

3.2.1. Choice of the Physical Model

As mentioned earlier, the physical model must be such that the boundary-value problem involving the field equations may be conveniently solved. In this regard, the slotted primary structure with its windings presents a formidable difficulty in applying and solving the field equations. To overcome this difficulty, the actual slotted structure is replaced by a smooth surface and the current-carrying windings are replaced by fictitious, infinitely thin current elements, called current sheets, having linear current densities (A/m). The current density distribution of the current sheet is the same as that of the slot-embedded conductor configurations (or windings), such that the field in the airgap remains unchanged. In replacing the actual slotted structure by an ideal smooth structure, the actual airgap of the machine is replaced by an effective airgap, which is about 1.020 to 1.2 times greater than the original airgap. A better estimate of the effective airgap is obtained by using the Carter coefficient as a multiplying factor.

The equivalence of an actual winding and its current is illustrated by means of Fig. 3.1, where a slot is shown to contain N conductors each carrying a current i. In this model, it is assumed that the permeability of iron is much greater than that of air ($\mu_{iron} \gg \mu_0$). In such a case, applying Ampere's law to the original structure of Fig. 3.1a and to the equivalent current sheet of Fig. 3.1b yields

$$\oint \overline{H} \cdot d\overline{l} = Ni = hw$$

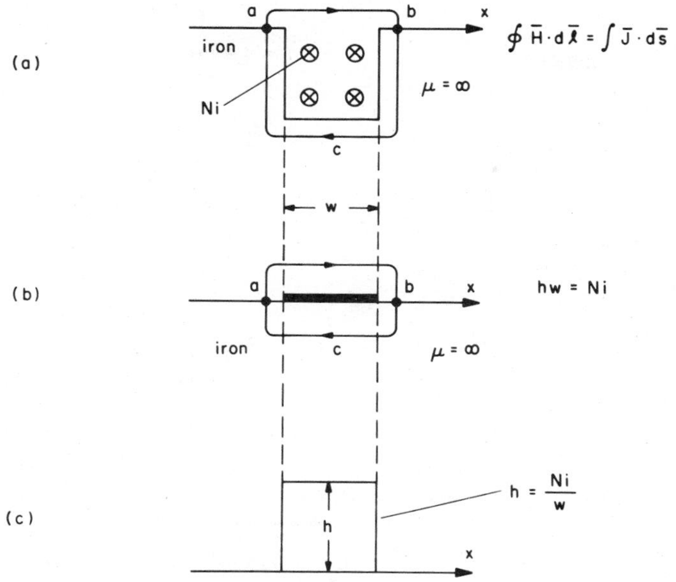

Fig. 3.1. (a) Original structure, (b) equivalent current sheet, (c) linear current density as pulse.

or

$$h = \frac{Ni}{w} \tag{1}$$

where h is the linear current density (A/m) and w is the width of the slot. This linear current density is shown as a pulse of width w and height h in Fig. 3.1c. Hence, a given actual slot-embedded winding can be replaced by a train of pulses of appropriate height. These pulses repeat periodically over every pair of poles, and can be Fourier-analyzed to find the predominant harmonic component. The details of finding equivalent current sheets for a given winding are given in Reference 1. For the present, taking a simplified approach, we assume that (ideally) the windings have predominant sinusoidal

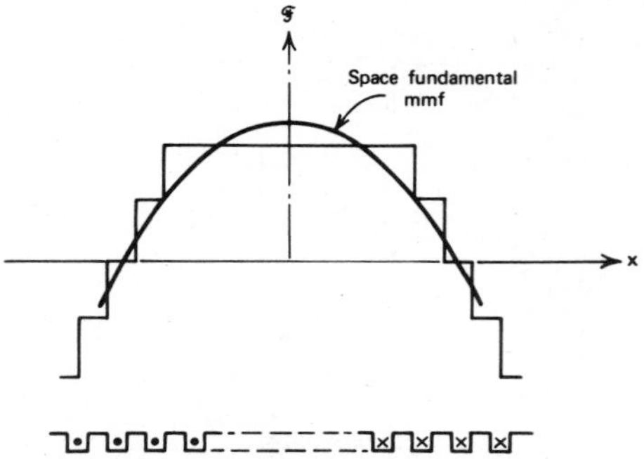

Fig. 3.2. Mmf distribution of a distributed winding and its resolution into space fundamental.

space distributions. For example, in Fig. 3.2 the stepped magnetomotive force (mmf) is resolved into its harmonics, and the fundamental component is then the equivalent current sheet. The amplitude of the current sheet J_m is found from the relationship

$$J_m = \frac{\text{maximum total current along machine periphery}}{\text{length of machine periphery}}$$

$$\times \text{ total winding factor} \qquad (2)$$

In this expression the total winding factor takes into account the deviation from a sinusoidal distribution because of chording, slots, and so on. We can also rewrite (2) as

$$J_m = \frac{m(2W\sqrt{2}I)}{2p\tau} k_w = \frac{\sqrt{2}mWIk_w}{p\tau} \qquad (3)$$

where m = number of phases, $2p\tau$ = length of machine (p being the number of pole pairs and τ the pole pitch), W = number of turns per phase, $\sqrt{2}$ I = maximum phase current, and k_w = total winding factor. The use of the current sheet concept is illustrated in the next section.

3.2.2. The Airgap Field Equation

In an electric machine, the current sheets result in electromagnetic fields in the various regions of the machine. Of these regions, airgap fields are the most important. Considering an idealized model of a LIM as shown in Fig. 3.3, let the primary and secondary currents, respectively, be replaced by their current sheets J^s and J^r having linear current densities. We

assume the currents to flow in the z-direction only and the permeability of the core material to be infinity. Then from Fig. 3.3 we have (assuming no relative motion between the primary and the secondary)

$$H_y' = \frac{\partial H_y}{\partial x} \Delta x + H_y \tag{4}$$

and

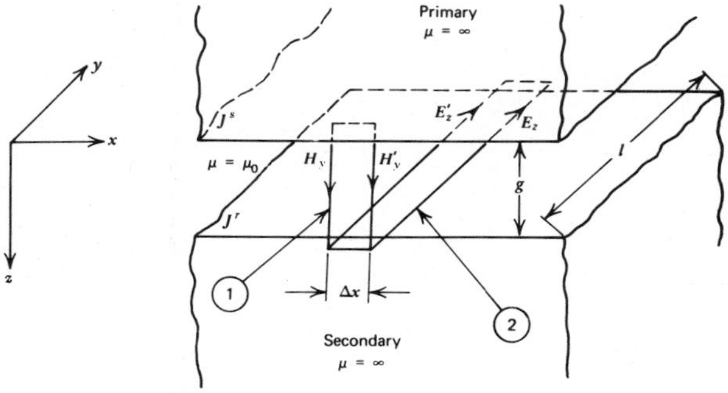

Fig. 3.3. An idealized model of a LIM and paths of integration.

$$E_z' = \frac{\partial E_z}{\partial \chi} \Delta x + E_z \tag{5}$$

From Ampere's law

$$\oint \overline{H} \cdot d\overline{l} = \oint_s \overline{J} \cdot d\overline{s} \tag{6}$$

we get

$$g \frac{\partial H_y}{\partial x} = J^s + J^r \tag{7}$$

and

$$\frac{\partial E_z}{\partial x} = \mu_0 \frac{\partial H_y}{\partial t} \tag{8}$$

But, from Ohm's law,

$$J^r = \sigma E_z \tag{9}$$

where σ is the surface conductivity. Thus (4) through (9) finally yield

$$g \frac{\partial^2 H_y}{\partial x^2} = \mu_0 \sigma \frac{\partial H_y}{\partial x} + \frac{\partial J^s}{\partial x}$$

Or, in terms of B_y, we have

$$\frac{\partial^2 B_y}{\partial x^2} - \frac{\mu_0 \sigma}{g} \frac{\partial B_y}{\partial t} = \frac{\mu_0}{g} \frac{\partial J^s}{\partial x} \tag{10}$$

which is known as the airgap field equation.

If we have a velocity U_x between the primary and the secondary, the modified airgap field equation becomes

$$\frac{\partial^2 B_y}{\partial x^2} - \frac{\mu_0 \sigma U_x}{g}\frac{\partial B_y}{\partial x} - \frac{\mu_0 \sigma}{g}\frac{\partial B_y}{\partial t} = \frac{\mu_0}{g}\frac{\partial J^s}{\partial x} \qquad (11)$$

For a LIM we may still use (10) and account for the relative motion by making use of the concept of slip frequency by putting $\partial/\partial t = j\omega s$, where ω is the primary angular frequency and s is the fractional slip. Thus for a given primary current sheet the airgap flux-density distribution can be determined.

In the beginning of this section we identified four general steps involved in solving the electromagnetic field problem in a LIM. Up to this point, we have considered the first two steps--(1) and (2). We illustrate the last two steps by means of two examples.

3.2.3. Solution to Field Equation of an Idealized LIM with an Infinitely Thick Secondary

The LIM modeled by Fig. 3.4 consists of three layers as follows:

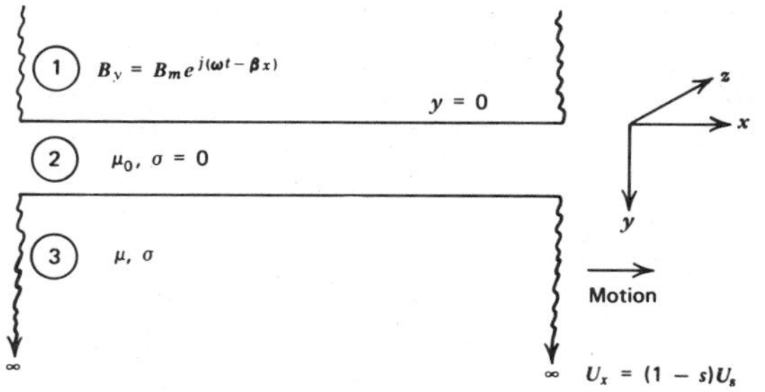

Fig. 3.4. Model of an idealized LIM with an infinitely thick secondary.

Layer 1: The primary;

Layer 2: The airgap;

Layer 3: The secondary.

The following assumptions are made to simplify the analysis:

1. All layers extend to infinity in the ±x-direction.

2. The secondary extends to infinity in the y-direction.

3. The excitation windings are located in the slotted primary structure. For convenience, the structure is "smoothed" to permit representation of the machine excitation as a current sheet of negligible thickness and finite width.

4. The motion of the secondary is in the x-direction only.

5. The physical constants of the layers are homogeneous, isotropic, and linear.

6. The ferromagnetic material does not saturate.

7. Variations in the z-direction are ignored.

8. All currents flow in the z-direction only.

9. The primary is constructed of laminated iron, to ensure that conductivity in the z-direction is negligible.

10. Time and space variations are sinusoidal.

A few comments about the assumptions are in order. Assumptions 1 and 2 form a starting point of the analysis. Assumption 3 makes the model amenable to mathematical analysis. Assumption 4 is an obvious one, since the secondary consists of a solid conductor moving in one direction only. Assumptions 5 and 6 are valid in the light of linearity assumption stated earlier. Assumptions 7 and 8 are made to reduce the problem to a two-dimensional field problem. The laminated primary core justifies Assumption 9. Assumption 10 is valid because the source voltage (or current) varies sinusoidally with time, and the ideal distribution of the primary winding assures a sinusoidal space distribution of the fields.

With the foregoing assumptions in mind, we state Maxwell's equations (in point form) as follows:

$$\overline{\nabla} \cdot \overline{B} = 0 \tag{12}$$

$$\overline{\nabla} \times \overline{H} = \overline{J} + \frac{\partial \overline{D}}{\partial t} \tag{13}$$

$$\overline{\nabla} \times \overline{E} = - \frac{\partial \overline{B}}{\partial t} \tag{14}$$

Ohm's law for a moving medium is given by

$$\overline{J} = \sigma(\overline{E} + \overline{U} \times \overline{B}) \tag{15}$$

From these equations we can derive the basic equations governing the electromagnetic phenomena in the LIM model of Fig. 3.4.

Since the displacement current density $\partial D/\partial t$ is negligible (at power frequencies), (1) becomes

$$\overline{\nabla} \times \overline{H} = \overline{J} \tag{16}$$

So

$$\overline{\nabla} \times \overline{B} = \mu\sigma(\overline{E} + \overline{U} \times \overline{B}) \quad (17)$$

The magnetic vector potential A is defined by

$$\overline{\nabla} \times \overline{A} = \overline{B} \quad (18)$$

Substitution of (18) into (17) gives

$$\overline{\nabla} \times (\overline{\nabla} \times \overline{A}) = \mu\sigma(\overline{E} + \overline{U} \times \overline{B}) \quad (19)$$

Expansion of (19) yields

$$\overline{\nabla}(\overline{\nabla} \cdot \overline{A}) - \nabla^2\overline{A} = \mu\sigma[- \frac{\partial\overline{A}}{\partial t} + \overline{U} \times (\overline{\nabla} \times \overline{A})] \quad (20)$$

But $\overline{\nabla} \cdot \overline{D} = 0$ (there being no free charges); also, $\overline{\nabla} \cdot \overline{A} = 0$ can be assumed. Hence

$$\nabla^2\overline{A} = \mu\sigma(\frac{\partial\overline{A}}{\partial t} - \overline{U} \times \overline{\nabla} \times \overline{A}) \quad (21)$$

When suitably excited, the primary creates a y-directed traveling field in the airgap given by

$$B_y = B_m e^{j(\omega t - \beta x)} \quad (22)$$

which implies $\partial/\partial x = -j\beta$ and $\partial/\partial t = j\omega$.

Since $\overline{A}$ is assumed to be z-directed and is not a function of z,

$$\overline{A}(x,y,t) = A(y)e^{j(\omega t - \beta x)}\hat{a}_z \quad (23)$$

where $\hat{a}_z$ is the z-directed unit vector. Now (21) can be rewritten:

$$\frac{\partial^2 A_z}{\partial x^2} + \frac{\partial^2 A_z}{\partial y^2} = \mu\sigma(j\omega A_z + \frac{U_x \partial A_z}{\partial x}) \tag{24}$$

where

$$A(y)e^{j(\omega t-\beta x)} = A_z \tag{25}$$

and

$$\overline{U} = U_x \hat{a}_x \tag{26}$$

Equation (24) is the basic governing equation. The solution to this equation, subject to the given boundary conditions, yields the quantitative information regarding the electromagnetic phenomena in the machine. For the model under consideration, we recall that the airgap field, produced by the primary, travels at a synchronous speed U_s, which is related to the slip s and the speed of the secondary U_x by

$$U_x = (1 - s)U_s \tag{27}$$

Because $\partial^2 A_z/\partial x^2 = \beta^2 A_z$, (24) becomes

$$\frac{\partial^2 A_z}{\partial y_2} = A_z \beta^2 (1 + \frac{j\mu\sigma s U_s}{\beta}) \tag{28}$$

If we put

$$\beta^2 (1 + \frac{j\mu\sigma s U_s}{\beta}) = \alpha^2 \tag{29}$$

for region 2, the airgap where $\sigma = 0$, then (28) reduces to

$$\frac{d^2A_z}{dy^2} - \beta^2 A_z = 0 \tag{30}$$

For region 3, the secondary, (28) becomes

$$\frac{d^2A_z}{dy^2} - \alpha^2 A_z = 0 \tag{31}$$

The solutions for (30) and (31) can be written as

$$A_{z_2} = (C_3 e^{\beta y} + C_4 e^{-\beta y}) e^{j(\omega t - \beta x)}$$

and

$$A_{z3} = (C_1 e^{\alpha y} + C_2 e^{-\alpha y}) e^{j(\omega t - \beta x)}$$

where the subscript numbers identify the region under consideration.

To evaluate the constants, the following <u>boundary conditions</u> are used:

$$y = 0, \quad B_y = B_m e^{j(\omega t - \beta x)}$$

$$y = g, \quad B_{y2} = B_{y3} \quad \text{and} \quad H_{x2} = H_{x3}$$

$$y \to \infty, \quad A_{z3} = 0$$

resulting in the following:

$$C_2 e^{-\alpha g} = C_3 e^{\beta g} + C_4 e^{-\beta g} \tag{32}$$

$$C_3 e^{\beta g} - C_4 e^{-\beta g} = -\frac{\alpha\mu_0}{\beta\mu} C_2 e^{-\alpha g} \tag{33}$$

$$C_3 = \frac{B_m}{j\beta} - C_4 \tag{34}$$

and

$$C_1 = 0 \tag{35}$$

From (32) through (35),

$$C_2 = \frac{B_m e^{\alpha g}}{j\beta\Delta} \tag{36}$$

$$C_3 = B_m e^{-\beta g} \frac{1 - \alpha\mu_0/\beta\mu}{2j\beta\Delta} \tag{37}$$

and

$$C_4 = B_m e^{\beta g} \frac{1 + \alpha\mu_0/\beta\mu}{2j\beta\Delta} \tag{38}$$

where

$$\Delta = \cosh \beta g + \frac{\alpha\mu_0}{\beta\mu} \sinh \beta g \tag{39}$$

The force density, F, is given by the <u>Lorentz force equation</u>

$$\overline{F} = \overline{J} \times \overline{B} \tag{40}$$

where

$$\overline{J} = \sigma(\overline{E} + \overline{U} \times \overline{B}) \tag{41}$$

Since $\sigma = 0$ in region 2, it is obvious that no force can develop there. In region 3

$$\overline{A}_3 = C_2 e^{-\alpha y} e^{j(\omega t - \beta x)} \hat{a}_z$$

$$\overline{B}_3 = \alpha C_2 e^{-\alpha y} e^{j(\omega t - \beta x)} \hat{a}_x + j\beta C_2 e^{-\alpha y} e^{j(\omega t - \beta x)} \hat{a}_y$$

$$\overline{E}_3 = - j\omega C_2 e^{-\alpha y} e^{j(\omega t - \beta x)} \hat{a}_z \qquad (42)$$

Again

$$\overline{F}_3 = \overline{J}_3 \times \overline{B}_3 \qquad (43)$$

Substitution of (41) into (43) gives

$$\overline{F}_3 = \sigma(\overline{E}_3 \times \overline{B}_3 + \overline{U} \times \overline{B}_3 \times \overline{B}_3) \qquad (44)$$

It can be noted that $\overline{E}_3 = -j\omega\overline{A}_3$ and $\overline{B}_3 = j\beta\overline{A}_3$. Substituting these into (44) yields

$$\overline{F}_3 = - \sigma(E_{z3}B_{y3} + U_x B_{y3} B_{y3})\hat{a}_x$$

$$+ \sigma(E_{z3}B_{x3} + U_x B_{x3} B_{x3})\hat{a}_y \qquad (45)$$

Hence

$$F_{x3} = - \sigma E_{z3} B_{y3} - \sigma U_x B_{y3} B_{y3}$$

So the time-average force density in the x-direction is

$$\langle F_{x3}\rangle = \frac{-\sigma}{2}\,\mathrm{Re}[(-j\omega\overline{A}_3)(-j\beta\overline{A}_3^*)]$$

$$-\ \frac{\sigma}{2}\,\mathrm{Re}[(1 - s)U_s(j\beta\overline{A}_3)(-j\beta\overline{A}_3^*)]$$

$$= \frac{(\sigma\beta^2 sU_s)\mathrm{Re}(\overline{A}_3\overline{A}_3^*)}{2} \tag{46}$$

where the asterisk indicates complex conjugation. From (36) through (42) we get

$$A_3 = \frac{B_m}{j\beta\Delta}\, e^{-\alpha(y-g)} e^{j(\omega t-\beta x)}\hat{a}_z$$

and

$$\langle F_{x3}\rangle = \frac{\sigma B_m^2 sU_s}{2|\Delta|^2}\,\mathrm{Re}[e^{-\alpha(y-g)} e^{-\alpha^*(y-g)}]$$

$$= \frac{\sigma B_m^2 sU_s}{2|\Delta|^2}\,[e^{-2(y-g)\mathrm{Re}(\alpha)}]$$

Force on the secondary in the x-direction per wavelength, for the entire width, l, and the entire depth of the secondary plate, in newtons, is

$$\langle F_x \rangle = \frac{\sigma B_m^2 s U_s}{2|\Delta|^2} \lambda \int_{-l/2}^{l/2} \int_g^{\infty} e^{-2(y-g)Re(\alpha)} dy\, dz$$

$$= \frac{\sigma B_m^2 s U_s \lambda l}{4\, |\Delta|^2\, Re(\alpha)} \qquad (47)$$

3.3. An Idealized LIM with a Secondary of Finite Thickness

Figure 3.5 shows the arrangement of a model having a secondary of finite thickness, h. The coordinate origin is now located on the secondary surface. The assumptions listed earlier are applicable also. In addition, the airgap is assumed to be very small with no fringing or decay of the B-field in the airgap. The following layers are shown in Fig. 3.5:

Layer 1: The primary.

Layer 2: The air gap.

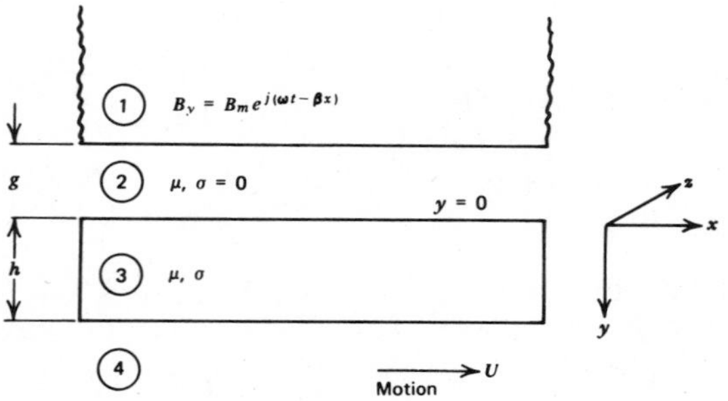

Fig. 3.5. Model of an idealized LIM with a secondary of finite thickness.

Layer 3: The secondary.

Layer 4: The air below the secondary.

As in the previous section, we have for region 3

$$\frac{d^2A_z}{dy^2} - \alpha^2 A_z = 0$$

and for region 4

$$\beta = \alpha$$

So for region 3 the solution

$$A_{z3} = (C_1 e^{\alpha y} + C_2 e^{-\alpha y}) e^{j(\omega t - \beta x)}$$

may be assumed. For region 4,

$$A_{z4} = (C_3 e^{\beta y} + C_4 e^{-\beta y}) e^{j(\omega t - \beta x)}$$

Using the defining equation for vector potential A,

$$\overline{\nabla} \times \overline{A} = \overline{B}$$

gives for region 3

$$\overline{B}_3 = \alpha(C_1 e^{\alpha y} - C_2 e^{-\alpha y})\, e^{j(\omega t - \beta x)} \hat{a}_x$$

$$+ j\beta(C_1 e^{\alpha y} + C_2 e^{-\alpha y})\, e^{j(\omega t - \beta x)} \hat{a}_y$$

and for region 4

$$\overline{B}_4 = \beta(C_3e^{\beta y} - C_4e^{-\beta y})\, e^{j(\omega t-\beta' x)}\hat{a}_x$$

$$+ j\beta(C_3e^{\beta y} + C_4e^{-\beta y})e^{j(\omega t-\beta x)}\hat{a}_y$$

The following boundary conditions now are used:

$$y = 0, \qquad B_{y3} = B_m e^{j(\omega t-\beta x)}$$

$$y = h, \qquad H_{x3} = H_{x4}, \qquad \text{and} \qquad B_{y3} = B_{y4}$$

$$y \to \infty, \qquad \overline{A}_4 = 0$$

The following results are obtained:

$$C_1 + C_2 = \frac{B_m}{j\beta}$$

$$C_3 = 0$$

$$C_1e^{\alpha h} + C_2e^{-\alpha h} = C_4e^{-\beta h}$$

$$C_1e^{\alpha h} - C_2e^{-\alpha h} = -\frac{\mu\beta}{\mu_0\alpha}C_4e^{-\beta h}$$

Manipulation of these four equations gives

$$C_1 = \frac{B_m e^{-\alpha h}}{2j\beta(\cosh \alpha h + (\mu\beta/\mu_0\alpha)\sinh \alpha h)} (1 - \frac{\mu\beta}{\mu_0\alpha})$$

$$C_2 = \frac{B_m e^{\alpha h}}{2j\beta(\cosh \alpha h + (\mu\beta/\mu_0\alpha)\sinh \alpha h)} (1 + \frac{\mu\beta}{\mu_0\alpha})$$

$$C_4 = \frac{B_m e^{\beta h}}{j\beta} (\frac{1}{\cosh \alpha h + (\mu\beta/\mu_0\alpha)\sinh \alpha h})$$

With these values, expressions for A and B can be written:

$$\overline{A}_3 = \frac{B_m}{j\beta\Delta} [\cosh \alpha(y - h)$$

$$- \frac{\mu\beta}{\mu_0\alpha} \sinh \alpha(y - h)] e^{j(\omega t - \beta x)} \hat{a}_z$$

and

$$\overline{B}_3 = \frac{B_m \alpha}{j\beta\Delta} [\sinh \alpha(y - h)$$

$$- \frac{\mu\beta}{\mu_0\alpha} \cosh \alpha(y - h)] e^{j(\omega t - \beta x)} \hat{a}_x$$

$$+ \frac{B_m}{\Delta} [\cosh \alpha(y - h) - \frac{\mu\beta}{\mu_0\alpha} \sinh \alpha(y - h)] e^{j(\omega t - \beta x)} \hat{a}_y$$

where $\Delta = \cosh \alpha h + (\mu\beta/\mu_0\alpha)\sinh \alpha h$. The average force density, (46), is

$$\langle F_{x3}\rangle = \frac{\sigma\beta^2 sU_s \ \mathrm{Re}(\overline{A}_3\overline{A}_3^*)}{2}$$

Now

$$\frac{\mathrm{Re}}{2}(\overline{A}_3\overline{A}_3^*) = \frac{\mathrm{Re}}{2} \ \frac{B_m}{j\beta\Delta}[\cosh\alpha(y-h)$$

$$-\frac{\mu\beta}{\mu_0\alpha}\sinh\alpha(y-h)]$$

$$\times \quad \frac{B_m}{-j\beta\Delta^*}[\cosh^*\alpha(y-h) - \frac{\mu\beta}{\mu_0\alpha}\sinh^*\alpha(y-h)]$$

After considerable algebraic manipulation, this reduces to

$$\frac{\mathrm{Re}}{2}(\overline{A}_3\overline{A}_3^*) = \frac{B_m^2}{4\beta^2|\Delta|^2}[\cosh 2\, m_1(y-h)$$

$$+\cos 2m_2(y-h) - \frac{2\mu\beta}{\mu|\alpha|^2} m_1 \ \sinh 2m_1(y-h)$$

$$+ m_2 \sin 2m_2(y-h) + \frac{\mu^2\beta^2}{\mu_0^2|\alpha|^2}\cosh 2m_1(y-h)$$

$$-\cos 2m_2(y-h)]$$

Thus the average force density can now be written as

$$\langle F_{x3}\rangle = \frac{B_m^2 \sigma s U_s}{4|\Delta|^2}[\cosh 2m_1(y - h) + \cos 2m_2(y - h)$$

$$- \frac{2\mu\beta}{\mu_0|\alpha|^2}\, m_1 \sinh 2m_1(y - h) + m_2 \sin 2m_2(y - h)$$

$$+ \frac{\mu^2\beta^2}{\mu_0^2|\alpha|^2} \cosh 2m_1(y - h) - \cos 2m_2(y - h)] \qquad (48)$$

Hence the force developed by the secondary in the x-direction per wavelength λ, for width l and thickness h, is obtained by multiplying the average force density expression by l and integrating over thickness h. Thus

$$F_x = \frac{B_m^2 \sigma s U_s l\lambda}{4\Delta^2\alpha^2}[\alpha^2(\frac{1}{2m_1} \sinh 2m_1 h + \frac{1}{2m_2} \sin 2m_2 h)$$

$$+ a^2(\frac{1}{2m_1} \sinh 2m_1 h - \frac{1}{2m_2} \sin 2m_2 h)$$

$$+ a(\cosh 2m_1 h - \cos 2m_2 h)] \quad \text{(newtons)} \qquad (49)$$

where $a = \mu\beta/\mu_0$ and $\alpha = m_1 + jm_2$.

Figure 3.6 plots (49) for a LIM having a secondary 3 mm thick. Before closing this section we wish to point out that the LIM models considered also develop a normal force in the y-direction, as given by (45).

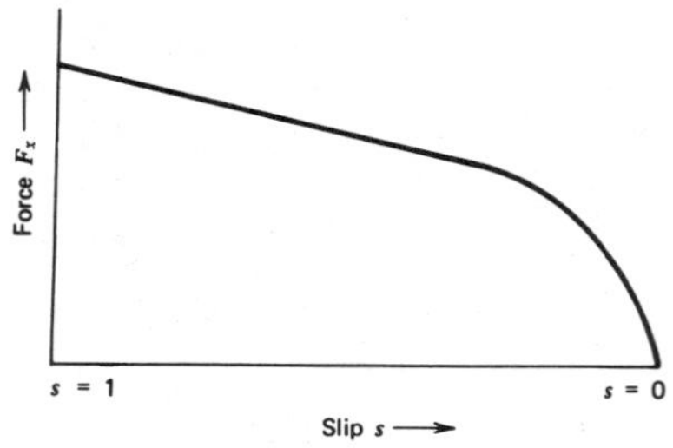

Fig. 3.6. Plot of F_x versus s as obtained from (49).

Second, in many realistic cases $m_1 >> 1$, and in such cases (49) simplifies to

$$F_x = \frac{B_m^2 \sigma s U_s \lambda l(m_2 \sinh 2m_1 h - m_1 \sin 2m_2 h)}{4m_1 m_2(\cosh 2m_1 h - \cos 2m_2 h)} \quad N \qquad (50)$$

Also, often $m_1 = m_2$. For example, for a LIM for which $f = 60$ Hz, $\lambda = 0.2$ m, $\mu = 1000\mu_0$, $\sigma = 4 \times 10^6$ mho/m, at $s = 0.1$, we have $\alpha = 436\underline{/45^\circ}$ and $\mu\beta/\mu_0\alpha = 72\underline{/-45^\circ}$.

3.4. A LIM Performance Criterion

There are a number of criteria that may be used to assess the design leading to the performance of a LIM. These criteria include: the goodness factor, G_2; the optimum goodness factor, G_0; and the magnetic Reynold's number, M. The details pertinent to these are available in References 2 through 4. To introduce the concept of goodness factor, consider an idealized LIM and let the primary current sheet be given by

$$J^s = J_m e^{j(\omega t - \beta x)}$$

This current sheet may be transformed to the secondary

coordinates by putting $\partial/\partial t = j\omega s$ so that the B_y at the secondary surface, from (10), is given by

$$(j\beta)^2 B_y - \frac{1}{g}(j\omega s \mu_0 \sigma) B_y = \frac{1}{g}\mu_0(-j\beta) J_m$$

from which we obtain

$$B_y = \frac{(j\mu_0 \beta/g) J_m}{\beta^2 + j\omega s \mu_0 \sigma/g} = \frac{\mu_0 J_m}{g\beta(sG - j)}$$

where

$$G = \frac{\omega \mu_0 \sigma}{g\beta^2} \qquad (51)$$

We define the ratio given by (51) as the goodness factor because it is related to the real part of the field, B_y, and denotes the active component or the force-producing component, in contrast to the reactive component of the field.

The definition given by (51) agrees with the alternative form given in Reference 4 as

$$G = \frac{2f\mu_0 \sigma \tau^2}{\pi g} \qquad (52)$$

where τ is pole pitch. Finally, the fundamental definition of the goodness factor for the secondary, in terms of an equivalent circuit, is given as

$$G = \frac{X_m}{R_2} \qquad (53)$$

where X_m is the magnetizing reactance and R_2 is the

secondary resistance. The details of equivalence of (51) and (53) are available in Reference 1.

The goodness factor is a useful index in preliminary designs of linear electric machines (LEMs). However, large goodness factor does not necessarily ensure maximum thrust and efficiency for a high-speed LIM. Certain modifications to the goodness factor given by (51) through (53) must be made in such cases, and it may be preferable to use the optimum goodness factor.

3.5. An Approximate Equivalent Circuit Analysis

Up to this point we have analyzed the LIM using the field equations. However, for an approximate analysis and design, it is advantageous to use the approximate equivalent circuit having known parameters. To determine the parameters of the circuit, we adapt the design formulas of the rotary induction motor for application to the LIM. We know that often the secondary of a LIM is made of a conducting sheet. The concept of surface resistivity, ρ_r, is very useful in finding the resistance of such a secondary. The surface restivity is defined by

$$\rho_r = \frac{\rho}{d} \tag{54}$$

where ρ = resistivity of the secondary material and d = thickness of the secondary sheet.

Although our analysis neglects the longitudinal end effect, a correction factor to determine the effective surface resistivity, as modified by end effects at standstill, may be used. Figure 3.7 graphically represents the effective resistivity variation due to end effect for a parallel connected short-primary LIM at standstill.

Likewise, the secondary overhangs also change the surface resistivity. The effective surface resistivity, approximately accounting for the secondary overhangs, is given by

$$\rho_r' = \frac{\rho_r}{1 - K} \tag{55}$$

where K is the Russell-Norsworthy overhang correction factor

$$K = \frac{\tanh(\pi l/2\tau)}{(\pi l/2\tau)\,[1 + \tanh(\pi l/2\tau)\tanh(\pi c/2\tau)]} \tag{56}$$

where l = stack width, τ = pole pitch, and c = secondary overhang. Thus we can account for the overhangs and end effects empirically, using (55) and (56) and Fig. 3.7.

Noting the above-named differences introduced by the presence of a sheet rotor and a short primary, we now represent a LIM by the approximate equivalent circuit of Fig. 3.8. This circuit is on a per-phase basis,

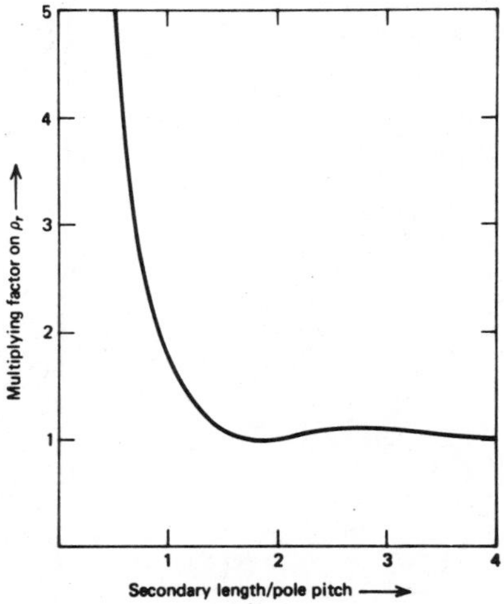

Fig. 3.7. Effective resistivity variation due to end effect.

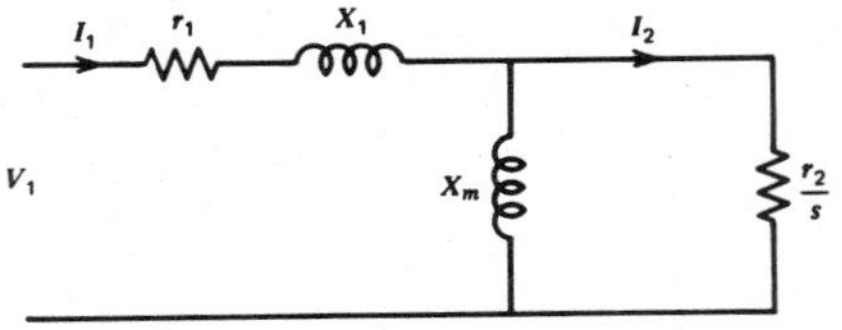

Fig. 3.8. An approximate equivalent circuit.

and for a double-sided LIM (DSLIM) the representation is for one side of the primary. With $X_2 \simeq 0$, adapting the standard design formulas, the parameters of the equivalent circuit of the LIM are given as follows.

Primary resistance:

$$r_1 = \frac{2\rho_c K_p^2 q m^2 l N^2 (1 + K_1 \tau / l)}{K_f K_d p \tau^2} \quad (\Omega/\text{phase}) \qquad (57)$$

with $K_p = t_c/w = 1.5$ to 2 and $K_d = t/w = 3$ to 5.

Primary leakage reactance, X_1 (for an open-slot primary core):

$$X_1 \simeq \frac{2\mu_0 \omega}{p} \left[\frac{(\lambda_c + \lambda_d) l}{q} + \lambda_e \tau K_1\right] N^2 \qquad (58)$$

with

$$\lambda_c \simeq \frac{1}{12} K_d (1 + 3\beta), \quad \lambda_d \simeq \frac{5g/w}{5 + 4g/w}$$

and

$$\lambda_e \simeq 0.3(3\beta' - 1) \qquad (59)$$

Secondary resistance, r_2:

$$r_2 = \frac{61}{\tau p} \rho_r'(K_w N)^2 \tag{60}$$

Magnetizing reactance, X_m:

$$X_m = \frac{6\mu_0 \omega}{\pi^2} \frac{\tau l}{pg}(K_w N)^2 \tag{61}$$

Goodness factor, G:

$$G = \frac{X_m}{r_2} = \frac{2\mu_0 f \tau^2}{\rho_r' \pi g} \tag{62}$$

In (57) through (62) the following symbols are used: t = slot depth, m; g = total airgap, m; t_c = slot pitch, m; w = slot width, m; ρ_c = volume resistivity of copper, Ω-m; K_1 = 1.2 to 1.8 = ratio of the mean length of coil end connection to the pole pitch; K_f = 0.5 to 0.6 = slot-filling factor; β' = chording factor; m = number of phases; q = number of slots per pole per phase; τ = pole pitch, m; N = number of turns per phase (for primary); K_w = primary winding distribution factor $\simeq$ 0.9, and p = pole pairs.

3.5.1. Performance Calculations

Once the parameters of the equivalent circuit are known, the performance of the LIM can be calculated in a routine manner. In summary, on a per-phase-per-side basis, for a three-phase DS LIM, the following relationships give the performance characteristics of the motor:

input power = $V_1 I_1 \cos \Phi_1$; primary copper loss = $I_1^2 r_1$;

developed power = $(1 - s)(V_1 I_{11}, \cos \Phi_1 - I_1^2 r_1)$; developed force = $1/U$ (developed power); secondary copper loss = $s(V_1 I_1 \cos \Phi_1 - I_1^2 r_1)$. In the preceding, s = slip, U = LIM speed, $\cos \Phi$ = input power factor, and other symbols appear in Fig. 3.8.

3.5.2. Design Considerations

Having presented a discussion of the performance calculations of a low-speed LIM, we now summarize some design aspects of LIMs. The design of a LIM involves many parameters that can be varied to affect the performance of the machine. The effects of varying some of the parameters are given in Table 3.1.

References

1. E. R. Laithwaite, Induction Machines for Special Purposes, Chemical Publishing Co., 1966.

2. "Study of linear induction motor and its feasibility for high speed ground transportation," Ai Research Manufacturing Co., U.S. Department of Transportation Study, Contract C-145-66 P.B. 174866, January 1967.

3. E. R. Laithwaite, D. Tipping, and D. E. Hesmondhalgh, "The application of induction motors to converyors," Proc. IEE, Vol. 107A, 1060, pp. 284-294.

4. G. F. Nix and E. R. Laithwaite, "Linear induction motors for low speed and standstill application," Proc. IEE, Vol. 113, No. 6, June 1966, pp. 1044-1056.

5. M. W. Davis, "Development of concentric linear induction motor," IEEE Trans. Power App. Syst., Vol. PAS-91, No. 4, 1972, pp. 1506-1513.

Table 3.1 Effects of Parameter Variation

Parameter	Increase	Decrease
Airgap, g	Larger magnetizing current	Larger goodness factor
	Larger exit-end losses	Larger output force Larger efficiency
Pole pitch, τ	Larger goodness factor	Larger number of poles
	Increase back iron thickness	
Number of poles, $2p$	Smaller end effects	Larger secondary leakage reactance
Secondary thickness, d	Larger goodness factor	Larger secondary leakage reactance
Secondary resistivity, ρ	Smaller end effects	Larger goodness factor
Tooth width, w	Larger leakage reactances	Larger force
		Larger efficiency

6. J. F. Eastham and J. H. Alwash, "Transverse flux tubular motors," Proc. IEE, Vol. 119, No. 10, October 1972, pp. 1709-1718.

CHAPTER 4

LOW SPEED LINEAR INDUCTION MACHINES

4.1. Flat Linear Induction Motors (LIMs) with Short Primary

Low-speed LIMs have characteristics quite distinct from those of high-speed LIMs. Therefore, we treat low-speed LIMs separately. By low speeds we mean speeds less than 10 m/sec. In most well-designed low-speed LIMs the longitudinal end effect can be neglected. There are many configurations of practical interest suitable for the various existing and future applications. Three of these configurations, considered representative, are treated here. These are:

1. Single- or double-sided flat LIM with short primary--used for various industrial applications.

2. Single-sided flat LIM with fixed short primary and short secondary--used for conveyors.

3. Tubular LIM with short primary--used for short-stroke actuators.

Beginning with a review of construction guidelines, in the absence of longitudinal end effect, the theory concentrates on transverse edge effect, skin effect, secondary solid back-iron (if any) saturation and eddy currents, and normal forces. All these phenomena are finally included in the expressions of lumped parameters of equivalent circuits in order to determine the LIM performance. Such an approach is design oriented, and consequently design guidelines and numerical examples are also given.

There is a considerable amount of published material[1,2,5,6] on LIMs for standstill and low-speed applications, and the interested reader may consult these for further studies.

4.2. Construction Guidelines of Flat LIMs

The primary and secondary configurations are considered individually.

4.2.1. Primary Construction

The LIM primary is made of a laminated core with open slots (Fig. 4.1). A three- or two-phase winding is embedded in the primary slots. Many winding configurations have been proposed. They resemble, to some extent, those of rotary induction motors. Among them four are considered here to be of practical interest:

1. One-layer winding with even number of poles (Fig. 4.2).

2. Three-layer winding with even number of poles (Fig. 4.3).

3. Two-layer winding with odd number of poles and half-filled end slots (Fig. 4.4).

4. "Economic" winding for very-low-power LIMs (Fig. 4.5).

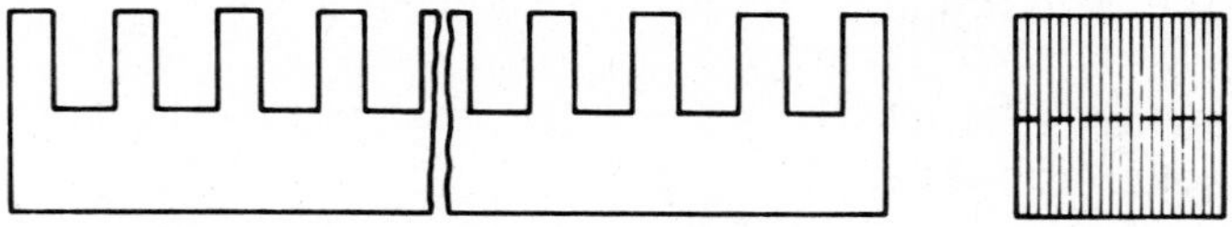

Fig. 4.1. LIM primary core.

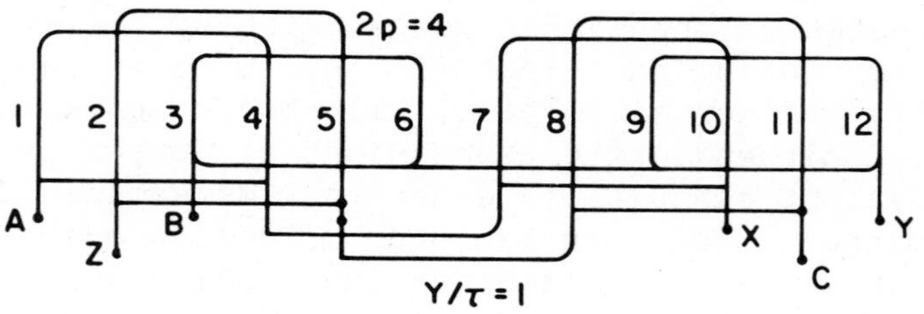

Fig. 4.2. One-layer winding with even number of poles.

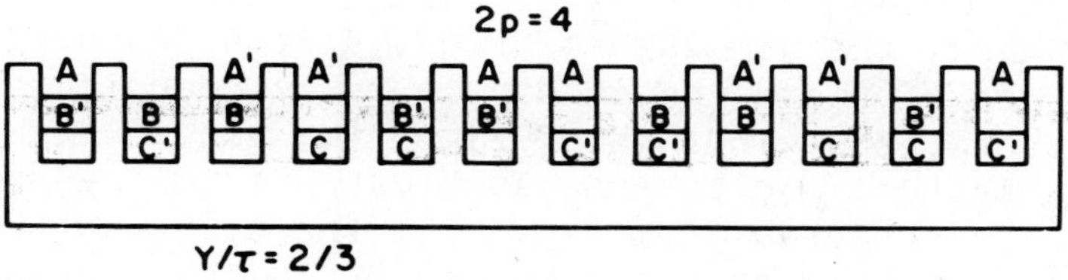

Fig. 4.3. Three-layer winding with even number of poles.

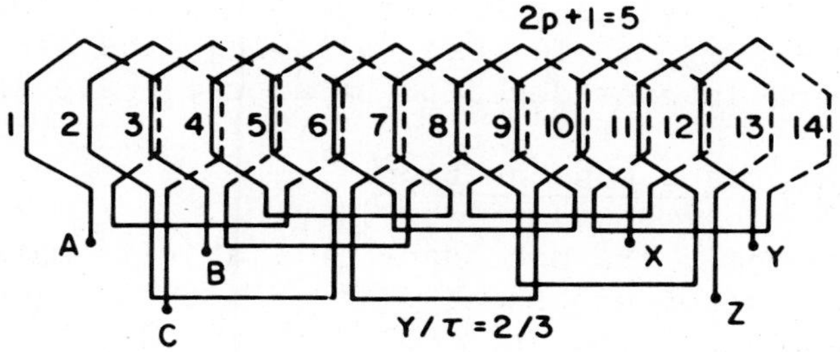

Fig. 4.4. Two-layer winding with odd number of poles and half-filled end slots.

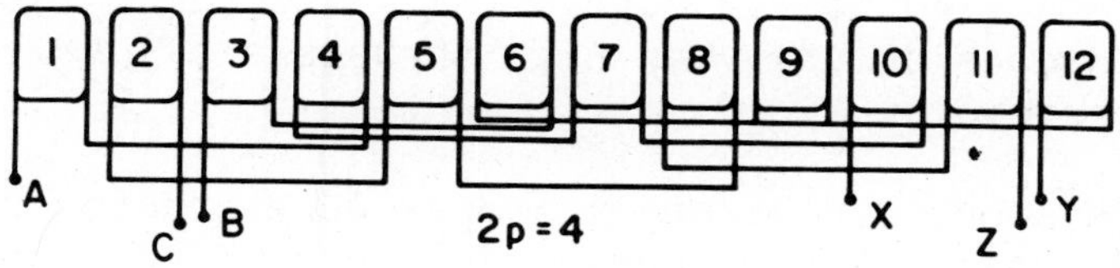

Fig. 4.5. "Economic" winding.

The advantages and disadvantages of these windings are related to manufacturing costs and the capacity of producing an airgap field distribution approaching a purely forward traveling wave. It is well established that the airgap field of the windings shown in Figs. 4.2, 4.3, and 4.5, with open secondary, exhibits notable pulsating components besides the traveling wave because of the open character of the magnetic circuit. Also, it has been found that the winding of Fig. 4.4 develops a purely traveling-wave airgap field in the central zone

[(2p - 1) poles in length] whereas pulsating components occur only along the marginal, half-wound, poles. Thus in high-thrust applications this winding is most suitable. Also, the windings of Figs. 4.2, 4.3 and 4.4 make better use of the magnetic core but, by reducing the end connections, the windings of Figs. 4.3 and 4.4 utilize copper better, and in some applications this asset prevails. As expected from Fig. 4.5, the economic winding produces pronounced space harmonics in the airgap field and its winding factor is rather poor, but for very low thrust levels (up to 20 to 30 N), the copper weight, mounting room savings prevail, promoting the use of this winding. However, in the following, the airgap field distribution is considered to be a pure traveling wave.

4.2.2. Secondary Construction

In general, the secondary of flat LIMs may be of an aluminum (or copper) sheet with or without a solid back iron plate (Fig. 4.6a, b). Also in special cases a laminated slotted core with a ladder secondary may be used (Fig. 4.6c). Configurations (a) and (b) are less expensive but have poorer energy conversion performance in comparison with configuration (c). The ladder secondary is worth considering when high thrusts are involved such as in conveyors.

LIM short-stroke oscillators were proposed three decades ago but they did not find industrial applications because of poor performance and dynamic instabilities. Consequently they are not treated here.

Let us now investigate specific phenomena in flat LIMs with short primaries.

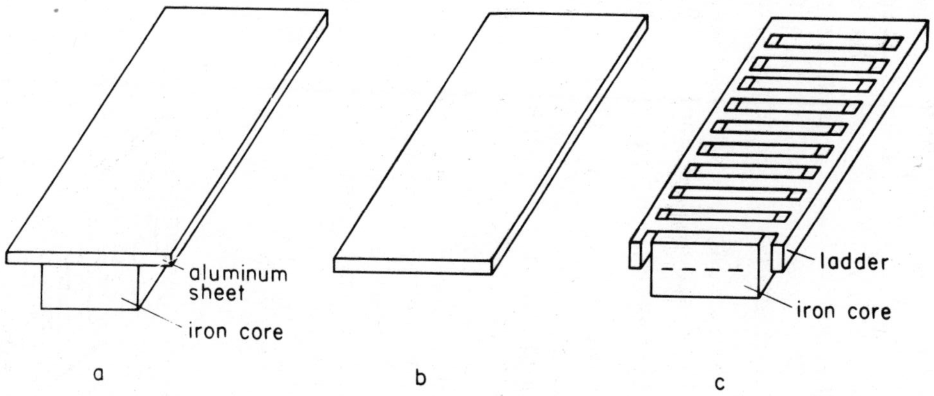

Fig. 4.6. LIM secondary (a) aluminum sheet on iron; (b) aluminum sheet; (c) ladder.

4.3. Specific Phenomena and Performances

The induced currents in the solid structure of the secondary (Fig. 4.7) are found by solving the field equation. Notice that the induced current density varies along the secondary thickness because of the skin effect. Finally the solid back iron secondary often gets saturated. The presence of the longitudinal component (J_x) of induced current density in the active zone ($|y| \leq a$) causes an apparent increase of secondary equivalent resistivity since the useful (J_y) component of current density is lowered by this effect. Also J_x causes a demagnetizing effect by its reaction field. These main aspects are called transverse end effect. A simultaneous treatment of transverse and skin effects together with secondary back iron saturation can be accomplished by a two dimensional analytical or numerical approach. Such approaches are time consuming and should be applied for refined studies. A superposition approach with transverse and skin effects treated separately has proved adequate for design purposes. This procedure is used here for the single-sided LIM (SLIM) with back iron in the secondary. The case of double-sided LIM (DSLIM) results is a simpler case of the former.

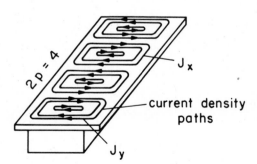

Fig. 4.7. Induced currents trajectory.

4.3.1. Saturation and Transverse End Effect

In the transverse view of a LIM shown in Fig. 4.8 the secondary back iron is made of three solid laminations. The influences of saturation and transverse effect are accounted for by calculating an equivalent conductivity, σ_e, and airgap, g_e, for an ideal motor such that the real and ideal motor are electromagnetically identical. Maxwell's equations applied to this ideal motor yield

$$g_e \frac{\partial H}{\partial x} = \underline{J}_1 + \underline{J}_2 d_a \qquad (1)$$

$$\frac{\partial \underline{J}_2}{\partial x} = -j\omega s \mu_0 \sigma_e \underline{H} \qquad (2)$$

where $\underline{H}$ = airgap field, and $\underline{J}_2$ = secondary current density, with $\underline{J}_1$ the primary current sheet fundamental in secondary coordinates

$$J_1 = J_{1m} e^{j(s\omega t - \pi x/\tau)}; \quad J_{1m} = \frac{3\sqrt{2}W_1 I_1 k_{w1}}{p\tau} \qquad (3)$$

In (1) through (3) we have s = motor slip, ω = angular frequency, τ = pole pitch. The solution to (1) becomes

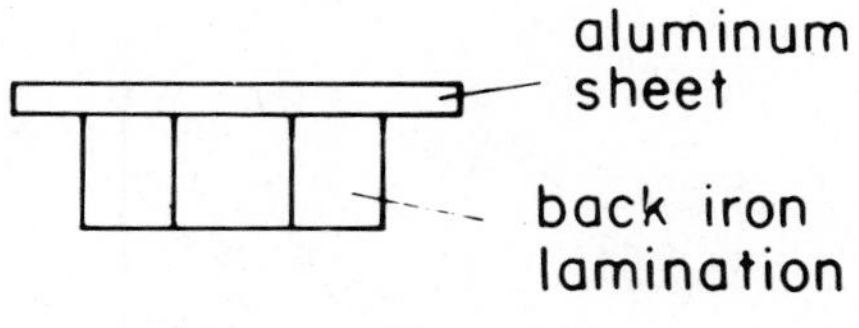

Fig. 4.8. LIM transverse view (three laminations in the back iron).

$$\underline{B}_g = \mu_0 \underline{H} = -j \frac{\tau \mu_0 J_{1m} e^{j(s\omega t - x\pi/\tau)}}{\pi g_e (1 + jsG_e)};$$

$$G_e = \frac{\mu_0 \omega \tau^2 \sigma_e d_a}{\pi^2 g_e} \qquad (4)$$

In (4) G_e is the realistic goodness factor of the ideal equivalent motor still to be determined from the following. The quantities σ_e and g_e must be such that the secondary back iron saturation and transverse effect are taken into account. First, only the transverse effect in secondary sheet is considered and the back iron influence in the magnetic reluctance is accounted for by an equivalent increase of airgap still to be determined. In this case the correction coefficients, K_{ta} and K_{tm}, are

$$K_{ta} = \frac{K_x^2 [1 + (sG_{ep} K_R / K_x)^2]}{K_R (1 + s^2 G_{ep}^2)} \qquad (5)$$

$$K_{ma} = \frac{K_R}{K_x} K_{ta} \qquad (6)$$

$$K_R = 1 - \mathrm{Re}[(1 - jsG_{ep}) \frac{\lambda}{\underline{\alpha} a_e} \tanh \underline{\alpha} a_e] \qquad (7)$$

$$K_x = 1 + \mathrm{Re}[(sG_{ep} + j) sG_{ep} \frac{\lambda}{\underline{\alpha} a_e} \tanh \underline{\alpha} a_e] \qquad (8)$$

$$\underline{\lambda} = \frac{1}{1 + (1 + jsG_{ep})^{1/2} \tanh \underline{\alpha} a_e \tanh (\pi/\tau)(c - a)} \tag{9}$$

with

$$G_{ep} = \frac{\mu_0 \omega \tau^2 \sigma_a' d_a}{\pi^2 g_0 K_c (1 + K_p)}; \quad \underline{\alpha} = 1 + jsG_{ep} \tag{10}$$

where a_e is the equivalent and semiwidth of primary stack accounting for the lateral fringing, and is given by

$$a_e \simeq a + g_0 \tag{11}$$

The coefficient K_p accounts for the back iron magnetic reluctances, still to be determined.

If the product $sG_{ep} << 1$, the transverse end-effect coefficients yield simplified expressions:

$$(K_{ta})sG_{ep} << 1 = 1/[1 - \tanh (\pi/\tau)a_e/\{(\pi/\tau)a_e[1 + \tanh (\pi/\tau)a_e \tanh(\pi/\tau)(c - a_e)]\}] \tag{12}$$

$$(K_{ma})sG_{ep} << 1 = 1 \quad \text{and} \quad (\alpha)sG_{ep} << 1 = \frac{\pi}{\tau} \tag{13}$$

The skin effect in the aluminum or copper plate secondary is moderate since the penetration depth in it is greater than its thickness. Thus the correction coefficient K_{sa} known from rotary machines may be used:

$$K_{sa} = \frac{da}{d_s} \frac{\sinh (2d_a/d_s) + \sin (2d_a/d_s)}{\cosh (2d_a/d_s) - \cos (2d_a/d_s)} \geq 1 \tag{14}$$

with the skin depth, d_s, given by

$$\frac{1}{d_s} = \text{Re}\ [\ \sqrt{(\frac{\pi}{\tau})^2 + js\omega\mu_0\sigma}\ ] \tag{15}$$

Thus the aluminum conductivity σ is corrected for skin effect by K_{sa}:

$$\sigma_a' = \frac{\sigma}{K_{sa}} \tag{16}$$

There is also a transverse effect in the secondary back iron. Since the iron conductivity, σ_i, is smaller than that of aluminum, the correction coefficients given by (12) and (13) hold, with $(c - a_e) = 0$, since the back iron and primary stack widths are considered equal to each other. Thus the transverse effect correction coefficient, K_{ti}, for iron is

$$K_{ti} = \frac{1}{1 - \dfrac{\tanh(\pi/\tau)a_e/i}{(\pi/\tau)a_e/i}} \tag{17}$$

where i is the number of secondary back iron laminations (in general i = 1, 2, 3).

Again the depth of field penetration in the back iron, δ_i, is

$$\delta_i^{-1} = \text{Real}\ [\ \sqrt{(\frac{\pi}{\tau})^2 + js\omega\mu\frac{\sigma_i}{K_{ti}}}\ ] \tag{18}$$

In (10), the back iron average permeability, μ, is still to be determined. Now considering the average length of magnetic flux paths in back iron to be τ/π and the core thickness to be δ_i, we obtain the ratio between airgap and back iron magnetic reluctances as

$$K_p = \frac{\tau^2}{\pi^2} \frac{\mu_0}{2g_0K_c\delta_i\mu} \tag{19}$$

Provided the μ of iron is known, the equivalent airgap g_e of a machine without transverse effect, and with ideal iron back core, is

$$g_e = (1 + K_p)g_0 \frac{K_c}{K_{ma}} \tag{20}$$

The total goodness factor of the machine G_e should account for the eddy currents in the back iron. Thus (10) becomes

$$G_e = \frac{\mu_0\omega\tau^2}{\pi^2 g_e} (\sigma_a' \frac{d_a}{K_{ta}} + \sigma_i \frac{\delta_i}{K_{ti}}) \tag{21}$$

Finally we replace the aluminum sheet conductivity by an equivalent one, σ_e, which yields from (21)

$$\sigma_e = \sigma_a'(\frac{1}{K_{ta}} + \frac{\sigma_i\delta_i}{\sigma_a' K_{ti} d_a}) \tag{22}$$

The goodness factor of the equivalent machine is

$$G_e = \frac{\mu_0\omega\tau^2}{\pi^2} \frac{\sigma_e d_a}{g_e} \tag{23}$$

But μ and thus δ_i and K_p are still to be determined. Assuming an exponential distribution of field along the iron depth, the amplitude of the longitudinal flux density on the back iron surface, B_{xi}, is given by

$$B_{xi} = \frac{\tau}{\pi} \frac{B_g}{\delta_i} K_{pf}; \qquad K_{pf} = 1.1 \text{ to } 1.5 \tag{24}$$

The factor K_{pf} accounts for the pulsating component of the core field in a LIM and B_g is computed from (4). An arbitrary initial value is assigned to the back iron average permeability μ. Calculating B_{xi}, a new value of μ is obtained from the magnetization curve of the back iron core. The average value of μ may be chosen in a number of ways but that corresponding to $0.9B_{xi}$ proved to yield good results.

In the above approach the secondary leakage reactance has been implicitly neglected, which is a realistic assumption for most practical applications. For the case of ladder secondary with laminated core, this approximation does not hold any more. However, we can proceed in this case as for a conventional rotary motor.

Due to the transverse effect (when $sG_{ep} > 1$) and also due to the back iron pole pitch, τ, saturation and eddy currents, the parameters should depend on airgap density, B_g, and slip angular frequency, $s\omega$. In other words, for given geometrical data the circuit parameters are dependent on $s\omega$ and J_{1m} (current sheet amplitude) and could be considered constant only if $sG_{ep} < 0.1$ and if there were no solid back iron in the secondary (for DSLIM no back iron is provided in most cases).

The equivalent circuit of the rotary machine may be used (Fig. 4.9), but its parameters, X_m and R_2', are in

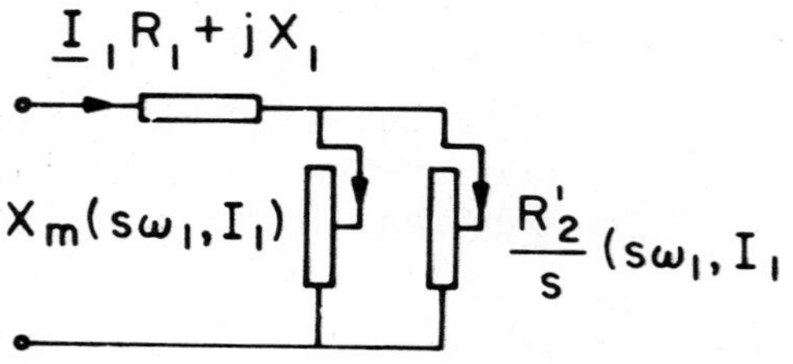

Fig. 4.9. Equivalent rotary scheme.

general dependent on $s\omega$ and I_1 for a given motor. Iron losses are neglected in Fig. 4.9 since the airgap flux density, B_g, is hardly greater than 0.3 to 0.4 while the primary core depth is larger than magnetically necessary. To define X_m we can use the rotary machine theory:

$$X_m(s\omega,\ I_1) = \frac{6\mu_0\omega 2a_e(K_{w1}W_1)^2}{\pi^2 pg_e(s\omega,\ I_1)}\tau = K_m W_1^2;$$

$$K_{w1} = \frac{\sin(\pi/6)}{q\ \sin(\pi/6q)}\sin\frac{\beta\pi}{2} \tag{25}$$

and the secondary equivalent resistance, $R_2'(s\omega,\ I_1)$, is

$$R_2'(s\omega,\ I_1) = \frac{X_m}{G_e}$$

$$= \frac{12a_e}{d_a\tau p\sigma_e(s\omega,\ I_1)}(K_{w1}W_1)^2 = K_{R_2'}W_1^2 \tag{26}$$

We still have to develop expressions for primary phase resistance, R_1, and leakage reactance, X_1. To obtain expressions useful for design purposes we define

$K_{slot} = \frac{h_s}{b_s} = 3$ to 6; $q = 1$ to 3 slots/pole/phase; h_s = slot depth; 10 b_s = slot width; K_f = 0.5 to 0.6 = total fill factor of the slot J_{co} = design current density; σ_{co} = conductivity of copper l_{ec} = the primary coil end connection length, $\simeq K_1\tau$; $K_1 = 0.6$ to 1.6

These definitions depend on the type of the winding, and $\beta' = y/\tau$.

$$K_d = \frac{3qb_s}{\tau} = 0.5 \text{ to } 0.75 \text{ (slot area/pole area)} \quad (27)$$

Now the primary phase resistance, R_1, can be expressed as

$$R_1 \simeq \frac{1}{\sigma_{co}} \frac{(4a + 2l_{ec})}{W_1 I_1} W_1^2 J_{co} = K_{R1} W_1^2 \quad (28)$$

Finally

$$W_1 I_1 = pq \frac{\tau^2}{9q^2} K_d^2 K_{slot} K_f J_{co} \quad (29)$$

and

$$W_1 I_1 = pq(2n_b I_1) = pqb_s h_s K_f J_{co} \quad (30)$$

$$X_1 = \frac{2\mu_0 \omega}{p} [(\frac{\lambda_c + \lambda_d}{q}) 2a + \lambda_e l_{ec}] W_1^2 = K_{\sigma 1} W_1^2 \quad (31)$$

with

$$\lambda_c \simeq \frac{1}{12} K_{slot}(1 + 3\beta'); \quad \lambda_d \simeq \frac{5(g_0 q/k_d \tau)}{5 + 4(g_0 q/K_d \tau)} \quad (32)$$

$$\lambda_e \simeq 0.3(3\beta' - 1); \quad \beta' = \frac{y}{\tau} = \text{chording factor} \quad (33)$$

When an odd number of poles is used, the end poles have

the slots half filled. To account for this aspect X_1 becomes

$$X_1 = \frac{2\mu_0\omega}{p}\left[\frac{\{\lambda_c(1 + 3/2p) + \lambda_d\}2a}{q} + \lambda_e l_{ec}\right]W_1^2 \qquad (34)$$

Performance calculations

To calculate the LIM performance the conventional equivalent circuit (Fig. 4.9) may be used taking into account the fact that the parameters X_m and R_2' are dependent on I_1 and $s\omega$.

Thus the thrust F_x is

$$F_x = \frac{3I_2'^2 R_w'}{s2\tau f_1} = \frac{3I_1^2 R_2'}{s2\tau f_1\ [1/(sG_e)^2 + 1]} \qquad (35)$$

For efficiency and power factor we have the following expressions:

$$\eta_1 = \frac{F_x 2\tau f_1(1 - s)}{F_x 2\tau f_1 + 3R_1 I_1^2} \qquad (36)$$

$$\cos\phi_1 = \frac{F_x 2\tau f_1 + 3R_1 I_1^2}{3V_{1f} I_1} \qquad (37)$$

4.3.2. Normal Force in a SLIM

The normal force, F_n, between LIM primary and secondary has two components:

1. An attraction force, F_{na}, between the primary core and secondary back iron.

2. A repulsion force, F_{nr}, between the primary and secondary currents.

In the absence of longitudinal effect the attraction force, F_{na}, is given by

$$F_{na} \simeq \frac{2a_e |\underline{B}_g^2| 2p\tau}{2\mu_0} \tag{38}$$

Making use of (4) we get

$$F_n = \frac{a_e L \mu_0 J_{1m}^2 \tau^2}{\pi^2 g_e^2 (1 + s^2 G_e^2)}; \quad L = 2p\tau \tag{39}$$

To calculate the repulsion force, F_{nr}, we need to know the tangential component of the airgap magnetic field; H_x on the primary core surface equals the primary current sheet. Hence,

$$H_x = J_{1m} \quad \text{and} \quad B_x = \mu_0 J_{1m} \tag{40}$$

Now the repulsion force, F_{nr}, is

$$F_{nr} \simeq 4 \frac{a_e}{2} \tau p d_a \ \mathrm{Real}(\underline{J}_2^* \mu_0 J_{1m}) \tag{41}$$

where $\underline{J}_2$ is the secondary current density. And, from (2),

$$\underline{J}_2 = s\omega\sigma_e \frac{\tau}{\pi} \underline{B}_g \tag{42}$$

Finally with (4)

$$F_{nr} = -2a_e \tau s\omega\sigma_e d_a \left(\frac{\tau}{\pi}\right)^2 \frac{(\mu_0 J_{1m})^2 sG_e}{g_e(1 + s^2G_e^2)} \qquad (43)$$

The net normal force, F_n, is

$$F_n = F_{na} + F_{nr}$$

$$= 2a_e p\tau \frac{\mu_0 J_{1m}^2 \tau^2}{\pi^2 g_e^2(1 + s^2G_e^2)} (1 - g_e s^2 \omega\sigma_e d_a \mu_0 G_e) \qquad (44)$$

As expected from (39), the net normal force becomes repulsive if

$$g_e s^2 \omega\sigma_e d_a \mu_0 G_e > 1 \qquad (45)$$

By using (23), the definition of G_e, (45) becomes

$$sG_e > \frac{\tau}{\pi g_e} \qquad (46)$$

In the low slip region the normal force of SLIM is expected to be attractive, while in the high slip region it could become repulsive if condition (46) is satisfied.

In general in low-speed LIMs condition (46) is not satisfied since G_e is rather small and is decreasing when s increases, as shown above, due to secondary back iron saturation. Thus the normal force at a low speed is in general an attraction one. Note again that the primary iron losses have been neglected. They could easily be added as in rotary induction machines.

4.3.3. Numerical Results

In order to show the influence of the transverse effect and of back iron saturation and current upon LIM parameters, a numerical example is worked out here.

Initial data are pole pitch $\tau = 0.084$ m; $2a = 0.08$ m; number of back iron laminations $i = 3$; number of poles $(2p + 1) = 7$; slots per pole per phase $q = 2$; turns per phase $W_1 = 480$; airgap $g_0 = 0.012$ m; aluminum thickness $d_a = 0.006$ m; aluminum width $2c = 0.120$ m; $\sigma_a = 3.5 \times 10^7 \, (\Omega m)^{-1}$; $\sigma_i = 3.55 \times 10^6 \, (\Omega m)^{-1}$; slot depth $h_s = 0.045$ m; slot width $b_s = 0.009$ m; $\beta' = 5/6$; iron magnetizing curve of solid back iron is given in Fig. 4.10.

For given values of primary current and frequency at standstill the equivalent relative airgap, g_e/g_0, secondary conductivity, σ_e/σ_a, realistic goodness fac-

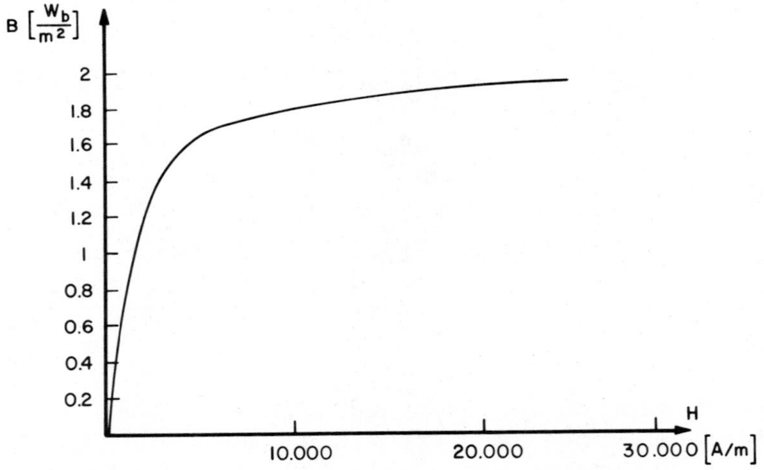

Fig. 4.10. Typical solid iron magnetization curve.

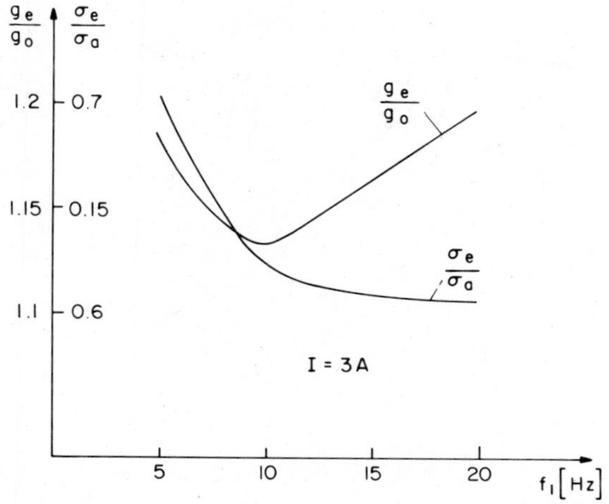

Fig. 4.11. Equivalent relative airgap and secondary conductivity for standstill performance.

tor, G_e, and thrust, $F_{\hat{x}}$, have been calculated. The numerical results are presented in Figs. 4.11 and 4.12. The influence of back iron saturation, eddy currents, and transverse effect are included in σ_e/σ_a and g_e/g_0. Running characteristics are obtained in the same manner as above. For a constant voltage supply an initial value is assigned to the primary current I_1, and thus σ_e and g_e are obtained. Finally from the equivalent circuit a new value of I_1 is obtained. After a few iterations the value of I_1 should stabilize.

Typical experimental thrust-speed and apparent input power curves of the above motor, at constant primary voltage and frequency, are given in Fig. 4.13. The critical slip is high, indicating a high-resistance secondary.

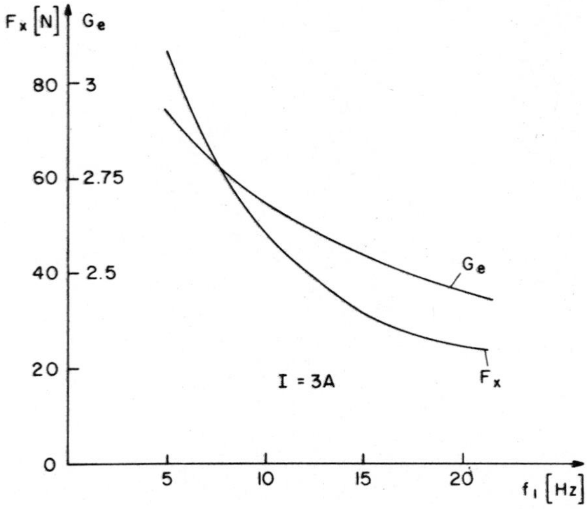

Fig. 4.12. Realistic goodness factor and thrust for standstill performance.

4.4. Design Criteria and Guidelines

In a LIM the back iron in the secondary is necessary to provide a return path for the magnetic flux. The inevitable saturation and eddy currents in the back iron of secondary render the SLIM parameters dependent on primary current and secondary frequency for a given motor as shown above. Thus any realistic design requires iterative procedures. Efficiency and power factor maximization are not the best overall design criteria any more, since both the airgap and the secondary equivalent resistance are higher than in rotary machines.

To start the design procedure we need some initial data to comply with some others obtained from past experience. The initial data considered here are: primary rated voltage, V_{1n}; rated thrust, F_{xn}; rated speed, U_n; and rated primary frequency, f_{1n}.

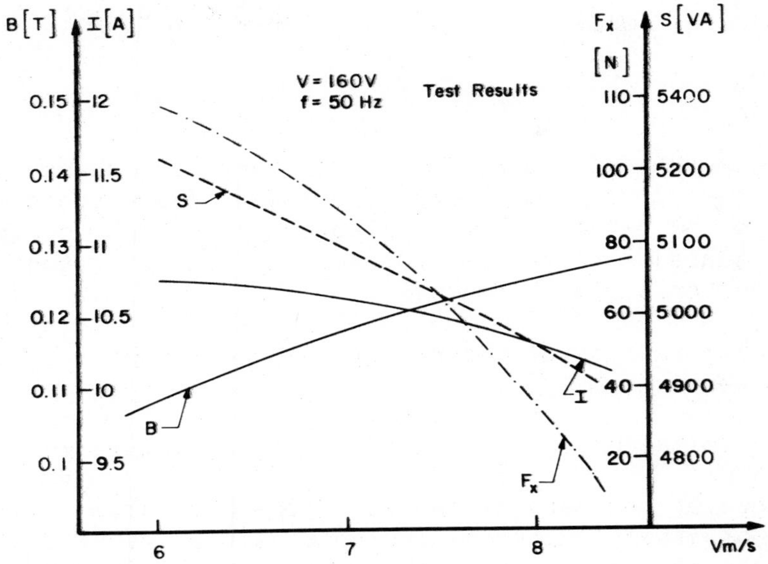

Fig. 4.13. Running test performance for f_1 = 50 Hz.

As data from past experience we take: mechanical airgap, $g_0 - d_a$ = 2 to 6 mm; airgap flux density range, B_g = 0.15 to 0.30 T; due to the high values of magnetic airgap, g_o = (2 to 12) mm; slot depth/slot width, K_{slot} = h_s/b_s = 3 to 6.

High values of K_{slot} are also due to the high values of magnetic airgap g_0 in comparison with rotary induction machines.

To avoid significant asymmetries between the phase currents the number of poles 2p (2p + 1) should be greater than 5.

The secondary overhangs should not exceed τ/π.

It is practical to fix the overhang length at the value $(c - a) = \tau/\pi$.

4.4.1. Design Criteria

Though difficult, it is necessary to choose some criteria toward an optimal design of low-speed LIMs. Due to the rather low power factor and high secondary resistance, a compromise between the following conflicting criteria must be made:

1. Maximization of mechanical power/input apparent power.

2. Maximization of mechanical power/primary weight.

Because of the many variables of the problem, a direct use of these criteria, though possible, requires large amounts of computation time. Thus a simpler criterion is used first: namely, maximum obtainable thrust at rated speed. The above criteria are then used for comparison between different solutions in order to choose the most appropriate one.

The maximum thrust for the rated speed requires, from (35), that

$$s_{nk} G_e = 1 \tag{47}$$

This maximum thrust corresponds to the critical slip, s_{nk}, at constant current supply. In general a low-speed LIM is supplied at a constant voltage (and frequency) and thus the critical slip at constant voltage, s_k, is given by

$$s_k = \frac{C_1 R_2'}{(R_1^2 + C_1^2 X_1^2)^{1/2}}$$

$$C_1 = 1 + \frac{X_1}{X_m} = 1 + \frac{X_1}{R_2'} G_e^{-1}; \quad X_2' = 0 \tag{48}$$

In general $s_{nk} < s_k$ and thus designing the motor at s_{nk} still provides a margin of stability when fed at a constant voltage.

4.4.2. Design Algorithm

With the initial data from (4), (26), and (35), the following equations are obtained:

$$(F_{xn})_{s_{nk}G_e = 1} = \frac{3I_1^2(W_1K_{w1})^2 12a_e}{(2\tau f_1 - U_n)d_a\tau p\sigma_e} \frac{1}{2} \tag{49}$$

$$s_{nk}G_e = \frac{2\mu_0\tau^2 f_1 s_{nk} d_a \sigma_e}{\pi g_e} = 1 \tag{50}$$

$$U_n = 2f_1(1 - s_{nk})$$

From (4), for a given airgap flux density and $s_{nk}G_e = 1$, we have

$$(B_{gn})_{s_{nk}G_e = 1} = \frac{3\sqrt{2}I_1W_1K_{w1}\mu_0}{g_e\pi p\sqrt{2}} \tag{51}$$

Also from (24), by limiting the longitudinal flux density on the solid back iron surface, we obtain

$$\delta_{iR} = \frac{\tau B_{gn}K_{pf}}{\pi B_{xin}} \tag{52}$$

Finally, by choosing the primary design current density, J_{co}, q slot per pole per phase, slot aspect ratio, K_{slot}, and B_{gn} (and K_d) in (30), we obtain

$$W_1 I_{1n} = C_s p\tau^2; \qquad C_s = \frac{1}{9q} K_d^2 K_{slot} K_f J_{co} \tag{53}$$

with

$$K_d = \frac{3b_s q}{\tau} = 1 - \frac{B_{gn}}{B_{tn}}$$

where B_{tn} is the design flux density in primary teeth. The unknowns of the problem are a_e/τ, τ, p, $(s_{nk}\omega)$, and d_a. Only iterative procedures can be used to solve this problem. However, we could use a_e/τ as a parameter and consequently from (52) with (18) we find

$$\tau^2 \omega s_{nk} = \frac{K_{ti}\pi^2}{\mu_0 \sigma_i} \sqrt{\left(\frac{B_{xin}}{B_{gn}}\right)^4 - 4} = C_\tau \tag{54}$$

The rated thrust (49) with (41) and (53) may be written as

$$F_{xn} = 6C_s B_{gn} p\tau^2 a_e K_{w1} \tag{55}$$

The primary frequency may also be a parameter. Then a new parameter has to be introduced:

$$p = 2, 3, 4, 5, 6, 7$$

Thus the pole pitch becomes

$$\tau = 3\sqrt{\frac{F_x}{6C_s B_{gn} p(a_e/\tau)}} \tag{56}$$

Now (ωs_{nk}) is obtained from (54). By considering (53), (49), (50), and (51) we get

$$\sigma_e d_a = \frac{3C_s \tau\pi}{\omega s_{nk} B_g} \qquad (57)$$

and

$$\sigma_a' d_a = \left(\frac{3\tau C_s \pi}{\omega s_{nk} B_g} - \frac{\sigma_i \delta_i}{K_{ti}}\right) K_{ta} \qquad (58)$$

Unfortunately d_a enters also in the expression of G_{ep}, that is in (5) through (10), from which we calculate K_{ta} (the aluminum overhangs are considered $c - a_e = \tau/\pi$). Thus only by iteration, $\sigma_a' d_a$ may be calculated from (58) and (5) through (10). Finally, knowing $\sigma_a' d_a$ from (5) through (10), we can find iteratively d_a, the aluminum sheet thickness. The solutions for d_a should still provide the mechanical airgap ($d_a < g_e$).

If the primary frequency ω is known, then for a given rated speed and $s_{nk}\omega$, from (54), we get directly the corresponding pole pitch:

$$\tau = \frac{\pi U_n}{\omega - \omega s_{nk}}; \qquad \frac{\omega\tau}{\pi} = U_n + \frac{C_\tau}{\pi\tau} \qquad (59)$$

For this case there is no need to take the number of pole pairs, p, as a parameter since, from (55), p can be calculated and then adjusted by changing C_s in (55), that is the slot aspect ratio, K_{slot}, in (53). Thus for rated speed all the terms entering the equivalent parameters X_m, (25), R_2' (26), R_1 (28), X_1 (31) can be calculated, since $W_1 I_{1n}$ is known from (53), except for W_1 (turns per phase).

Now if the primary voltage is fixed, W_1 is really obtainable from the equivalent circuit (Fig. 4.9), and is given by

$$W_1 = V_{1f}/\{(W_1 I_{1n})[(K_{R1} + \frac{K_{R'2}}{s_{nk}(1 + 1/(s_{nk}^2 G_e^2)})^2 + K_{\sigma 1} + \frac{K_m}{1 + s_n^2 G_e^2})^2]^{1/2}\} \quad (60)$$

All LIM performance for constant voltage supply can now be calculated iteratively if for each value of I_1 and $s\omega$ the saturation and eddy currents in the solid back iron of the secondary are evaluated as indicated in Section 4.2. It is evident that the degree of saturation (B_{xi}) and the depth of field penetration in the back iron (δ_i) vary for a given motor with I_1 and $s\omega$. Also the airgap flux density, B_g, varies with I_1 and $s\omega$. Up to this point, the ratio a_e/τ is still a parameter, and if the primary frequency is not fixed, the number of pole pairs is also a parameter. To choose the most suitable solution for this purpose, two more criteria are used. Thus we have to determine the apparent power for rated speed and also the motor cost (or weight). Because in some applications the cost of the secondary is an important portion of overall drive costs, it should also be accounted for. The apparent power, S_n (for rated speed), is

$$S_n = 3V_{1f}I_{1n} = 3(W_1 I_{1n})^2$$

$$[(K_{R_1} + \frac{K_{R'_2}}{s_{nk}[1 + 1/(s_{nk}G_e)^2]})^2$$

$$+ (K_{\sigma 1} + \frac{K_m}{1 + s_{nk}^2 G_e^2})^2]^{1/2} \qquad (61)$$

with $s_{nk}G_e = 1$.

The cost of the primary, C_p, contains mainly the iron core and the copper costs. Hence,

$$C_p = \frac{6W_1 I_{1n}(2a_e + K_1\tau)}{J_{co}} \gamma_{co} C_{co} +$$

$$[2p\tau(1 - K_d)2aK_{slot} \frac{K_d\tau}{3q} + 3\delta_i 2p\tau 2a]\gamma_i C_{ii} \qquad (62)$$

where γ_{co}, γ_i and C_{co}, C_{ii} are the specific weights and, respectively, specific prices of copper and core laminations.

The secondary materials costs are

$$C_{sec} = [(2a + \frac{2\tau}{\pi})\gamma_{Al} c_{Al}$$

$$+ 3\delta_i (2a(2a_e + \frac{2\tau}{\pi})\gamma_i C_i]L_1 \qquad (63)$$

with γ_{Al} and c_{Al} the aluminum specific weight and specific cost, respectively and L_1 is the total LIM excursion.

When $L_1 >> 2p\tau$ the secondary cost may prove to be a significant part of the total cost of active materials, C_t; that is,

$$C_t = C_p + C_{sec} \tag{64}$$

and (64) should be accounted for.

Thus, a high ratio a_e/τ leads to a smaller length of primary but the secondary gets wider and thus the cost of the secondary increases. A minimum cost, C_t, is expected to be obtained at a certain value of a_e/τ. The choice between the two criteria is a difficult task left to the designer.

A numerical example follows.

4.4.3. A Numerical Design Example

Let us design a low speed SLIM from the following data: LIM excursion length L_1 = 50 m; rated thrust F_{xn} = 1500 N; rated speed U_n = 6 m/sec; primary frequency f_1 = 50 Hz; primary phase voltage V_1 = 220 V.

Additionally the following data from past experience are admitted: airgap flux density at rated speed B_{gn} = 0.3 T; primary tooth design flux density B_{tn} = 1.6 T; back iron tangential flux density B_{xin} = 1.7 T; slot aspect ratio K_{slot} = 4 to 6; slot fill factor K_f = 0.6; design current density J_{co} = 4 A/mm^2; .nr 99 10 chording factor β = 1 (no chording); mechanical airgap g_m = 0.002 m.

Following the procedure described above we obtain the main unknowns: $s_{nk}\omega$, τ, δ_i, d_a as functions of a_e/τ

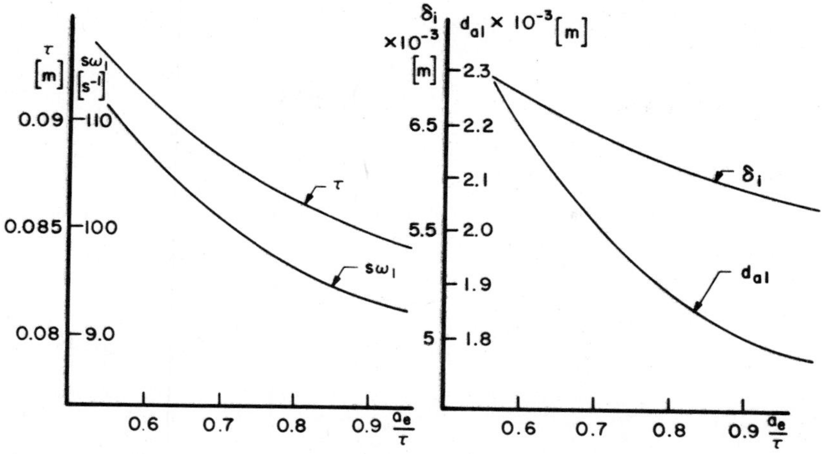

Fig. 4.14. (a) Dependance of $s_{nk}\omega$ and τ on a_e/τ,
(b) dependance of δ_i and d_{1a} on a_e/τ.

as shown in Fig. 4.14a and b. Also the apparent power S_{1n}, specific thrust $f_x = (F_x/2ap\tau)$, primary plus secondary costs in equivalent units of solid iron specific price, c_i (with $c_{ii}/c_i = 1.5$; $c_{co}/c_i = 10$ and $c_{Al}/c_i = 3$) are shown in Fig. 4.15a and b.

Considering the two criteria (apparent power and overall costs) we finally choose: $a_e/\tau = 0.9$; $s_{nk}\omega = 93\ \text{sec}^{-1}$; $\tau = 0.085$ m; $d_a = 1.8 \times 10^{-3}$ m; $W_1 = 384$ turns/phase; $I_{1n} = 32.757$ A; $S_1 = 21.9198$ kVA.

The entire thrust-speed and current-speed curves can be obtained by simply programming the computation of parameters on a computer. It would be interesting to lower the primary frequency and look for its consequences in apparent power. It should be noted that for a DSLIM with aluminum sheet secondary the design procedure simplifies notably because of the absence of solid back iron in secondary. However the algorithm remains the same in principle and thus is not detailed here again.

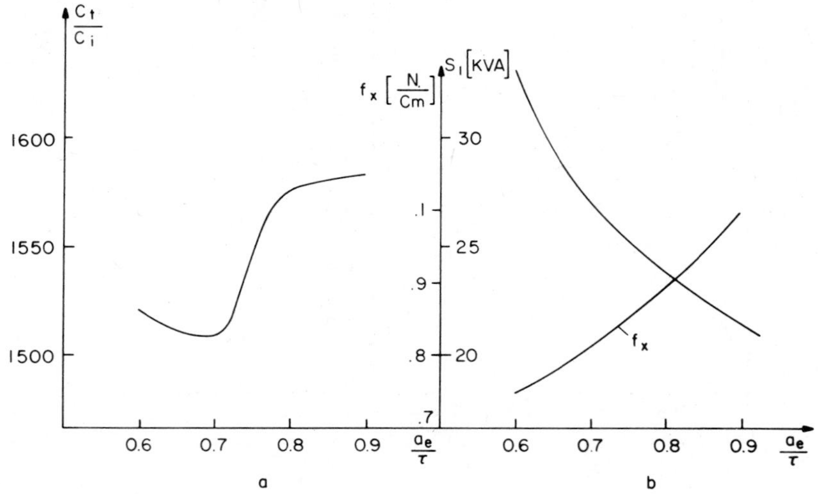

Fig. 4.15. (a) Apparent power S_1(VA), specific thrust f_x(N/cm^2), (b) motor costs C_t/C_i in in relative units against a_c/τ.

4.5. Short Primary-Short Secondary Flat LIMs

In ore transportation or material handling, LIM-activated cars or car-conveyors have been proposed. In this case the short primaries are fixed to the ground and the short secondaries are placed horizontally on the lower parts of the cars (Fig. 4.16). The secondary could be made of an aluminum (copper) sheet fixed to a solid back iron plate (Fig. 4.17a) or a copper (aluminum) ladder placed in a laminated core as in rotary induction motors (Fig. 4.17b).

Whereas the first solution is less expensive, the second is more energy efficient and its overheating limitations are less severe. The primary is utilized for a short duration and the secondary heating-cooling aspects

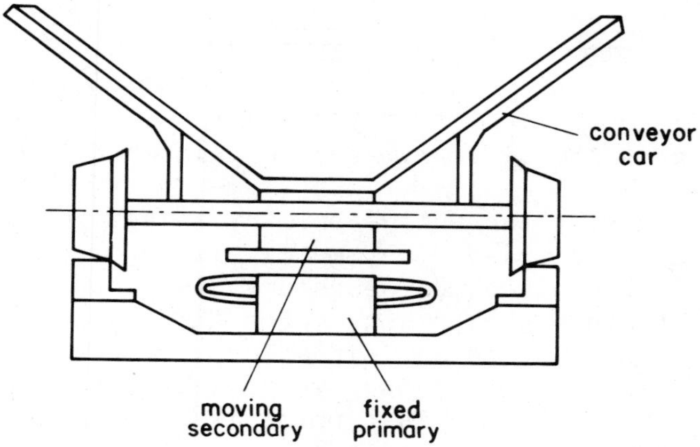

Fig. 4.16. Short-primary-short-secondary LIM.

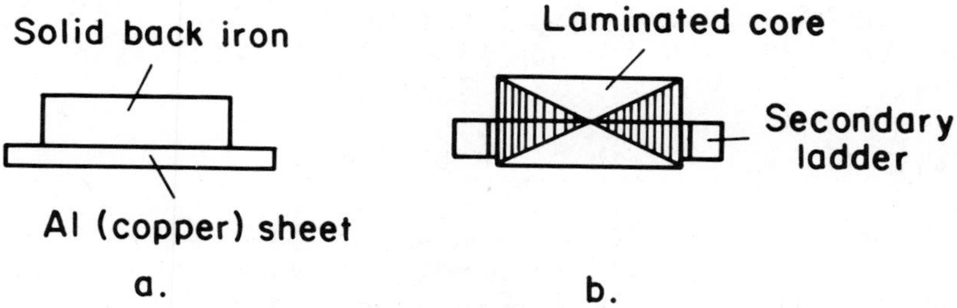

Fig. 4.17. Two possible secondary structures.

become predominant. The speed range is in general between 1 and 10 m/sec. Thus, it seems appropriate to choose the ladder laminated secondary solution to obtain an energy-efficient drive. Between two secondaries placed on adjacent cars there is a considerable distance. Hence, the entire primary length is not active when two adjacent secondaries are placed above the

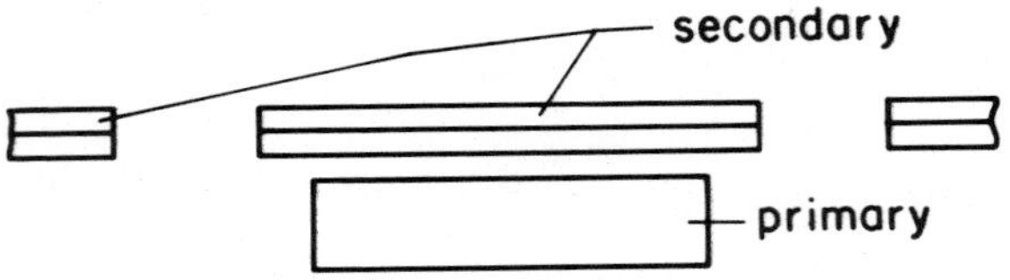

Fig. 4.18. Primary-secondary interactions.

primary, whereas the entire primary is active when one single car is placed symmetrically above the primary (Fig. 4.18). It becomes necessary to investigate the magnetic field distribution, motor thrust, power factor, and efficiency in these situations. Because of the low-speed operating range we neglect the longitudinal effect. The first and last primaries placed under the train cars are in a different situation in the sense that at the beginning they have no secondary above while being supplied with three phase currents. The train is considered sufficiently long to neglect this aspect. Thus each primary in time experiences a time-dependent presence of secondaries. The LIM performance is time-dependent and an average value characterizes the drive overall performance.

4.5.1. Magnetic Field Distribution

The secondary may be longer or shorter than the primary. We consider here the secondary that is slightly longer than the primary such that at most two secondaries are placed above one primary (Fig. 4.19). To solve the field problem the ladder secondary is replaced by an equivalent, infinitely thin, sheet whose electrical conductivity, $\underline{\sigma}_{es}$, is considered as a complex number to account for ladder leakage reactance. The equivalent airgap remains equal to the real airgap. A series connect of primary coils is considered. Thus in the zones of secondaries (active zone) the airgap field may be considered to exist along OZ only (the airgap being small). Between the secondaries the airgap field is much smaller and has two components, one along OZ and

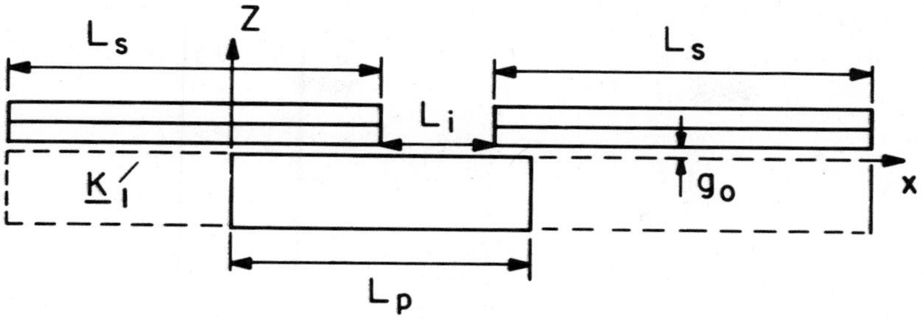

Fig. 4.19. Primary-secondary relative position.

the other along OX. The primary current sheet, J_s, when expressed in secondary coordinates, may be written as

$$\underline{J}_s = J_{1m} \exp[j(\frac{\pi x_1}{\tau} - s\omega t - \frac{L_{1s}\pi}{\tau})] \qquad (65)$$

Active zones

Neglecting the longitudinal end effect due to speed U, the airgap flux density, B_g, in the active zone, in secondary coordinates, is given by

$$\frac{\partial \underline{B}_g}{\partial t} = j\omega s \underline{B}_g \qquad (66)$$

Ampere's and Faraday's laws yield

$$\frac{g}{\mu_0} \frac{\partial^2 B_g}{\partial x^2} + j\omega s d_e \sigma_{es} B_g$$

$$= j \frac{\pi}{\tau} J_{1m} \exp[j(\frac{\pi}{\tau} x_1 - \frac{L_{1s}\pi}{\tau})] \qquad (67)$$

where d_e is the thickness of an equivalent sheet replacing the ladder secondary. The complete solution to (67) is

$$\underline{B}_g = \underline{B}_0 \exp[j(\frac{x_1 \pi}{\tau} - \frac{L_{1s} \pi}{\tau})]$$

$$+ \underline{B}_1 \exp[\underline{\lambda}_1 x] + B_2 \exp[\underline{\lambda}_2 x] \qquad (68)$$

where

$$\underline{\lambda}_{1,2} = \sqrt{- \frac{j\omega s d_e \sigma_{es} \mu_0}{g_e}} = \sqrt{- \frac{js\underline{G}_e \pi^2}{\tau^2}} \qquad (69)$$

In (69), σ_{es} represents a complex fictitious conductivity of the ladder secondary accounting for the secondary leakage, and σ_{es} is given by

$$\underline{\sigma}_{es} = \sigma_e[\frac{1}{1 + (X_2'/R_2')^2} - j \frac{1}{R_2'/X_2' + X_2'/R_2'}] \qquad (70)$$

The goodness factor, $\underline{G}_e$, in this case is a complex number such that

$$\underline{G}_e = \frac{X_m}{R_2' + jX_2'} \qquad (71)$$

From (67) and (68) $\underline{B}_0$ can be written as

$$B_0 = j \frac{\tau}{\pi g_e} \frac{\mu_0 J_{1m}}{1 + js\underline{G}_e} \qquad (72)$$

with

$$\underline{G}_e = \frac{\omega\tau^2}{\pi^2} \mu_0 \underline{\sigma}_{es} (\frac{d_e}{g_e}) \tag{73}$$

Evidently (71) and (73) are equivalent.

Intersecondary spacing

Here the Laplace equation holds and the airgap field has two components, $\underline{B}_z$ and $\underline{B}_x$. The pertinent equations are

$$\frac{\partial^2 \underline{B}_z}{\partial x^2} + \frac{\partial^2 \underline{B}_z}{\partial z^2} = 0;$$

$$\frac{\partial \underline{B}_z}{\partial z} + \frac{\partial \underline{B}_x}{\partial x} = 0; \qquad (\underline{B}_x)_{z=0} = \underline{J}_s \mu_0 \tag{74}$$

Finally $\underline{B}_z$ is

$$\underline{B}_z = j J_{1m} \mu_0 \exp[j(\frac{x_1 \pi}{\tau}$$

$$- s\omega t - \frac{L_{1s}\pi}{\tau})] \exp[- \frac{z\pi}{\tau}] \tag{75}$$

Outside the primary

As shown in Fig. 4.19 a fictitious primary sheet, K_1, flowing into an ideally reactive material accounts for the real airgap in this zone. Consequently,

$$g \frac{\partial \underline{H}_{i,0}}{\partial x} = \underline{K}_{i,0} + \underline{J}_2 d_e \tag{76}$$

$$\underline{K}_1 = j\sigma_r\underline{E}_1; \quad \frac{\partial \underline{E}_1}{\partial x} = -j\omega\mu_0\underline{H}_{i,0} \tag{77}$$

$$\frac{\partial \underline{J}_2}{\partial x} = -j\omega s\underline{\sigma}_{es}\mu_0 H_{i,0} \tag{78}$$

The solution to (76) through (78) is

$$\underline{H}_{i,0} = \underline{A}_{i,0} e^{\underline{\gamma}_{i,0}(x_1 + x_{1,0})};$$

$$x_1 = L_{1s}; \quad x_0 = L_{1s} + L_p \tag{79}$$

With

$$\underline{\gamma}_{i,0} = \sqrt{-js\underline{G}_e(\frac{\pi}{\tau})^2 + \frac{\sigma_r\omega\mu_0}{g}};$$

$$G_e = G_{er} - jG_{ei} \tag{80}$$

or

$$\underline{\gamma}_{i,0} = +-(\gamma_r + j\gamma_i) \tag{81}$$

with

$$\gamma_r = \sqrt{a + \frac{\sqrt{a^2 + 4b^2}}{2}}; \quad \gamma_i = \frac{b}{\gamma_r} \tag{82}$$

$$a = \frac{\sigma_r\omega\mu_0}{g} - sG_{ei}(\frac{\pi}{\tau})^2; \quad b = -sG_{er}(\frac{\pi}{\tau})^2 \tag{83}$$

The value of σ_r is that which in the absence of the conducting secondary makes the field distribution fit the results obtained by tests.

In the entry and exit zones we have the same situation since the longitudinal (speed-dependent) effect has been neglected:

$$\frac{\sigma_r \omega \mu_0}{g} = \frac{0.5}{g^2} \tag{84}$$

We now have four integration constants for the active zones and two for the entry and exit zones. The boundary conditions are as follows.

The resultant flux density on the primary surface is continuous for $x_1 = L_{1s}, L_s, L_s + L_i, L_{1s} + L_p$, and the current density integrals are zero along the two secondaries:

$$\int_0^{L_{1s}} \underline{J}_i \, dx_1 + \int_{L_{1s}}^{L_s} \underline{J}_2 \, dx_1 = 0 \tag{85}$$

$$\int_{L_{1s}-L_i}^{L_{1s}+L_p} \underline{J}_2 \, dx_1 + \int_{L_{1s}+L_p}^{2L_s+L_i} \underline{J}_0 \, dx_1 = 0 \tag{86}$$

4.5.2. Thrust and Power

The thrust for both active zones is

$$F_x = 2a \, \text{Real}\Big[\int_{L_{1s}}^{L_s} (\underline{J}_2^* \underline{H}_i) \, dx_1$$

$$+ \int_{L_s+L_i}^{L_{1s}+L_p} (\underline{J}_2^* \underline{H}_i)\, dx_1] \tag{87}$$

Secondary power loss, P_2 (as in conventional rotary induction motors), is

$$P_2 = sF_x 2\tau f_1 \tag{88}$$

The airgap magnetic energy, W_m, is

$$W_m = \frac{2ag_0}{2\mu_0}[\int_{L_{1s}}^{L_s} |\mu_0 \underline{H}_i|^2\, dx_1$$

$$+ \int_{L_s+L_i}^{L_{1s}+L_p} |\mu_0 \underline{H}_i|^2\, dx_1] \tag{89}$$

Thus the reactive power, Q_m, corresponding to airgap secondary (the secondary leakage is included in the equivalent secondary complex conductivity, σ_{es}) is

$$Q_m = 2\omega W_m$$

We include the primary winding losses P_{co1} and leakage Q_1

$$P_{co1} = 3R_1 I_1^2; \quad Q_1 = 3X_1 I_1^2 \tag{90}$$

as well as the iron losses, P_i. Thus the total power balance is given by well as the iron losses, P_i. The total power balance is given by

$$3V_1 I_1$$

$$= \sqrt{(F_x v_s + P_{co1} + P_i)^2 + (Q_m + Q_1)^2} \qquad (91)$$

The normal attraction force, F_n, between primary and secondary is

$$F_n = \frac{W_m}{g_0} \qquad (92)$$

Thus the performance, including efficiency and power factor, can be calculated.

4.6. Tubular LIMs

The tubular LIM does not have end connections in the primary winding coils and in the secondary current paths. (There are, however, transverse flux tubular LIMs where these effects do occur.) Furthermore, the zero net normal force (as in rotary machines) facilitates easy sliding on linear bearings. However the stroke length could not go 1 to 1.5 m beyond the primary length due to mechanical alignment problems. For a specific low-speed or standstill application the abovementioned assets and limitations should be considered carefully before a flat or a tubular LIM configuration is used.

4.6.1. Some Further Construction Guidelines

When building a tubular LIM the primary and secondary magnetic and electric circuits should be considered first. Due to the tubular aspect, the primary could use longitudinal laminations embedded in four to six separate cores framed together after the primary winding has been inserted in their slots (Fig. 4.20) or made of transverse laminations (Fig. 4.21).

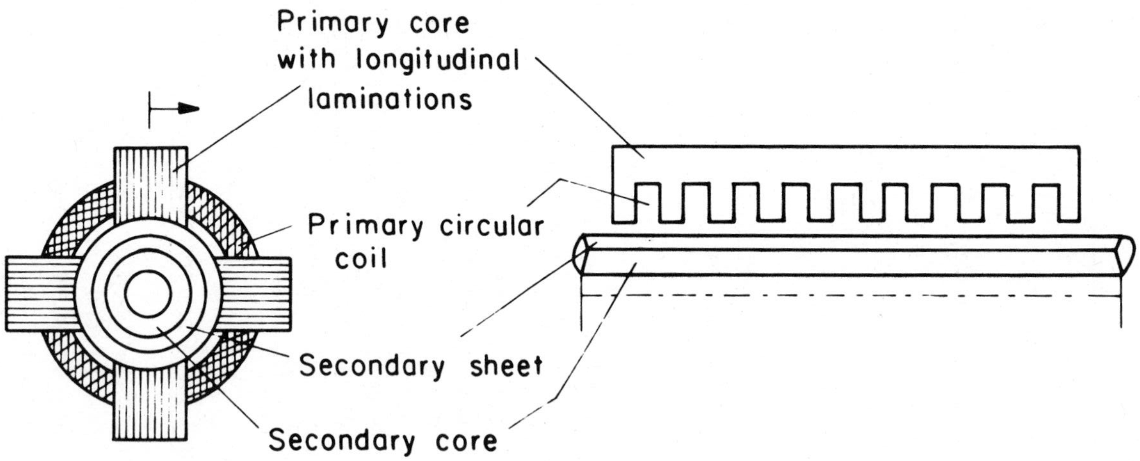

Fig. 4.20. Tubular LIM with longitudinal laminations.

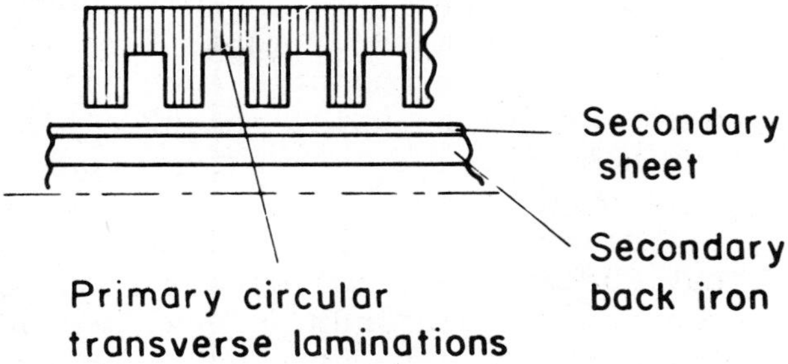

Fig. 4.21. Tubular LIM with transverse (circular) laminations.

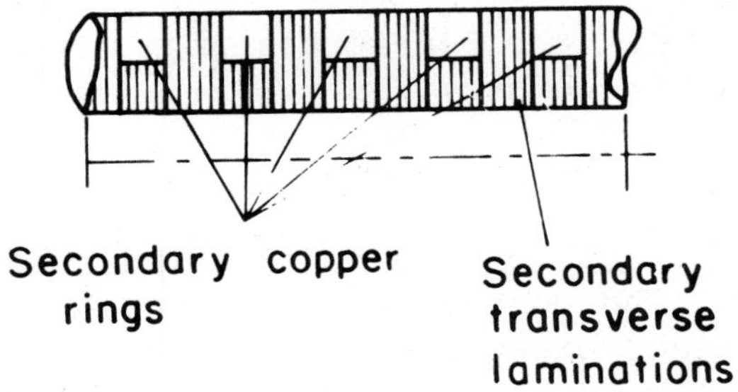

Fig. 4.22. Tubular secondary with transvese laminations.

In the first case the precision in aligning the four to six cores along a circumference and the difficulties in inserting the primary winding in the slots are the main drawbacks.

In the second solution (transverse laminations), the inevitable interspaces occupied by lamination insulation between the outer laminations (behind the slots) add to the motor magnetic airgap. As long as the pole pitch is not greater than 0.1 to 0.15 m, the influence of this phenomenon is not expected to adversely affect the performance. However, the higher the LIM magnetic airgap (mechanical airgap + conductive sheet thickness), the smaller is the effect of interspaces between transverse laminations. The primary winding circular coils may be connected (as usual in flat LIMs) to form an even or an odd number of poles. Allowing a high level of saturation in the magnetic core, it seems that an even number of poles, while making a better use of LIM cores, does not require an extra core thickness since the pulsating component of primary airgap field is rather small. The secondary could be made of copper or aluminum sheet backed by a solid back iron rod or of a series of copper or aluminum rings separated by transverse laminations all mounted on a central iron rod (Fig. 4.22). The transverse laminations may be replaced by solid iron rings, but in this case full advantage of conducting rings cannot be taken, due to severe skin effect in solid iron. Finally the secondary should be coated with a thin layer (0.1 to 0.2 mm) of tough material on which the linear bearings may slide without wearing out the secondary.

4.6.2. Specific Phenomena and Lumped Parameters

Because of low speeds, the longitudinal end effect is neglected. The conducting sheet on solid iron secondary is the least expensive solution and has been used frequently for short duty purposes. However, the overheating, and consequently the thermal deformation, of aluminum (or copper) sheet in any short-stroke application adds to low energy conversion performance, resulting in an uneconomical solution. The treatment of this type of motor involves the skin effect and saturation in the secondary and also the curvature influence. A rather complete treatment of this problem has been carried out for linear cylindrical induction magnetohydrodynamic (MHD) pumps. But we can use the previous analysis carried out for flat LIMs if:

1. The transverse effect is neglected ($K_{tr} = K_{tm} = K_{ti} = 1$).

2. The width of the machine $2a_e = D\pi$ (D = external diameter of secondary).

3. The effect of secondary curvature is neglected.

4. The primary phase resistance, R_1, and leakage reactance, X_1, should account for the circular aspect of the primary winding coils:

$$R_{1t} = \rho_{co} \frac{\pi D_{av} J_{co} W_1^2}{I_1 W_1} \tag{93}$$

$$X_{1c} = 2\mu_0 \omega \frac{\pi D_{av}}{pq} (\lambda_s + \lambda_d) W_1^2 \tag{94}$$

5. The slot and differential specific permeance expression, (27), is invalid.

It should be noted that an average diameter has been adopted for the primary coils length computation. Also the leakage reactance expression implies that the primary coils are totally included in the iron structure. This condition is only partially fulfilled for the motor with longitudinal laminations and entirely fulfilled for the motor with transverse laminations. The average diameter is chosen to ascertain that the total length of the turns of a coil in one slot is kept the same as in reality, and thus D_{av} depends mainly on the external diameter of the secondary and the slot depth. The magnetizing reactance, X_m, and secondary resistance, R_2', can be expressed as

$$X_m = \frac{6\mu_0 \omega}{\pi^2} (K_{w1} W_1)^2 \frac{\tau \pi D_0}{p g_e} \tag{95}$$

$$R_2' = \frac{6\pi D (K_{w1} W_1)^2}{p \tau d_a \sigma} \tag{96}$$

Observing this particular aspect, the performance computation procedure is similar to that developed for flat LIMs. Tubular LIMs with transverse laminations and secondary conducting rings are treated next.

4.6.3. Secondary with Conducting Rings

The expressions for the primary parameters R_1 and X_1 remain the same as in (93) and (94). Expressions for the secondary parameters should be developed especially for this case. First, it should be noted that this secondary resembles the cage-type rotor of rotary induction machines and that the secondary leakage reactance should be accounted for. Expressions of secondary resistance, R_2', and leakage 10 reactance, X_2', may be derived from the formulas used for rotary induction motors:

$$X_2' = 2\mu_0\omega(\pi D_{avs})12\frac{(W_1K_{w1})^2}{N_{s2}}(\lambda_{s2} + \lambda_{d2}) \tag{97}$$

$$R_2' = \rho_{cs}\frac{\pi D_{avs}}{S_{ring}}12\frac{(W_1K_{w1})^2}{N_{s2}} \tag{98}$$

where N_{s2} = number of slots of secondary per primary length; D_{avs} = average diameter of conducting rings and S_{ring} = cross section of rings. Again to avoid high parasitic thrusts the number of primary and secondary slots obeys the rules valid for rotary induction machine design. Now the equivalent circuit of rotary induction machines may be used and the thrust and performance can be calculated accordingly. The skin effect in primary winding, especially the secondary rings, may be accounted for as in rotary induction machines.

4.6.4. Design Aspects of Tubular LIMs

The tubular LIM is used for relatively short stroke applications and thus, at very low speeds, almost at standstill. Again two main criteria should be used: an energy conversion one and one that accounts for the overall cost of the drive. The cost criterion remains the same as for flat LIMs but the expressions to evaluate it are slightly altered to account for the tubular structure. Given the very low speed, the energy conversion criterion should now aim to obtain the maximum thrust at standstill per unit apparent power. The primary frequency can be chosen below the industrial one to obtain good performance. The costs of cycloconverters or other electronics such as dc link inverters used as frequency changers should also be included in the overall cost of the drive. Finally, when computing different solutions the energy and maintenance costs should also be accounted for in a complete design study.

However, as a starting point, we choose the criterion of maximum thrust at standstill. The presence of secondary leakage reactance warrants a new definition of an equivalent goodness factor. We should separate the thrust-producing (real) part of the equivalent impedance made of jX_m and $R_2'/s + jX_2'$ in the standard equivalent circuit (Fig. 4.23). Thus,

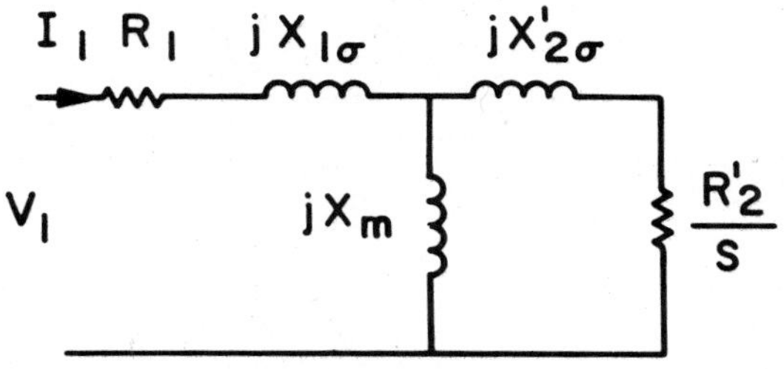

Fig. 4.23. Equivalent circuit of tubular LIM with secondary conducting rings.

$$R_e = \frac{X_m^2 R_2'/s}{(R_2'/s)^2 + (X_m + X_{2\sigma}')^2}$$

or

$$R_e = \frac{R_2'}{s[1/(sG_e)^2 + (1 + X_{2\sigma}'/X_m)^2]} \qquad (99)$$

instead of

$$R_{e0} = \frac{X_m^2 R_2'/s}{(R_2'/s)^2 + X_m^2} = \frac{R_2'}{s[1/(sG_e)^2 + 1]} \qquad (100)$$

which holds when $X_2' = 0$

Keeping the expression of the goodness factor, G_e,

$$G_e = \frac{X_m}{R_2'} \qquad (101)$$

the maximum thrust is obtained for

$$(sG_e)_{opt} = \frac{1}{(1 + X_2'/X_m)} \qquad (102)$$

The maximum thrust should be reached for standstill (s = 1) and thus

$$(G_e)_{opt} = \frac{1}{1 + (X_{2\sigma}'/X_m)} \qquad (103)$$

Evidently for the sheet on iron secondary $X_2' \simeq 0$ and

$(G_e)_{opt,} = 1$. When initiating the design procedure the ratio X_2/X_m is not known and can only be assigned a value and then adjusted iteratively.

Now the design procedure developed for flat LIMs can be used here. Since the secondary is active almost all the time, its heating problems become severe. This aspect could be accounted for by limiting the current density in the secondary.

4.6.5. Some Numerical Results

Consider a tubular LIM with transverse laminations in primary and secondary and copper secondary rings with the following data: primary internal diameter $D_i = 0.15$ m; primary external diameter $D_e = 0.27$ m; primary slot depth $d_s = 0.04$ m; pole pitch $\tau = 0.06$ m; number of poles $2p = 8$; $q = 2$ slots per pole per phase; slots in secondary per primary length $N_{s2} = 72$; airgap $g_0 = 3 \times 10^{-3}$ m; primary slots width $b_{s1} = 1.25 \times 10^{-2}$ m; secondary slot pitch $\tau_{s2} = 10^{-2}$ m; secondary slot width $b_{s2} = 0.6 \times 10^{-2}$ m; secondary slot depth $c_{s2} = 0.4 \times 10^{-2}$ m; $J_{co} = 3 \times 10^6$ A/m^2; $K_f = 0.6$;

Let us calculate the primary frequency at which the maximum thrust condition at standstill is fulfilled. Thus first we calculate the secondary leakage reactance $X'_{2\sigma}$, at standstill (from (88))

$$X'_{2\sigma} = 2\mu_0 \pi D_{avs} \frac{12}{N_{s2}} (\lambda_{s2} + \lambda_{d2})(W_1 K_{w1})^2 \qquad (104)$$

with

$$\lambda_{s2} = \frac{h_{s2}}{3b_{s2}}; \quad \lambda_{d2} = 5g_0/b_{s2}\text{Over}5 + 4g_0/b_{s2}$$

From the above data

$$X'_{2\sigma} = 0.985 \times 10^{-6} f_1 W_1^2 K_{w1}^2 \tag{105}$$

The secondary resistance referred to the primary, from (103), is given by

$$R'_2 = \rho_{cs} \frac{\pi D_{avs} 12}{h_{se} b_{se}} W_1^2 K_{w1}^2 = 0.876 \times 10^{-4} W_1^2 K_{w1}^2 \tag{106}$$

The magnetizing reactance from (95) becomes

$$X_m = \frac{6\mu_0 \omega}{\pi^2} (W_1 K_{w1})^2 \frac{\tau \pi D_i}{p g_0 K_c} \simeq 10^{-5} f_1 W_1^2 K_{w1}^2 \tag{107}$$

From (105) through (107), the maximum thrust condition at standstill becomes

$$G_e = \frac{X_m}{R'_2} = \frac{1}{1 + (X'_{2\sigma}/R'_2)};$$

$$0.114 f_1 = \frac{1}{1 + (1.12 \times 10^{-2} f_1)} \tag{108}$$

The primary frequency f_1 is found from (108):

$$f_1 = 8.046 \text{ Hz} \tag{109}$$

In order to calculate the performance, we should know the primary resistance, R_1, and leakage reactance X_1. They are found as follows:

$$R_1 = \rho_{co} \frac{\pi D_{av}}{(I_1 W_1)} J_{co} W_1^2 \tag{110}$$

$$I_1W_1 = pq2n_bI_1 = pqh_sb_sK_fJ_{co}$$

$$= 4 \times 1.4 \times 1.2 \times 10^{-4} \times 0.6 \times 3 \times 10^6 = 3456\text{At} \qquad (111)$$

Hence,

$$R_1 = 2 \times 10^{-8} \times \frac{\pi \times 0.19}{3456} \times 3 \times 10^6 \times W_1^2$$

$$= 1.035 \times 10^{-5} W_1^2 \qquad (112)$$

$$X_1 = 2\mu_0\omega \frac{\pi D_{av}}{pq} (\lambda_{s1} + \lambda_d) W_1^2 = 2.483 \times 10^{-5} W_1^2 \qquad (113)$$

Now the voltage, and consequently the number of turns per phase, is one choice. The apparent power is not dependent on the voltage since I_1W_1 is already fixed. Thus the apparent power found from the equivalent circuit with $K_{w1} = 0.96$ is

$$S_1 = 3I_1^2[R_1 + jX_1 + \frac{jX_{1m}(R_2' + jX_2')}{jX_m + R_2' + jX_2'}] \qquad (114)$$

For the above data, we obtain

$$S_1 = 2970 \text{ VA} \qquad (115)$$

Now the electromagnetic power, P_e, is given by

$$P_e = F_{xs}U_s = F_{xs}2\tau f_1 = 1311.4 \text{ W} \qquad (116)$$

Thus the thrust at standstill is

$$F_{xs} = 1358.27 \text{ N} \tag{117}$$

The power factor at standstill is $\cos \phi = 0.566$.

It is to be observed that, because of the good quality of the secondary, moderate airgap, and low frequency, a high thrust at standstill is obtained for a moderate apparent power, that is 0.457 N/VA. The above performance is quite satisfactory for short-stroke applications.

References

1. E. R. Laithwaite, Induction Machines for Special Purposes, George Newness, London, 1966.

2. S. A. Nasar and I. Boldea, Linear-Motion Electric Machines, Wiley-Interscience, New York, 1976.

3. P. K. Budig, A.C. Linear Motors (in German), VEB VERLAG Technik, Berlin, 1978.

4. M. Poloujadoff, Theory of Linear Induction Machines, Clarendon Press, Oxford England, 1980.

5. H. Bolton, "Transverse edge effect in sheet rotor induction motors," Proc. IEE, Vol. 116, No. 5, 1969, pp. 723-731.

6. N. Umezu and S. Nonaka, "Characteristics of low speed linear induction machines," Elec. Eng. Jap., Vol. 97, No. 1, 1977, p. 59.

7. G. Rojat, A. Nicholas, and A. Foggia, "Analysis and realization of an inverter-fed linear induction motor for material handling system," International Conference on Electrical Machines, Brussels, Belgium, 1979.

8. E. R. Laithwaite, D. Tipping, and D. E. Hesmondhalgh, "The application of induction motors to conveyors," Proc. IEE, Vol. 107A, 1960, pp. 284-297.

9. E. R. Laithwaite and S. A. Nasar, "Linear-motion electrical machines," Proc. IEEE, Vol. 58, No. 4, 1970, pp. 531-542.

10. M. A. Saleh, M. F. Sakr, S. M. Ali, and A. Fachim, "Electric tubular actuator with composite rotor," L3/6, International Conference on Electric Machines, Brussels, Belgium, 1979.

11. M. A. Saleh, S. M. Ali, and A. A. Fahim, "Numerical analysis of tubular induction actuator with composite rotor," L3/7, International Conference on Electric Machines, Brussels, Belgium, 1979.

CHAPTER 5

LINEAR MOTION MAGNETO-HYDRODYNAMIC MACHINES

By replacing the solid conducting secondary of linear motion induction and dc homopolar motors with an ionized gas (or plasma) or a liquid metal, we obtain a linear motion magneto-hydrodynamic (MHD) machine. The linear induction and dc homopolar conduction MHD machines are used in metallurgy and nuclear plants as liquid-metal pumps, whereas the homopolar dc MHD machines with plasma as the working fluid are intended as high power topping generators for the conventional fuel or nuclear power plants to improve their efficiency from 35-40% to 50-55%. Research efforts and developments in the field of linear motion MHD systems have grown steadily in the last two decades and consequently many large induction and conduction dc pumps are used in metallurgy and nuclear reactors. Experimental plasma dc generators of up to 25 MW have already been connected to the electric power grid for hundreds of hours and demonstration MHD power plants of up to 300 MW have been designed and are planned to be in operation in the late 1980s. A theoretical study of linear motion MHD machines is very complex, involving electromagnetic and hydrodynamic problems. The resemblance of operating principles, primary windings, and electromagnetic specific phenomena to those of linear motion electric machines have led us to include this topic in the present work. We stress mainly electromagnetic aspects--only in a simplified form--and tackle special problems of the fluid flow disturbance in the magnetic fields.

5.1. Linear Induction Pumps

5.1.1. Introduction

Pumping or stirring liquid metals in metallurgical and nuclear power applications by conventional hydraulic pumps poses significant difficulties such as:

1. Bearing wearout at high temperatures (above 400 C).

2. Leak-proof sealing of the liquid metal.

3. Cavitation-caused defects.

The absence of moving parts, construction simplicity, perfect sealing, and easy maintenance are the main assets of linear electromagnetic pumps--both of induction and conduction type. Because of the absence of electrodes and the ability to work at conventional ac voltage levels, linear induction pumps are simpler and less expensive than conduction (dc) pumps. The latter require a low voltage dc source and electrodes, but the energy conversion efficiency is in general higher than that of linear induction pumps, mainly because of high reactive currents necessary for ac magnetizing of large airgaps. In general linear induction pumps are to be used for higher conductivity liquid metals and the conduction pumps for those of low conductivity.

5.1.2. Construction Guidelines

Numerous physical configurations for linear induction pumps and stirrers are feasible.[1-4] However, in the following only three representative structures are considered. These are:

1. Linear flat induction pumps with double-sided primary, with or without lateral end plates of high conductivity, as shown in Fig. 5.1.

2. Linear flat induction pumps, shown in Fig. 5.2, with single-sided primary and no secondary back iron.

3. Linear cylindrical, or tubular, pumps, as shown in Fig. 5.3.

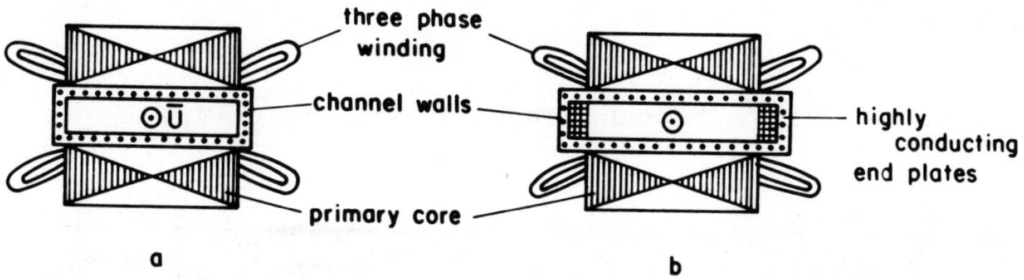

Fig. 5.1. Linear induction pumps a) No end plates; b) End plates.

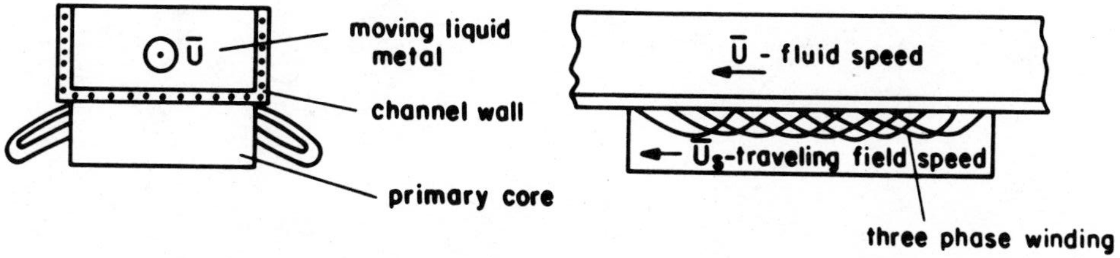

Fig. 5.2. Linear induction pumps without secondary back iron.

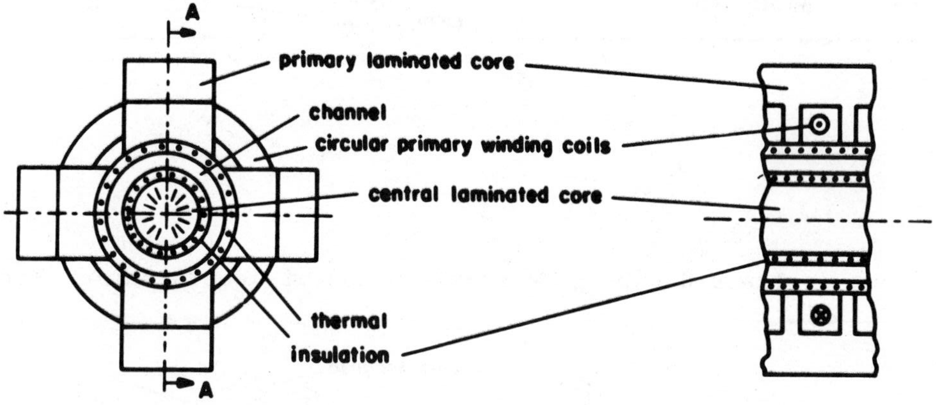

Fig. 5.3. Linear tubular pumps.

The primary core and windings of linear electromagnetic pumps resemble those of linear flat and tubular induction motors,[5] and consequently they are not treated here again. There are auxiliary aspects that are unique to pumps, such as duct materials, thermal isolation, and others; however, these are beyond our scope here.

5.1.3. Electromagnetic Phenomena at a Constant Speed

The complex interaction of hydrodynamic and electromagnetic phenomena makes an exact and complete mathematical treatment of linear induction pumps extremely difficult. In order to circumvent this difficulty the liquid metal speed is first considered the same throughout of the channel; that is, laminar flow at constant speed is assumed. And, thereby a linear induction pump is reduced to a linear induction motor (LIM). Thus, the electromagnetic phenomena are isolated and their treatment can be approached by using the analytical and numerical approaches commonly used for LIMs.[5]

The most important electromagnetic phenomena to be considered are transverse effect, skin effect, airgap leakage, and the longitudinal effect. In what follows, a superposition approach is used to account for these phenomena. Thus, the transverse edge and skin effects are treated for the case of flat pumps and skin effect for cylindrical induction pumps, yielding finally correction coefficients that enter the realistic goodness factor, G_e, known from LIMs.[5] The longitudinal effect could then be treated by a quasi-dimensional approach known from LIMs[5] and is not developed again here.

A separate treatment of transverse and skin effects as they occur in the flat and cylindrical pumps, respectively, follows.

5.1.4. Double-Sided Flat Induction Pumps--Transverse and Skin Effects

A transverse view of a flat linear induction pump with end plates is shown in Fig. 5.4. It is well known from the transverse effect study in LIMs[7] that the transverse fringing of primary flux lines could be approximated satisfactorily by simply increasing the stack width, 2a, by roughly one airgap length, g_0. Thus the channel width, $2a_e$, becomes

$$2a_e = 2a + g_0 \tag{1}$$

Now the equivalent stack width should also be considered to be $2a_e$ where no fringing beyond $2a_e$ exists anymore. The following assumptions are also considered to hold:

1. The airgap magnetic field has only one component, along the channel depth, and consequently, the induced currents density exhibits two components: J_x and J_y.

2. The primary current sheet is

$$J_{1m} = J_{1m}e^{j\omega_1 t - \pi/\tau x} \tag{2}$$

3. There is no longitudinal end effect, and thus the airgap magnetic field and the induced currents components are also traveling waves similar to that of (2).

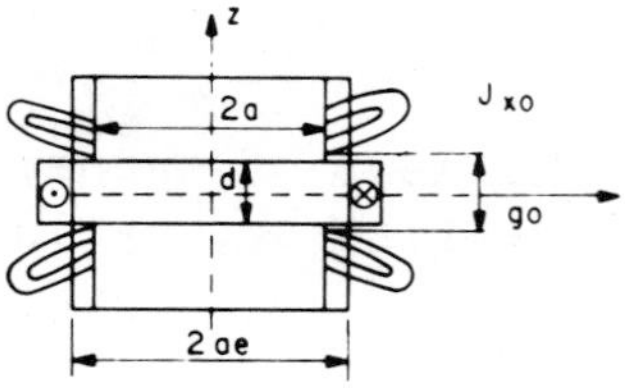

Fig. 5.4. Linear induction pump (transverse view).

4. The end plate thickness $(c - a_e)$ is so small that the current density in it could be considered to have one component, $J_{x0}(x, t)$, constant along OY.

5. The airgap field is constant along OZ but the skin effect is accounted for by correcting the conductivity of metal (liquid), σ, and of end plates, σ_0:

$$\sigma_e = \frac{\sigma}{K_{skin}} \; ; \quad \sigma_{0e} = \frac{\sigma_0}{K_{skin,0}} \tag{3}$$

with:

$$K_{skin,0} = \frac{d}{2d_{s,0}} \left[\frac{\sinh (d/d_{s,0}) + \sin(d/d_{s,0})}{\cosh (d/d_{s,0}) - \cos(d/d_{s,0})}\right] \tag{4}$$

$$d_s = \sqrt{\frac{2}{\mu_0 s\omega_1 \sigma}} \; ; \quad d_{s,0} = \sqrt{\frac{2}{\mu_0 \omega_1 \sigma_0}} \tag{5}$$

where s is the slip, as given by

$$s = 1 - \frac{u\pi}{\tau\omega_1} \; . \tag{6}$$

In (5) it should be noticed that within the end plates the primary angular frequency is ω_1, since the plates are not in motion. Now, the Poisson-type equation that holds in this case, for the reaction airgap field H_r, as known from LIMs theory[5] is

$$\frac{\partial^2 \underline{H}_r}{\partial y^2} - \left[\left(\frac{\pi}{\tau}\right)^2 + j\omega_1 s\sigma_e\mu_0 \frac{d}{g_0}\right]\underline{H}_r = jJ_{1m} s\omega_1 \sigma_e \mu_0 \frac{d}{g_0} \tag{7}$$

The factor $\exp[j(\omega_1 t - (\pi/\tau)x)]$ is implicitly present in (7). The solution to (7) is

$$\underline{H}_r = \underline{C}_1 \sinh \underline{\gamma} y + \underline{C}_2 \cosh \underline{\gamma} y + \frac{sG\tau J_{1m}}{\pi(1 + jsG)g_0} \qquad (8)$$

with

$$\underline{\gamma} = \frac{\pi}{\tau} \sqrt{1 + jsG} \ ; \quad G = \frac{\omega_1 \tau^2}{\pi^2} \mu_0 \sigma_e \frac{d}{g_0} \qquad (9)$$

where G is the so called realistic goodness factor as used for LIMs.

The current density components in the liquid metal are

$$\underline{J}_y = - \frac{g_0}{d} \frac{\partial \underline{H}_r}{\partial x} \qquad (10)$$

$$\underline{J}_x = \frac{g_0}{d} \frac{\partial \underline{H}_r}{\partial y} \qquad (11)$$

The resultant airgap field, $\underline{B}_z$, is given by

$$\underline{B}_z = \mu_0 \underline{H}_r + \frac{j\mu_0 J_{1m}\tau}{\pi g_0} \qquad (12)$$

The boundary conditions are

$$(\underline{J}_{2y})_{y=0} = 0 \qquad (13)$$

$$(c - a_e) \frac{\partial \underline{J}_{x0}}{\partial x} = (J_y)_{y=a_e} \qquad (14)$$

$$\frac{\underline{J}_{x0}}{\sigma_{0e}} = \left(\frac{\underline{J}_x}{\sigma_e}\right)_{y=a_e} \qquad (15)$$

Considering that the factor exp $[j(\omega_1 t - (\pi/\tau)x)]$ is also present in J_{x0}, from (10) through (15), $\underline{C}_1$, $\underline{C}_2$, and $\underline{J}_{x0}$ are obtained as

$$\underline{C}_1 = 0$$

$$\underline{J}_{x0} = -\frac{sGJ_{1m}}{\pi d(1 + jsG)} \frac{1}{\left[(c - a_e) + \dfrac{d}{\underline{\gamma} g_0} \dfrac{\sigma_e}{\sigma_{0e}} \dfrac{1}{\tanh \underline{\gamma} a_e}\right]} \qquad (16)$$

$$\underline{C}_2 = \frac{\underline{J}_{x0}\sigma_e d}{\sigma_{0e} g_0 \underline{\gamma} \sinh \underline{\gamma} a_e} \qquad (17)$$

<u>Case of end plates</u>

In this case there is only one boundary condition for determining $\underline{C}_2'$; that is,

$$(\underline{J}_{yi})_{y=a_e} = 0 \qquad (18)$$

Thus $\underline{C}_2'$ is given by

$$\underline{C}_2' = -\frac{sGJ_{1m}\tau}{\pi g_0(1 + jsG)\cosh \underline{\gamma} a_e} \qquad (19)$$

<u>Case of no transverse effect</u>

In this case simply $\underline{C}_2 = 0$, and therefore,

$$\underline{J}_{yi} = -\frac{jsGJ_{1m}}{d(1} + jsG) \; ; \quad \underline{B}_{zi} = \frac{j\mu_0 J_{1m}\tau}{\pi g_0(1 + jsG)} \qquad (20)$$

where $\underline{B}_{zi}$ is the resultant field in the airgap in the absence of transverse effect. To account globally for the transverse effect two correction coefficients defined for LIMs are used. These factors are

$$K_{tr} = \frac{\mathrm{Real}\left(\int_0^{a_e} \underline{J}_y \, dy\right)}{a_e \, \mathrm{Real}(\underline{J}_{yi})} \leq 1 \tag{21}$$

$$K_{tm} = \frac{\mathrm{Real}\left(\int_0^{a_e} \underline{B}_z \, dy\right)}{a_e \, \mathrm{Real}(\underline{B}_z)_{y=0}} \leq 1 \tag{22}$$

The first coefficient, K_{tr}, accounts for the fact that the longitudinal components J_x and J_{x0} produce an apparent decrease of the liquid metal conductivity compared to the case when no such effect is present. The second one, K_{tm}, takes into account the apparent demagnetizing effect of the same longitudinal current component J_x under the primary stack. Thus, in fact, a realistic goodness factor G_e to fully account for the transverse effect could be defined as[5]

$$G_e = \frac{G K_{tm} K_{tr}}{K_{a1}} \tag{23}$$

The coefficient K_{a1} takes into account globally the airgap leakage, that is, the airgap magnetic field longitudinal components that were neglected in the above analysis[5]:

$$K_{a1} \simeq \frac{\tau}{\pi g_0} \sinh\left(\frac{\pi}{\tau} g_0\right) \tag{24}$$

In a two-dimensional numerical (finite difference or finite element) analysis, both transverse effect and skin effect (and, implicitly, airgap leakage) would be simultaneously treated but this is beyond our purpose here.[6] The pressure gradients, at constant speed, are obtained from Navier-Stokes flow equation with U = constant:

$$\frac{\partial p_x}{\partial x} = \text{Real}(\underline{J}_y \underline{B}_0^*) = \text{Real}(j\underline{J}_y \frac{\mu_0 J_{1m}}{\pi g_0 \tau}) \tag{25}$$

$$\frac{\partial p_y}{\partial y} = \text{Real}(\underline{J}_x \underline{B}_0^*) = \text{Real}(-j\underline{J}_x \frac{\mu_0 J_{1m}}{\pi g_0 \tau}) \tag{26}$$

Corresponding to the propulsion and lateral forces in LIMs[5] there are two magnetic pressure gradients, one along the direction of motion and the other along OY, tending to press the fluid toward the stack center ($y = 0$). Unless the transverse effect is strongly attenuated by the highly conducting end plates, when $(2a_e/\tau) < 1$, the lateral magnetic pressure gradient could cause notable fluid speed variations along the transverse direction.

A numerical example

Let us suppose that $J_{1m} = 10^5$ A/m; $\tau = 0.2$ m; $g_0 = 0.02$ m; $d = 0.012$ m; $f_1 = 50$ Hz; $s = 0.30$; $2a = 0.16$ m; $\sigma = 5.10^6\ (\Omega m)^{-1}$; $\sigma_0 = 4.10^7\ (\Omega m)^{-1}$; $(c - a_e) = 0.01$ m. Using the expressions developed above the transverse distribution of current densities and of

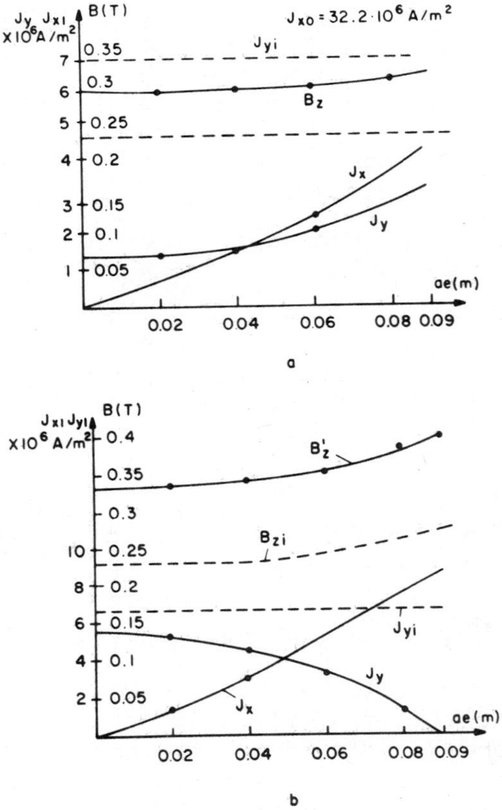

Fig. 5.5. Transverse effect. (a) With end plates, (b) without end plates.

the resultant airgap magnetic field with and without end plates for the ideal case (no transverse effect) are shown in Fig. 5.5. Also $K_{tr} = 0.738$, $K_{tm} = 0.804$; $G = 4.704$; $G_e = 2.82$. The highly conducting end plates (Fig. 5.5a) tend to reduce the transverse variation of the current densities, reducing the transverse effect.

5.1.5. Absence of secondary back iron skin effect

In the absence of secondary back iron (Fig. 5.6) the magnetic field exhibits a three dimensional distribution that is very difficult to calculate. Let us

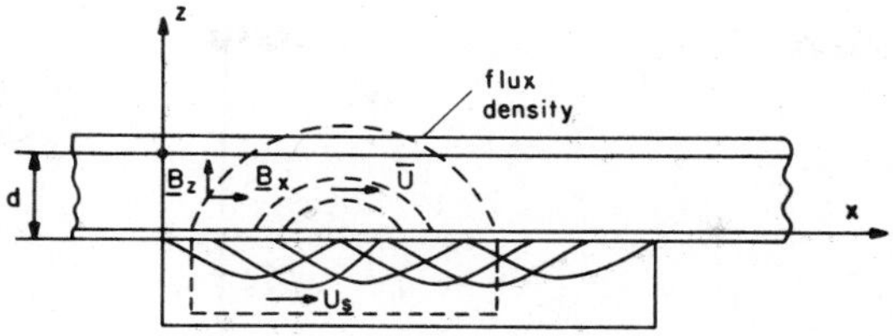

Fig. 5.6. No secondary back iron.

isolate only the distribution along OZ, which is most significant for this case. Thus, implicitly, the transverse and longitudinal effects are neglected. Also since the liquid height, d, is in general larger than the duct wall thickness, d_d ($d_d << d$), the latter is neglected in order to further simplify the final field distribution expressions. The problem is again two-dimensional, but in the ZOX plane. Hence, the flux density has two components, B_x and B_z, but the induced current density has only one, $-J_y$.

The Poisson equation in the liquid metal is now written for the resultant magnetic field components:

$$\frac{\partial^2 \underline{B}_x}{\partial x^2} + \frac{\partial^2 \underline{B}_x}{\partial z^2} - j\omega_1 s\sigma_e \mu \underline{B}_x = 0 \tag{27}$$

The solution to (27) is of the form

$$\underline{B}_x = \underline{B}_1' e^{-\underline{\gamma}' z} + \underline{B}_1 e^{\underline{\gamma}' z} \tag{28}$$

with

$$\underline{\gamma}' = \sqrt{\left(\frac{\pi}{\tau}\right)^2 + j\omega_1 s\sigma_e \mu} \tag{29}$$

According to the divergence law, the $\underline{B}_z$ component of airgap flux density is

$$\underline{B}_z = -\int \frac{\underline{B}_x}{\partial x}\, dz = j\,\frac{\pi}{\underline{\gamma}'\tau}(\underline{B}_1 e^{\underline{\gamma}'z} - \underline{B}_1' e^{-\underline{\gamma}'z}) \tag{30}$$

In the airspace above the liquid metal, the Laplace equation holds. Thus, the final expressions for the magnetic field components here are

$$B_{x0} = \underline{B}_1'' e^{-(\pi/\tau)(z-d)} \tag{31}$$

$$B_{z0} = -j\underline{B}_1'' e^{-(\pi/\tau)(z-d)} \tag{32}$$

Evidently (31) through (32) fulfill the condition of cancelling the magnetic field when $z \to \infty$. The following boundary conditions hold:

$$(\underline{B}_x)_{z=0} = \mu_0 J_{1m} \tag{33}$$

$$\left(\frac{\underline{B}_x}{\mu}\right)_{z=d} = \left(\frac{\underline{B}_{x0}}{\mu_0}\right)_{z=d} \; ; \quad (\underline{B}_z)_{z=d} = (\underline{B}_{z0})_{z=d} \tag{34}$$

In what follows the liquid metal is considered nonmagnetic, implying that $\mu = \mu_0$. The integration constants $\underline{B}_1$, $\underline{B}_1'$, and $\underline{B}_1''$ are then given by

$$\underline{B}_1 = \frac{\mu_0 J_{1m}}{1 - \dfrac{\tau\gamma' - \pi}{\tau\underline{\gamma}' + \pi}\, e^{-2\underline{\gamma}'d}} \tag{35}$$

$$\underline{B}_1' = \mu_0 J_{1m} - \underline{B}_1 \tag{36}$$

$$\underline{B}_1'' = \underline{B}_1 e^{-\underline{\gamma}'d} + \underline{B}_1' e^{\underline{\gamma}'d} \tag{37}$$

Specific cases

In the absence of the fluid, d → 0 and thus $\underline{B}_1 = \underline{B}_1'' = \mu_0 J_{1m}$ and $\underline{B}_1' = 0$. The same is valid for d → ∞, that is, when the liquid metal layer is very thick. Again considering the speed constant along OX and OZ the pressure gradients are

$$\frac{\partial p_x}{\partial x} = -\ \text{Real}\ (\underline{J}_y \underline{B}_z^*) = f_x \tag{38}$$

$$\frac{\partial p_z}{\partial y} = \text{Real}\ (\underline{J}_y \underline{B}_x^*) = f_y \tag{39}$$

where

$$\underline{J}_y = -\ j\omega_1 s\sigma_e \int \underline{B}_z\, dx = s\omega\sigma_e \underline{B}_z \frac{\tau}{\pi} \tag{40}$$

Thus, the liquid metal experiences two pressure gradients--one along the direction of motion, ∂p/∂x, which contributes to the pumping effect; and the other, $\partial p_z/\partial y$, which acts along the OZ producing a repulsion effect along OZ. This repulsion force could produce variations of liquid metal speed along the channel depth and thus a stirring effect. For this type of induction pump, or stirrer, the channel thickness, d, is of the order of magnitude of field penetration depth:

$$d_s = \frac{1}{\text{Real}(\underline{\gamma}')} \tag{41}$$

Therefore, this effect could not be neglected in any realistic performance assessment of this type of a pump.

A numerical example

The following initial data are considered: $\tau = 0.20$ m, $\sigma_e = 5.10^6\ (\Omega\ m)^{-1}$, $f_1 = 50$ Hz, s = 0.3, d = 0.2 m. The amplitude and phase of the current density J_y is plotted in Fig. 5.7a against z. The notable variation

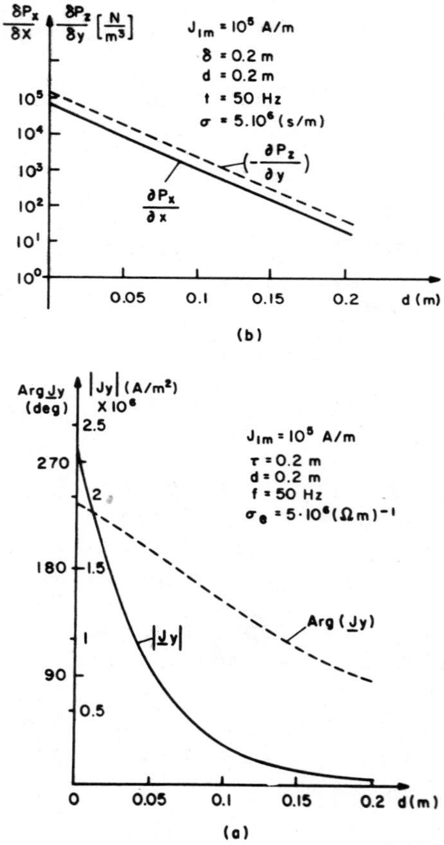

Fig. 5.7. (a) Current density $\underline{J}_y$, (b) magnetic pressure gradients along the channel thickness.

of the pressure gradients $\partial p_x/\partial x$ and $\partial p_z/\partial y$ along OZ shown in Fig. 5.7b indicates the importance of the phenomenon.

The active and reactive powers leaving the primary may be calculated as

$$\underline{S}_e = P_e + jQ_e = \frac{1}{2}\,(2a_e)\,\frac{2p\tau}{\mu_0\sigma}\,(\underline{J}_y B_x^*)_{z=0} \tag{42}$$

Adding the Joule losses, P_{co1}, and the reactive leakage power in the primary, Q_1, the full power balance of the machine is found:

$$S_1 = P_e + P_{co1} + j(Q_e + Q_1) \tag{43}$$

Hence the electrical efficiency and power factor of this pump can be determined. Of course, in complete performance assessment the hydraulic losses should also be considered but this aspect falls beyond our scope. The pressure of the channel wall, metallic or of thermal isolating material, would only have added two equations and two integration constants without affecting the nature of the problem. When used as an electromagnetic stirrer, or homogenizer, a single-phase supply of the primary could be used. In this case it is simpler to start the problem by assuming that the primary current sheet J_1 is given by

$$J_1 = \frac{2I_1 w_1 \sqrt{2}K_{w1}}{p\tau}\cos\omega_1 t\,\cos\frac{\pi}{\tau}x \tag{44}$$

The solution to the field problem is obtained in the time domain directly. As no special problem arises no further details are given in the following discussion.

5.1.6. Cylindrical Structure--Skin Effect and Curvature

A representative structure for a cylindrical linear induction pump is shown in Fig. 5.8. There are a few layers of media between the two magnetic cores in addition to the channel itself. Poisson and Laplace equations in polar coordinates are valid for these media. Thus for the channel the magnetic vector potential satisfies the following equation:

$$\frac{\partial^2 A}{\partial r^2} + \frac{1}{r}\frac{\partial A}{\partial r} + \left(\frac{1}{r^2} + \underline{\gamma}'^2\right)\underline{A} = 0 \qquad (45)$$

where

$$\underline{\gamma}'^2 = \alpha^2(1 + jsG) \; ; \quad \alpha = \frac{\pi}{\tau} \qquad (46)$$

In all other media if they are nonmagnetic and nonconducting $\underline{\gamma}'$ will be replaced by α in (45). The solution (45) is expressed in terms of Bessel functions of the first and second kind with complex argument:

$$\underline{A} = \underline{D}_1 I_1(j\underline{\gamma}'r) + \underline{D}_2 K_1(j\underline{\gamma}'r) \qquad (47)$$

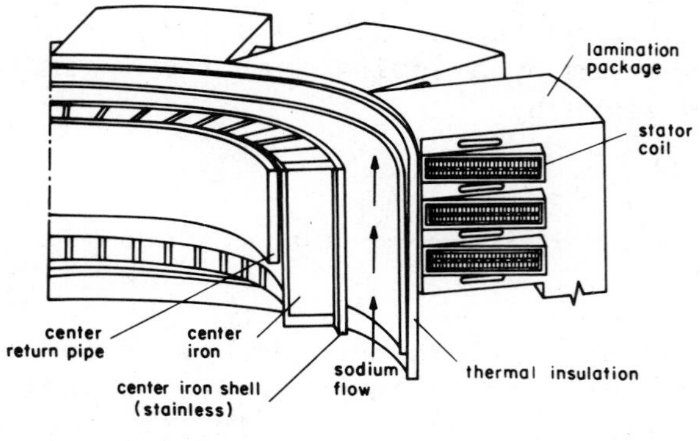

Fig. 5.8. Schematic structure of cylindrical induction pump.

The boundary conditions are standard, that is, continuity of tangential magnetic field and radial flux density at a source-free interface. At the surface of the current sheet, the tangential magnetic field equals the primary current sheet density:

$$\left[\frac{\partial \underline{A}_0(r)}{\partial r} \right]_{r=r_1} = \mu_0 J_{1m}; \quad \left[\frac{\partial \underline{A}_m(r)}{\partial r} \right]_{r=r_2} = 0 \qquad (48)$$

The magnetic field components and the current density are given by

$$\underline{B}_{ri}(r) = -\frac{\partial \underline{A}_i}{\partial x} = +j\frac{\pi}{\tau}\underline{A}_i(j\underline{\gamma}'r) \qquad (49)$$

$$\underline{B}_{xi}(r) = \frac{\partial \underline{A}_i}{\partial r}; \quad \underline{J}_{\theta i}(r) = -j\omega_1 s\sigma_e A_i(j\underline{\gamma}'r) \qquad (50)$$

Thus, the magnetic pressure gradients in the channel (the ith layer) are

$$\frac{\partial p}{\partial x} = -\mathrm{Real}[\underline{J}_{\theta i}(j\underline{\gamma}'r) \cdot \underline{B}^*_{ri}(j\underline{\gamma}'r)] = f_x \qquad (51)$$

$$\frac{\partial p}{\partial r} = +\mathrm{Real}[J_{\theta i}(j\underline{\gamma}'r) \cdot B^*_{xi}(j\underline{\gamma}'r)] = f_r \qquad (52)$$

Power balance

The electromagnetic power passing from the primary, S_e, is determined by evaluating the Poyting vector as

$$S_e = \frac{1}{2}\,\frac{2\pi r_1 L}{\mu_0}\,[E_0(j\alpha r_1)\, B^*_{x0}(j\alpha r_1)] \qquad (53)$$

where

$$E_0(j\alpha r_1) = -j\omega_1 A_0(j\alpha r_1) \tag{54}$$

The mechanical power, P_m, and secondary losses, P_2, are

$$P_m = (1 - s)\ \mathrm{Real}(\underline{S}_e) \ ; \quad P_2 = s\ \mathrm{Real}(\underline{S}_e) \tag{55}$$

The global power balance should include the primary Joule losses P_{co1} and the core losses P_{i1} as well as the leakage reactive power $Q_{\sigma 1}$. Also, the fluid flow losses, P_{flow}, must be included in the overall power balance equation.

$$P_{co1} = 3R_1 I_1^2 \ ; \quad Q_{\sigma 1} = 3X_{1\sigma} I_1^2 \tag{56}$$

The total apparent power, $\underline{S}1$, is therefore

$$\underline{S}_1 = P_{flow} + \underline{S}_e + P_{co1} + P_{i1} + jQ_{\sigma 1} = P_1 + jQ_1 \tag{57}$$

where P_{flow} denotes the fluid flow losses.

Now the electrical efficiency, η, and power factor, $\cos\phi$, are

$$\eta = \frac{P_m}{\mathrm{Real}(\underline{S}_1)} \ ; \quad \cos\phi = \frac{P_1}{S_1} \tag{58}$$

<u>Note</u>

So far the speed has been considered uniform. A typical turbulent profile could be considered along r, as a starting point, but then the channel should be divided into layers of constant speed and the same equations (45) hold for all of them. There will be more equations and more boundary conditions but the nature of the solution of the problem is as presented above.

<u>A numerical</u> example[8,9,10]

The initial data are: pole pitch $\tau = 0.4865$ m; frequency $f_1 = 20$ Hz; slip at channel center $s = 0.217$ (when nonuniform speed profile is considered); fluid velocity at center $U = 15.2$ m/sec; and number of poles $2p = 8$.

The computed results concerning the flux density components and pressure gradient, at uniform speed, are shown in Fig. 5.9. It is evident that the pressure gradient $\partial p/\partial x$ at uniform speed varies little with the radius. When a typical parabolic speed profile is considered the pressure gradient varies considerably with the radius, as shown in Fig. 5.10. The net effect will be a strong tendency to straighten the speed profile. By an iterative procedure using the flow equation, this new speed profile could be found. After a few iterations the speed profile should stabilize. The nominal overall performance of this large pump is shown in Fig. 5.11. It is very similar to that of a high rotor resistance, large airgap induction motor. For complete design data of this pump see Reference 9.

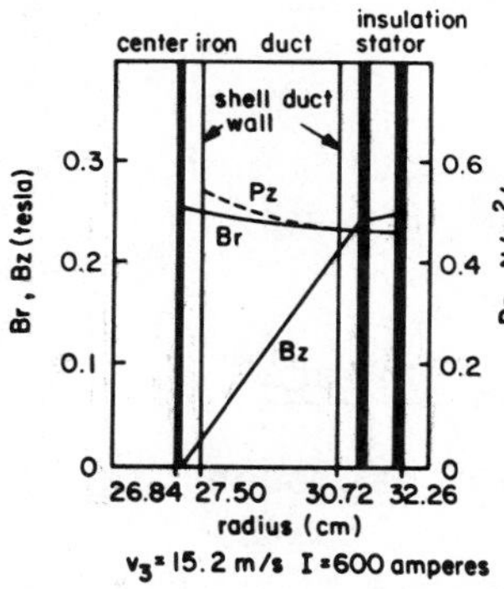

Fig. 5.9. Geometric parameter of the channel, flux density, and pressure gradient (constant speed).

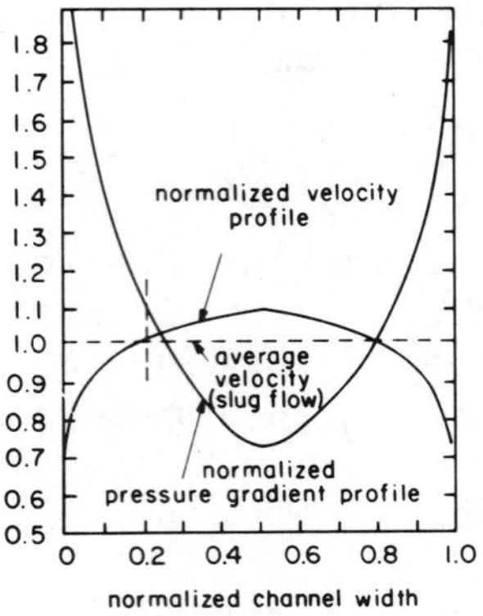

Fig. 5.10. Parabolic speed profile and corresponding pressure gradient.

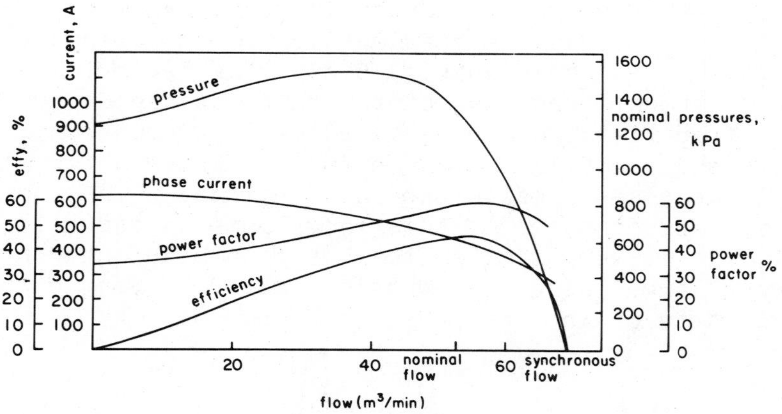

Fig. 5.11. Overall performance of a large linear induction pump.

Note on longitudinal effects

So far the speed has been considered constant along the direction of motion (OX). In this case if the goodness factor, G, is close to its optimum value, G_e (Fig 5.12), in the low slip region, the longitudinal effect, as known in LIMs, has notable consequences. Let us call these consequences the electromagnetic longitudinal

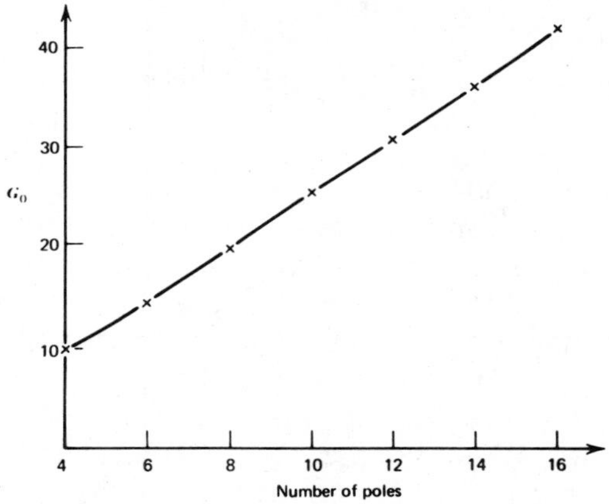

Fig. 5.12. Optimum goodness factor.

effects. The main effect is to make the thrust for unit volume (that is the magnetic pressure gradient, f_x) vary significantly along the motor length. For low-speed, low-conductivity liquid metals, G is considerably smaller than G_0. Thus the electromagnetic longitudinal effect is small. However, if the primary magnetic field has a significant pulsating component in the airgap, the magnetic pressure would pulsate accordingly along OX. For an odd number of poles with half-filled end slots the primary magnetic field is a traveling wave except for the end poles. In high-speed pumps of higher conductivity liquid metals, the electromagnetic longitudinal effect must be included (G close to G_0) in a theoretical assessment. The magnetic pressure variation along OX is expected to cause a speed variation along OX. This phenomena may be called the hydromagnetic longitudinal effect. The hydrodynamic interactions are modeled by the Navier-Stokes equations of flow in a magnetic field. For a one-dimensional flow (the speed has only one component along OX) these equations yield

$$\rho u \frac{du}{dx} + \frac{\partial p}{\partial x} = f_x(x, r) + \rho\nu\left(\frac{\partial^2 u}{\partial x^2} + \frac{1}{r}\frac{\partial u}{\partial r} + \frac{u}{r^2}\right) \quad (59)$$

$$\frac{\partial p}{\partial r} = f_z(x, r) \tag{60}$$

These equations can be solved within a reasonable number of iterations by the finite difference[11] or the Galerkin method if proper initial values are chosen. Initial values for pressure gradient $\partial p/\partial x$ may be obtained by considering only the electromagnetic longitudinal effect[5] ($u = ct$) while the radial initial profile of speed can be calculated as done previously (when neglecting the electromagnetic longitudinal effect). With these initial values equations (59) and (60) are solved numerically and new values for pressure gradients and speed distribution are found. After a few properly managed iterations satisfactory convergence for the speed profile is attained. It is beyond our scope here to elaborate further on the hydromagnetic longitudinal and radial effects.

5.2. Linear Motion DC Conduction Pumps

There exist several forms of dc pumps. However, three typical ones are considered here. These are

1. DC-fed conduction pumps (Fig. 5.13).

2. Transformer-fed conduction pumps (Fig. 5.14).

3. DC conduction pumps without field winding (Fig. 5.15).

The dc-fed conduction pump works like a dc motor and has a magnetic, and a hydraulic circuit and two electric circuits as shown in Fig. 5.13. Two electrodes are in general used to supply the pump at a low dc voltage and high current. The need for a low voltage high current dc source is a drawback of this pump. This difficulty is circumvented by using a transformer whose

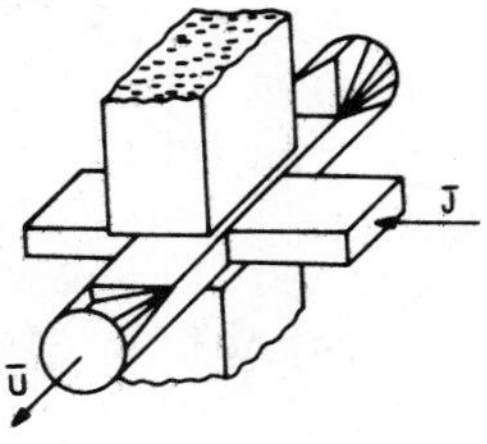

Fig. 5.13. DC conduction pump.

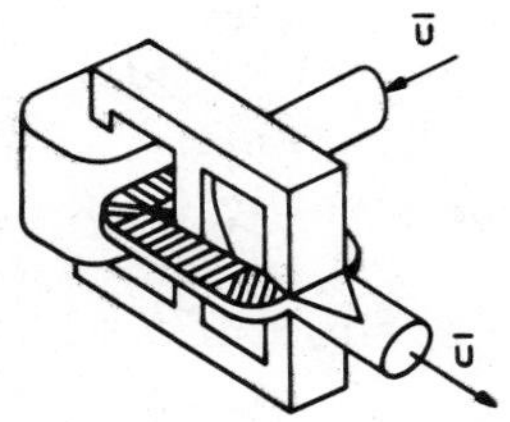

Fig. 5.14. Transformer-fed conduction pump.

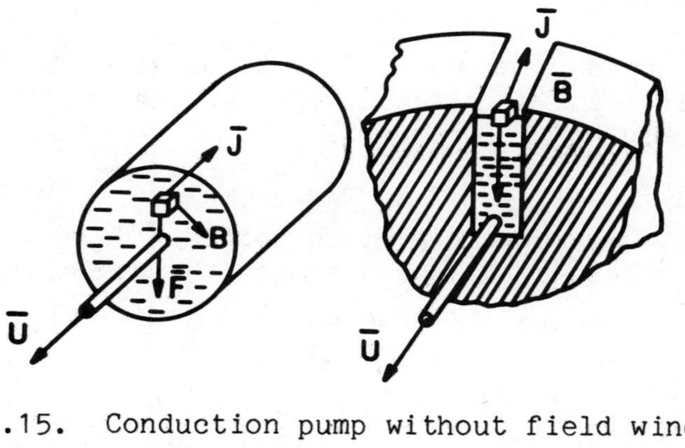

Fig. 5.15. Conduction pump without field winding.

primary produces directly the excitation field and also induces in its secondary a voltage. A part of its secondary represents the pump channel (Fig. 5.14). In fact both the excitation field and the current in the channel are time-varying, resulting in a pulsating force (pressure). The dc pump without field winding (Fig. 5.15) is based on magnetostriction effect of the current interaction with its own magnetic field. Only the dc conduction pump is considered here.

5.2.1. Theory of DC Pumps[12-15]

To account for most phenomena and still keep the analysis simple to use, an approximate field theory has been developed and is presented here. The following assumptions are made: The liquid metal flow is considered laminar and a uniform speed distribution all over the channel cross section is assumed.

1. The pump channel is supplied with a voltage, V_0.

2. The fluid conductivity, σ, and speed, u, are constant throughout the channel.

3. The excitation field has only one component and its distribution along the x- and y-directions are as shown in Fig. 5.16.

4. The airgap remains constant outside the channel, leading to a slight overestimation of longitudinal effect.

5. The skin effect in the channel due to longitudinal effect may be neglected since for most practical liquid metals and speed ranges the depth of penetration is greater than 30 mm.[15]

To account for the presence of metal electrodes we assume that the excitation field is constant in the transverse direction in the active zone and sinusoidal

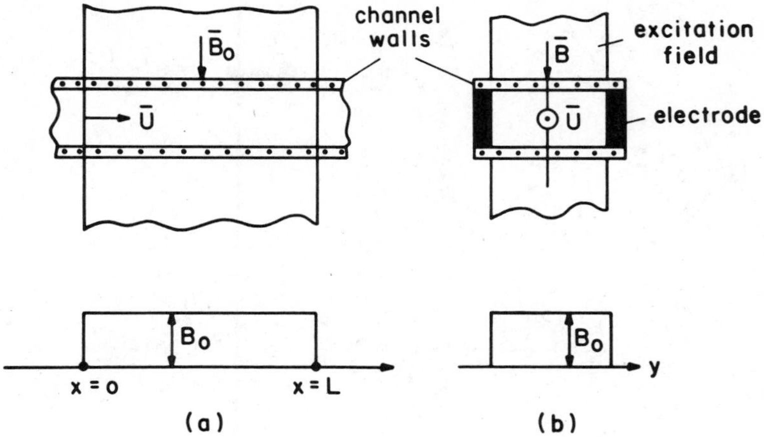

Fig. 5.16. Excitation field distribution along Ox and Oy.

outside the channel (0 => x and x => l). Thus the current density has only one component in the active zone, J_y, and two components in the entry and exit zones.

By applying Ampere's and Faraday's laws the corresponding equations for the reaction field are

$$g_0 \frac{\partial^2 H_r}{\partial x^2} - \mu_0 \sigma u d \frac{\partial H_r}{\partial x} = 0 \qquad \text{for } 0 \leqq x \leqq L \qquad (61)$$

$$g_0 \frac{\partial^2 H_{r0}}{\partial x^2} - \mu_0 \sigma u d \frac{\partial H_{r0}}{\partial x} - g_0 H_{r0} \left(\frac{\pi}{2a}\right)^2 = 0$$

$$0 \geqq x \text{ and } x \geqq L \qquad (62)$$

The conventional uniformly distributed induced current density, J_0, is yet to be considered. The corresponding solutions for the longitudinal effect induced current densities are

$$J_{ey} = -\frac{g_o}{d}\frac{\partial H_{r0}}{\partial x} = -A_e\,\gamma_1'\,\frac{g_0}{d}\,e^{\gamma_1' x}\quad \cos(\frac{\pi}{2a}y);$$

$$\text{for } x \leqq 0 \tag{63}$$

$$J_{ex} = \frac{g_o}{d}\frac{\partial H_{r0}}{\partial y} = -A_e\,\frac{\pi}{2a}\,\frac{g_0}{d}\,e^{\gamma_1' x}\quad \sin(\frac{\pi}{2a}y)$$

$$J_{yt} = -\frac{\partial H_r}{\partial x}\frac{g_0}{d} - J_0 = -\frac{g_0}{d}\,A_1\gamma_2 e^{\gamma_2 x} - J_0;$$

$$\text{for } 0 \leqq x \leqq L \tag{64}$$

$$J_{0y} = -A_0\frac{\gamma_2' g_0}{d}\,e^{\gamma_2'(x-L)}\quad \cos(\frac{\pi}{2a}y)\ ;\quad \text{for } x \geqq L \tag{65}$$

$$J_{0x} = -A_0\,\frac{\pi}{2a}\,\frac{g_0}{d}\,e^{\gamma_2'(x-L)}\quad \sin(\frac{\pi}{2a}x)$$

$$\gamma_{1,2}' = \frac{\mu_0 u\sigma d}{2g_0} \pm \sqrt{(\frac{\mu_0 u\sigma d}{2g_0})^2 + (\frac{\pi}{2a})^2}\ ;$$

$$\gamma_2 = \mu_0\sigma u\,\frac{d}{g_0} \tag{66}$$

Boundary conditions

The current density J_0 is to be found in the absence of longitudinal effect and thus fulfills the condition

$$\frac{J_0 L}{\sigma} = V_0 - 2uB_0 a \tag{67}$$

The condition for the two electrodes to be equipotential surface is implicitly satisfied by the solution to (64). The reaction field fulfills the conservation of flux law:

$$\int_{-\infty}^{\infty} H_r \, dx = 0 \tag{68}$$

Also the flux density should be continuous at x = 0 and x = L:

$$(H_{re})_{x=0} = \frac{B_0}{\mu_0} + (H_r)_{x=0} \tag{69}$$

$$(H_{r0})_{x=L} = \frac{B_0}{\mu_0} + (H_r)_{x=L}$$

Thus we find

$$A_1 = \frac{B_0}{\mu_0} \frac{(1/\gamma_2' - 1/\gamma_1')}{1/\gamma_1' - 1/\gamma_2 (1 - e^{\gamma_2 L}) - 1/\gamma_2'} \tag{70}$$

$$A_e = A_1 + \frac{B_0}{\mu_0}$$

$$A_0 = A_1 e^{\gamma_2 L} + \frac{B_0}{\mu_0}$$

A numerical example

To illustrate the influence of longitudinal effect we present here a numerical example having the following data: $u = 20$ m/sec; $\sigma = 10^6$ $(\Omega\ m)^{-1}$; $2a = 0.2$ m; $L = 0.2$ m; $B_0 = 0.5$ T; $J_0 = 5 \times 10^6$ A/m^2; $g_0 = 0.015$ m; $d = 0.01$ m. From (67) it follows that $V_0 = 3$ V. The values of A_1, A_e, and A_0 follow from (70) such that $A_1 = -3.206 \times 10^4$ A/m; $A_e = 3.65 \times 10^5$ A/m; $A_0 = -5.167 \times 10^5$ A/m. The resultant flux density and transverse current density, J_y, along the direction of motion for $y = 0$ are shown in Fig. 5.17. It is clear that the longitudinal effect could affect the performance importantly by producing additional Joule losses and a drag force.

5.2.2. Thrust (or Pressure) Reduction and Additional Losses

The thrust, F_x, is obtained from

$$F_x = -2ad \int_0^L J_{yt}(x)(B_0 + \mu_0 H_r)\, dx \qquad (71)$$

In general A_1 is negative. So the second term represents, as expected, a drag force caused by the longitudinal effect. Also the distribution of pressure

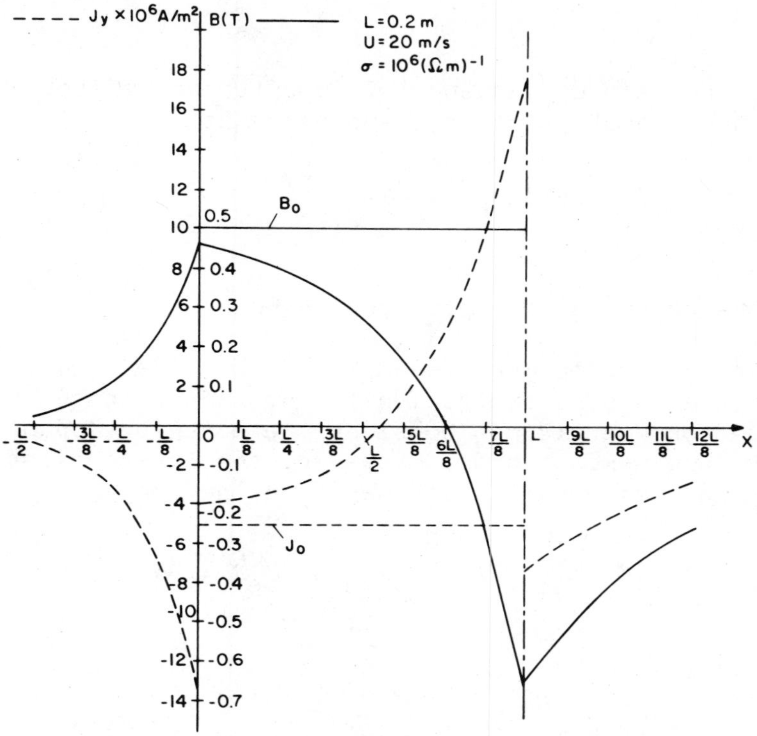

Fig. 5.17. Flux density and current density distribution (conduction pump).

along the channel length becomes nonuniform and thus turbulence may occur due to longitudinal effect.

The average pressure in the channel, p_{ae}, in the presence of longitudinal effect is given by

$$p_{ae} = \frac{F_x}{2ad} \qquad (72)$$

When the longitudinal end effect does not exist, the ideal (uniform) pressure is, simply,

$$p_i = B_0 J_0 L \tag{73}$$

Thus, the main consequence of longitudinal effect is a pressure reduction, $\Delta p/p_i$, expressed as

$$\frac{\Delta p}{p_i} = \frac{p_{ae} - p_i}{p_i} \tag{74}$$

For the numerical example $\Delta p/p_i = -0.41$

According to our assumption, outside the active zone there is no excitation field. Hence in the thrust evaluation, we did not consider the contribution of the entry and exit zones. However, the losses exist here also, and the total Joule losses in the pump are

$$P_j = \frac{d}{\sigma}\Big[\int_{-a}^{a}\int_{-\infty}^{0}(J_{xe}^2 + J_{ye}^2)\,dx\,dy$$

$$+ 2a\int_0^L J_{yt}^2\,dx + \int_{-a}^{a}\int_{L}^{\infty}(J_{x0}^2 + J_{y0}^2)\,dx\,dy\Big] \tag{75}$$

and finally

$$P_j = \frac{2ad}{\sigma}\Big[\frac{1}{\gamma_1'}\Big(\frac{A_e g_0}{2d}\Big)^2\Big(\gamma_1'^2 + \Big(\frac{\pi}{2a}\Big)^2\Big)$$

$$- \frac{1}{\gamma_2'}\Big(\frac{A_0 g_0}{2d}\Big)^2\Big(\gamma_2'^2 + \Big(\frac{\pi}{2a}\Big)^2\Big) + J_0^2 L$$

$$+ 2J_0 A_1 \frac{g_0}{d}\big(e^{\gamma_2 L} - 1\big) + \frac{1}{2}A_1^2 \gamma_2 \Big(\frac{g_0}{d}\Big)^2\big(e^{2\gamma_2 L} - 1\big)\Big] \tag{76}$$

Again the ideal losses

$$P_i = \frac{2adL}{\sigma} J_0^2 \tag{77}$$

Thus the increase in losses $\Delta P/P_i$ is

$$\frac{\Delta P}{P_i} = \frac{P_j - P_i}{P_i} \tag{78}$$

For our case $\Delta P/P_i = 1.55$.

As can be seen the longitudinal effect has an important influence on performance of the pump.

In the literature of dc pumps the longitudinal effect in the active zone is taken for reaction current influence[12-16] but since the phenomenon is speed dependent and becomes 0 for $L \to \infty$, it is a direct consequence of longitudinal effect. Its effect is, however, more important in the active zone because the electrodes provide a very highly conducting path for induced return currents so that the current density is considered purely transverse in the active zone. An exact treatment of the entry and exit zones should take into consideration the fact that the airgap increases sharply to infinity. Consequently an additional reluctance-type braking force is produced primarily as in a LIM. Only a numerical method can provide an exact treatment of this problem. However, the influence of these end zones is less important than that of the active zone.

5.2.3. Longitudinal Effect-Compensation Scheme

Under the name of reaction field compensation there are a number of solutions proposed to neutralize the longitudinal effect in the active zone ($0 \leq x \leq L$). Among them two are most popular. These methods are schematically shown in Fig. 5.18a and b.

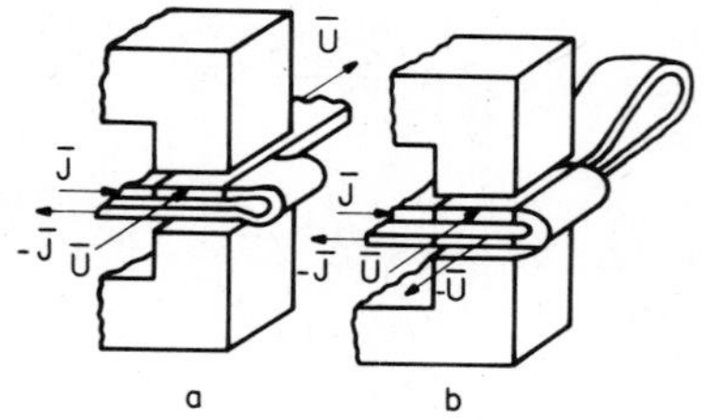

Fig. 5.18. Reaction field compensation (a) With back current bar, (b) with double channel.

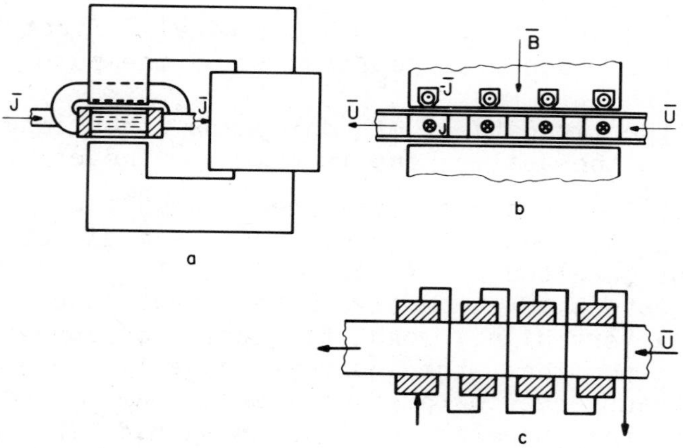

Fig. 5.19. Series compensation winding.

In both methods, the reaction field is neutralized. Thus the longitudinal effect in the active zone is also canceled. Unfortunately, in the first solution almost half of the airgap is not used, and in the second case the construction is complicated. The losses in the compensating conductors of the channel and additional losses in the field winding due to airgap increase should be smaller than the longitudinal effect losses. Since the excitation flux density in the airgap is smaller than 1 T, copper bars embedded in slots placed in the pole shoes of the stator core (Fig. 5.19) may be connected in series with the segments of the electrodes. Hence, the reaction field is fully compensated, but also a more uniform distribution of the pressure is obtained, while the airgap is not increased and the voltage to supply the pump could be increased e-fold (e = number of segments). Since the copper resistivity is much smaller than that of the liquid metal, the losses in the compensating bars are less than 10% of those in the liquid metal. If segmenting of electrodes is not possible, the copper bars in slots may be paralleled to form one resultant bar. It is felt that in this way the application of dc conduction pumps could be enlarged further taking advantage of the dc magnetization of large airgaps to obtain performances higher than those of linear induction pumps, at least for low conductivity liquid metals. However, the problems of voltage drop at the electrodes and the electrode liquid metal compatibility should not be underestimated when linear dc conduction pumps are compared with the linear induction pumps.

5.3. Linear Motion Dc MHD Generators

In principle a linear motion dc MHD generator converts the kinetic energy of an electrically conducting fluid into electrical energy by an interaction of the fluid with a dc homopolar magnetic field. Such an operation corresponds to the generator mode of linear motion dc pumps. However, in general the working fluid of a dc MHD generator is an ionized gas (plasma), that is, a compressible fluid. This fact and others result in many differences between the liquid metal dc pump and a dc MHD generator, justifying a separate treatment of the latter.

Proposed in principle by Faraday, the first MHD generator was built by Karlovitz during the period 1938 through 1946. Later, its feasibility was demonstrated by AVCO Laboratories in the early 1960s.[17] Since then this field has developed steadily in most industrialized countries.

By burning the fuel at above 3000 K more of its thermal enthalpy is extracted in the MHD generator and in the joint conventional thermal power plant. Thus, the overall efficiency of the thermal power plant increases from 35-40% to 45-55%. Since most electrical energy is produced in thermal power plants, it is expected that MHD generators will supplement thermal power plant operation.

A complete theory of a plasma MHD generator is very involved, due to the electromagnetic, hydraulic, and thermal interactions within a large channel where, in general, the temperature, speed, and static pressure are not constant throughout the channel. A numerical method is necessary to obtain a refined analysis of MHD generators. Also, there are several technical problems associated with a practical MHD generator. These would require a detailed treatment, which is beyond our scope here. Instead, we present a rather simplified study in order to gain an insight into most specific aspects and performance of MHD plasma generators. Construction alternatives, electromagnetic phenomena, electrical efficiency, one-dimensional flow equations, and adiabatic efficiency are treated here in some detail assuming that the plasma conductivity is the same throughout the channel.

5.3.1. Construction Guidelines

In its simplest form, the construction of a dc MHD generator (Fig. 5.20) resembles that of a homopolar dc pump with continuous electrodes. Power levels of MHD generators are generally high (of the order of hundreds of megawatts). Thus, the channel length is of the order of a few meters and its cross section of the order of 1 to 3 m^2. Also, the electrical conductivity of the plasma, σ, at 3000 K (when seeded with 1 to 2% of alkali compounds) is less than 100 $(\Omega m)^{-1}$. Due to this fact, unless very strong magnetic fields (3 to 7 T) are pro-

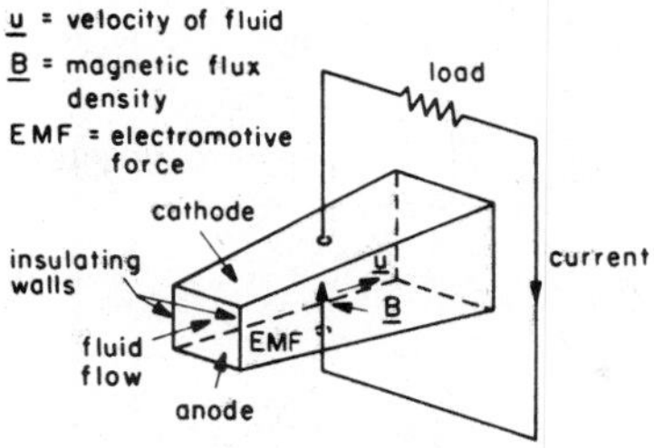

Fig. 5.20. Plasma dc MHD generator.

vided together with very high speeds of fluid (600 to 2000 m/sec), the electrical efficiency of such a generator is likely to be low. Therefore, only very large superconducting magnets are to be considered for such generators. Their optimal shape is a matter of design, but in any case a very high degree of magnetic field uniformity throughout the channel should be provided in order to avoid notable disturbances in the plasma. An adequate solution, the double-saddle magnet, is shown in Fig. 5.21.

The continuous-electrode solution, though the simplest, poses a few problems, especially concerning the low level of voltage obtained (like in dc pumps) and the reduction of performance due to the Hall effect, as explained later. Therefore, for practical purposes the electrodes are segmented (Fig. 5.22). The load could be

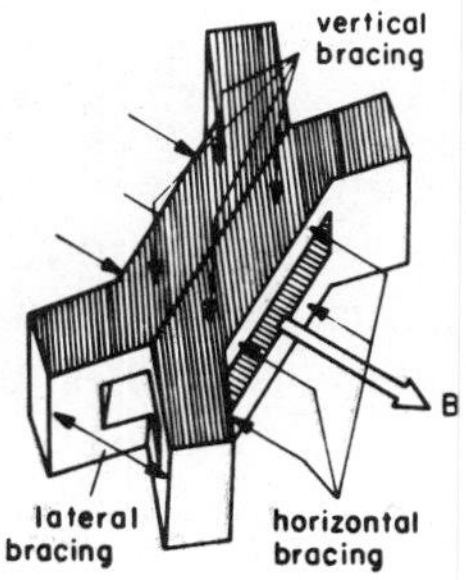

Fig. 5.21. Double-saddle MHD superconducting magnet.

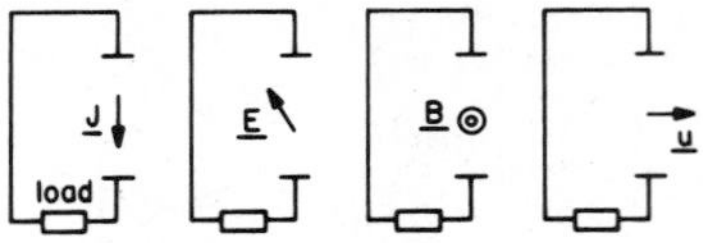

Fig. 5.22. Segmented electrode.

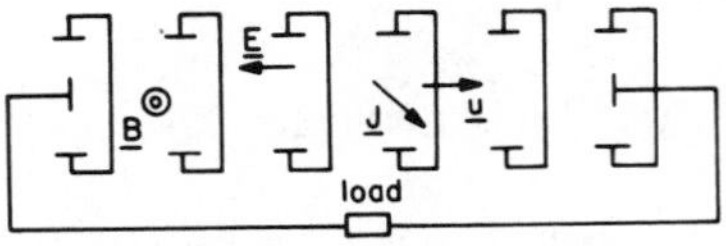

Fig. 5.23. Hall MHD generator.

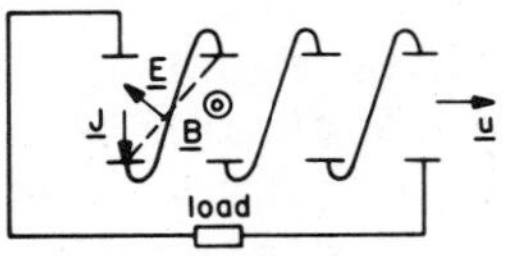

Fig. 5.24. Slant-duct MHD generator.

distributed to all of them (Fig. 5.22) or they could be short circuited and only one load connected (Hall generator, Fig. 5.23). In practice a combination of these two solutions is used (Fig. 5.24). It is called the slant-duct or diagonally connected MHD generator.

5.3.2. Simplified Treatment of Electromagnetic Phenomena

MHD generators may be designed to work at constant speeds along the channel.[18] In what follows the constant-speed case is considered. Considering the complexity of the problem, only a simplified treatment of electromagnetic phenomena is presented to obtain insight into specific phenomena. The following assumptions are made:

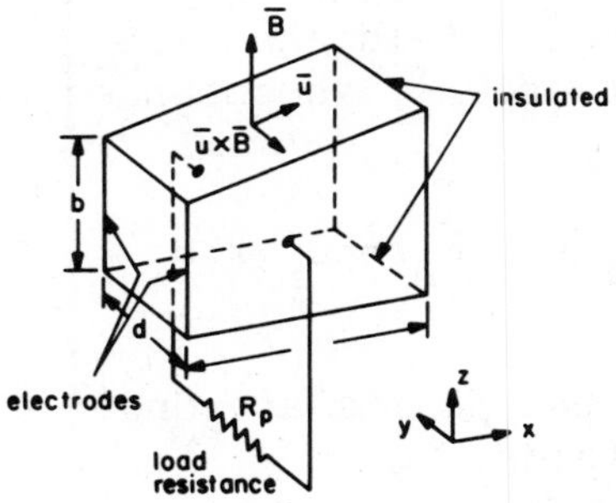

Fig. 5.25. MHD channel.

1. The plasma enters the MHD channel at the point 1 (Fig. 5.25) with the speed u, along OX, pressure P_1, and temperature T_1 and keeps a constant flow rate along OX.

2. The magnetic field B is uniform and acts along OZ; the channel walls at z = ±b/2 are electric insulators whereas those at y = ±d/2 are conductors representing the electrodes.

3. The plasma is a perfect gas, without friction, and is compressible; its properties are the same throughout the channel.

4. The reaction field of the induced currents is negligible both because the product σu is small and because B is large.

5. Heat transfer from the channel is neglected.

6. Additional Joule losses induced in the plasma at the entry and exit ends of the channel are neglected.

The electrons in the gas are considered to move at a relative speed u_n due to the combined interaction with the induced electric field and external magnetic field. Thus Hall effect occurs, accompanied by a Hall electric field, $\overline{E}_n$:

$$\overline{E}_n = \overline{U}_n \times \overline{B} \tag{79}$$

The electrons in the plasma are under the influence of three electric fields:

$\overline{E}$ = external (applied)

$\overline{E}_{ind}$ = induced by motion

$\overline{E}_n$ = Hall field

It is known that

$$\overline{E}_{ind} = \overline{U} \times \overline{B} \tag{80}$$

The resultant electric field $\overline{E}''$ is therefore

$$\overline{E}'' = \overline{E} + \overline{E}_{ind} + \overline{E}_n = \overline{E} + \overline{U} \times \overline{B} + \overline{U}_n \times \overline{B} \tag{81}$$

But the resultant electric field and the current density are related through the electric conductivity by

$$\overline{J}_n = \sigma\overline{E}'' \tag{82}$$

Also, $\overline{J}_n$ is

$$\overline{J}_n = -\overline{U}_n e n_n \; ; \quad u_n = \frac{|\overline{U}_n|}{|\overline{E}''|} \tag{83}$$

with e = the electron charge, n_n = concentration of electrons in the plasma, and u_n = electric mobility of electrons. Making use of (82) and (83) in (81) we obtain

$$\bar{J}_n = \sigma(\bar{E} + \bar{U} \times \bar{B}) - \mu_n(\bar{J}_n \times \bar{B}) \qquad (84)$$

In general the Hall angle, θ, and the Hall parameter, β_n, are defined by

$$\tan\theta = \frac{U_n B\sigma}{J_n} = u_n B = \beta_n \qquad (85)$$

It is well understood that when θ = 0, β = 0 (u_n = 0) there is no Hall effect. An additional, but in general negligible, Hall effect is experienced by plasma ions[17] (ion slipping). Equation (84) could be also expressed as Fig. 5.26. This representation of (84) may be decomposed along the two axes so that

$$J_{ny} = \sigma(E_y - UB) + \mu_n J_{nx} B$$

$$J_{nx} = \sigma E_x - \mu_n J_{ny} B \qquad (86)$$

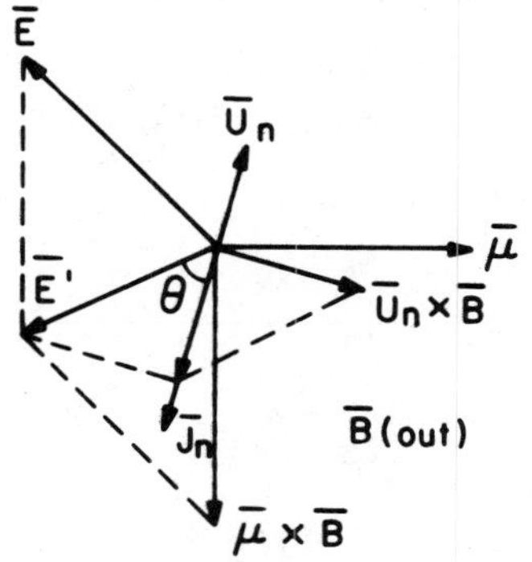

Fig. 5.26. General vector diagram.

$$J_{nz} = 0$$

Finally,

$$J_{nx} = \frac{\sigma[E_x - \beta_n(E_y - UB)]}{1 + \beta_n^2} \tag{87}$$

$$J_{ny} = \frac{\sigma(E_y - UB + \beta_n E_x)}{1 + \beta_n^2} \tag{88}$$

The Hall parameter, β_n, corresponds in fact to the relationship

$$\omega\tau = \beta_n \tag{89}$$

where ω is the Larmor frequency of an electron around a magnetic field, B, and τ is its relaxation time. We now consider the four major electrode configurations.

<u>Continuous electrode</u>

In this case there is one single electrode along the channel length, and thus the longitudinal component E_x of the applied electric field (or voltage) is zero ($E_x = 0$). The current densities (87) and (88) yield the simplified expressions

$$J_{nx} = - \frac{\sigma\beta_n(E_y - UB)}{1 + \beta_n^2} \tag{90}$$

$$J_{ny} = \frac{\sigma(E_y - UB)}{1 + \beta_n^2} \tag{91}$$

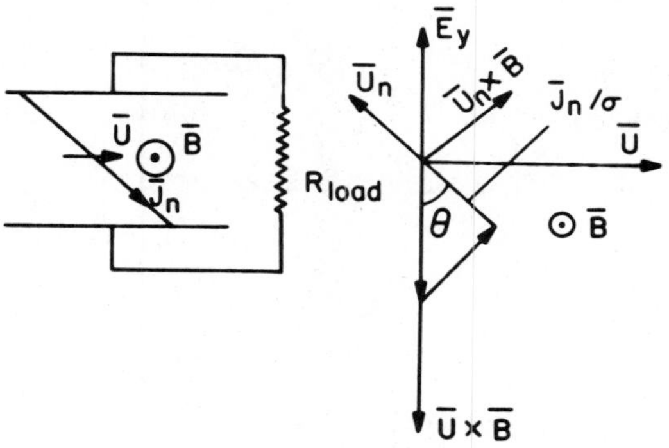

Fig. 5.27. Vector diagram--continuous electrode.

These currents are shown as vectors in Fig. 5.27. The longitudinal component of the current density, J_{nx}, does not produce electric power and contributes only to the Joule losses. If the Hall parameter $\beta >> 1$, $J_{nx} >> J_{ny}$, rendering this type of MHD generator less efficient.

On no-load ($J_{ny} = 0$) we get

$$E_{y0} = UB \tag{92}$$

The load factor is now defined as

$$K = \frac{E_y}{UB} \tag{93}$$

Thus the electric power (dP_e/dx) generated per unit length of channel is

$$\frac{dP_e}{dx} = -J_{ny}E_y bd = \frac{\sigma U^2 B^2 bdK(1 - K)}{1 + \beta_n^2} \tag{94}$$

and the mechanical (braking) power extracted from the plasma, dP_m/dx, is

$$\frac{dP_m}{dx} = -J_{ny}BUbd \tag{95}$$

The electrical efficiency of the generator, η_e, is

$$\eta_e = \frac{dP_e}{dP_m} = K \tag{96}$$

The electric power generated per unit volume of channel, p_e, is

$$p_e = \frac{dP_e}{bd}\,dx = \frac{\sigma U^2B^2K(1 - K)}{1 + \beta_n^2} \quad (W/m^3) \tag{97}$$

which is theoretically reduced by the Hall parameter, β_n, if $\beta_n >> 1$.

A high power/volume ratio is necessary to reduce the generator size, and thus when $\beta_n >> 1$ the Hall effect should be avoided or, on the contrary, utilized.

One way to reduce the Hall effect drastically is to segment the electrodes.

<u>Segmented electrodes</u>

For identical loads for each pair of segments it could be accepted that $J_{nx} = 0$. Thus (87) and (88) become

$$E_x = \beta_n(E_y - UB) \tag{98}$$

$$J_{ny} = \sigma(E_y - UB)$$

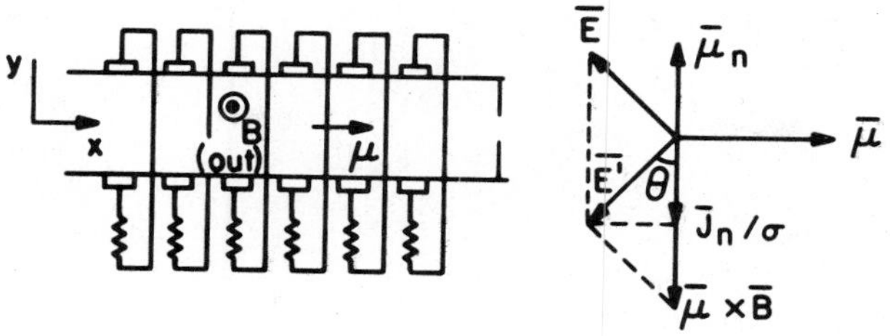

Fig. 5.28. Vectorial diagrams--segmented electrode.

the phase relationships of which are shown in Fig. 5.28. The value of p_e is now

$$p_e = \sigma U^2 B^2 K(1 - K) \quad (W/m^3)$$

$$\eta_e = K \tag{99}$$

Hence, the electrical efficiency is the same as above but the electric power per unit volume is considerably increased. However, the division of loads makes this solution hardly practical.

Hall type MHD generator

By short circuiting the opposing segments (Fig. 5.23), E_y becomes zero. Thus, again from (87) and (88), we obtain

$$J_{nx} = \frac{\sigma(E_x + \beta_n UB)}{1 + \beta_n^2}$$

$$J_{ny} = \frac{\sigma(-UB + \beta_n E_x)}{1 + \beta_n^2} \tag{100}$$

Now, for no-load conditions $J_{nx} = 0$ and thus (100) yields

$$E_{x0} = -\beta_n UB$$

$$J_{ny0} = -\sigma UB \tag{101}$$

Therefore, for no-load conditions Joule losses inevitably occur in the channel. The load factor, K_H, is now

$$K_H = \frac{E_x}{E_{x0}} = \frac{E_x}{-\beta_n UB} \tag{102}$$

Again the dP_e/dx and dP_m/dx are, respectively,

$$\frac{dP_e}{dx} = \frac{\sigma U^2 B^2}{1 + \beta_n^2} K_H(1 - K_H)bd \tag{103}$$

$$\frac{dP_m}{dx} = \frac{\sigma U^2 B^2}{1 + \beta_n^2} (1 + \beta_n^2 K_H)bd \tag{104}$$

The electrical efficiency is

$$\eta_{eH} = \frac{dP_e}{dP_m} = \frac{\beta_n^2 K_H(1 - K_H)}{1 + \beta_n^2 K_H} \tag{105}$$

The peak value of η_{eH} is obtained for

$$K_{H0} = \frac{1}{\beta_n}\sqrt{1 + \frac{1}{\beta_n^2}} - \frac{1}{\beta_n^2} \tag{106}$$

and thus, the value of η_{eHmax} is

$$\eta_{eHmax} = 1 + \frac{2}{\beta_n^2} - \frac{2}{\beta_n}\sqrt{1 + \frac{1}{\beta_n^2}} \qquad (107)$$

The maximum electrical efficiency increases with β_n. For $\beta_n = 7$, $K_{H0} = 0.124$ and $\eta_{eHmax} = 0.75$. The Hall-type MHD generator works efficiently up to almost short-circuited condition, but the other two work best close to no-load conditions.

The electric power density, p_{eH}, is

$$p_{eH} = \frac{dP_e}{(bd)dx} = \frac{\sigma U^2 B^2 \beta_n^2 K_H(1 - K_H)}{1 + \beta_n^2} \qquad (108)$$

For high values of β_n ($\beta_n > 7$) the ideal power density (of a segmented-electrode-type generator) is almost obtained with the Hall-type generator. Since in practice the value of β_n is in many cases greater than 0.7 but smaller than 7, none of the above solutions represents an optimum. Thus, the slant-wall duct generator has been proposed[21] to combine the advantages of segmented electrodes and of Hall-type connection by a diagonal connection of electrodes.

Slant-wall duct generator

A typical slant-wall duct is shown in Fig. 5.29. The inclination angle of the short-circuited electrodes is α_i and should be close to the Hall angle θ ($\tan \theta = \mu_n B$) corresponding to an electrical equipotential line. The electric field E'' in the fluid will be almost perpendicular to the short-circuited electrodes exploiting better, electrically, the duct volume. In this case,[18]:

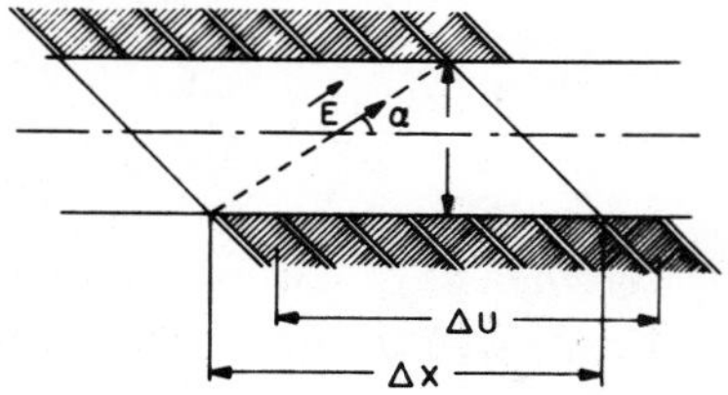

Fig. 5.29. Slant-wall duct.

$$\frac{E_x}{E_y} = -\tan \alpha_i \tag{109}$$

The open-circuit voltage per unit width of this generator, E_0, is

$$E_0 = UB(\tan \alpha_i + \beta_n) \tag{110}$$

Thus the loading factor K_D is

$$K_D = \frac{E_x[1 + (\tan \alpha_i)^{-2}]}{E_0} \tag{111}$$

From (87) and (88) with (109) and (111) we get

$$E_x = -\frac{K_D UB[(\tan \alpha_i)^{-1} - \beta_n]}{[(1 + (\tan \alpha_i)^{-2}]} \tag{112}$$

$$E_y = \frac{K_H UB[(\tan \alpha_i)^{-1} + \beta_n]}{[1 + (\tan \alpha_i)^{-2}]} \tag{113}$$

$$J_{nx} = \sigma UB \frac{\beta_n(1 - K)}{(1 + \beta_n^2) - K(\tan \alpha_i)^{-1}/[1 + (\tan \alpha_i)^{-2}]} \tag{114}$$

$$J_{ny} = \sigma UB \frac{1 - K}{1 + \beta_n^2 + K/[1 + (\tan \alpha_i)^{-2}]} \tag{115}$$

The electric power per unit volume, P_{eD}, is

$$p_{eD} = \frac{\sigma U^2 B^2 [(\tan \alpha_i)^{-1} + \beta_n]^2}{(1 + \beta_n^2)[1 + (\tan \alpha_i)^{-2}]} \tag{116}$$

As expected when $\alpha_i \to 0$ the case of Hall generator is obtained and $\alpha_i \to \infty$ corresponds to the continuous-electrode situation. The electrical efficiency of the generator is

$$\eta_{e\Delta} = \frac{(E_x J_{nx} + E_y J_{ny})}{J_{ny} UB} \tag{117}$$

With the above expressions it may be demonstrated that in general the slant-wall duct generator can produce higher power densities, p_{eD}, than the continuous-electrode and Hall generators at high electrical efficiences ($\eta_{e\Delta}$ = 0.7 to 0.8) for moderate values of Hall parameter β_n = (0.5 to 5). This range of β_n is most common in practice, and thus, the slant-wall duct generator has become the most practical solution so far. There are also some advantages in the technology of making the short-circuiting connections and the increasing of output voltage due to the series connection of electrode segments.

5.3.3. One Dimensional Magnetic Flow Equations

So far the speed and plasma properties have been considered constant throughout the channel. The one-dimensional flow equation will be developed in what follows to show the electromagnetic-flow interactions. The equation of motion including the Lorenz force, for a

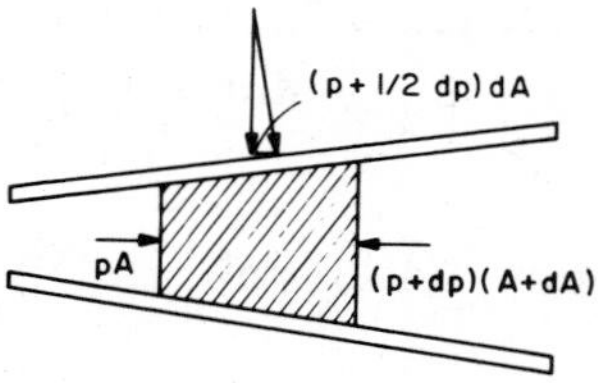

Fig. 5.30. Control volume of channel.

control volume of fluid (Fig. 5.30), yields the form

$$pdA + pA - (p + dp)(A + dA) - J_{ny}BA\ dx = d(\zeta AU^2) \qquad (118)$$

where ζ = the fluid density. Again, if the wall friction and viscosity losses are neglected, we obtain

$$\frac{d}{dx}(\zeta AU^2)dx = -A(\frac{dp}{dx})dx - J_yBA\ dx \qquad (119)$$

Denoting by m the flow rate $m = \zeta\ AU$, (119) becomes

$$\frac{m}{A}\frac{dU}{dx} + \frac{dp}{dx} = -J_{ny}B \qquad (120)$$

If the speed U is constant,

$$(\frac{dp}{dx})_{U=ct} = -J_{ny}B \qquad (121)$$

The total energy per unit weight of fluid, ε, is

$$\varepsilon = \varepsilon_{int} + \frac{p}{\zeta} + \frac{1}{2}U^2 = h + \frac{1}{2}U^2 \qquad (122)$$

where ε_{int} = the internal specific energy, p/ζ = specific energy of flow, and h = specific enthalpy. Applying now the law of energy conservation for the same constant volume of Fig. 5.30 we get

$$\frac{d}{dx}\zeta AU(h + \frac{1}{2}U^2) = -(J_xE_x + J_yE_y)A \qquad (123)$$

For constant speed and no Hall effect, (123) becomes

$$(\zeta U \frac{dh}{dx})_{U=ct} = (-J_{ny}E_y)_{J_{nx}=0} \qquad (124)$$

Equations (120) and (123) represent the basic equations of one-dimensional flow of the plasma in a uniform magnetic field. In practice the variations of the fluid density (due to temperature and pressure differences) and of magnetic field from point to point make the complete study of the plasma flow very difficult, and this problem is beyond our scope.

5.3.4. Thermodynamic Aspects (Adiabatic Efficiency)

It is evident that, when the speed is constant along the channel, by an adequate increase of channel cross section from entry to exit end, the flow rate can be kept constant. Electrical energy is obtained by decreasing the plasma pressure and temperature from entry to exit end of channel. Let us consider a constant-speed MHD generator. The speed of sound U_s in plasma (considered as a perfect gas) is

$$U_s = \frac{\gamma p}{\zeta} \; ; \quad \gamma = \frac{C_p}{C_v} \; ; \quad h = C_p T \; ; \quad p = \zeta RT \qquad (125)$$

where C_p and C_v = specific heat coefficients; R = the perfect gas constant. The Mach number M is the ratio between the actual speed, U, and the corresponding speed of sound:

$$M = \frac{U}{U_s} \qquad (126)$$

Let us consider that the stagnation temperature is T_0 at a point in the plasma (zero speed), T is the static temperature, and U is the speed. The energy conservation provides the equality:

$$h + \frac{U^2}{2} = h_0 \ ; \quad (h_0 - h) = C_p(T_0 - T) \tag{127}$$

Thus

$$\frac{T_0}{T} = 1 + \frac{U^2}{2C_pT} = 1 + \frac{\gamma - 1}{2\gamma RT} U^2 = 1 + (\frac{\gamma - 1}{2})M^2 \tag{128}$$

We also assume that in the absence of magnetic field the relation between pressures and temperatures is that corresponding to isentropic conditions:

$$\frac{p_0}{p} = (\frac{T_0}{T})^{(\gamma-1)/\gamma} = (1 + \frac{\gamma - 1}{2} M^2)^{\gamma/(\gamma-1)} \tag{129}$$

So far the plasma processes have been considered conservative. The electric energy produced is accompanied by Joule losses in the plasma, and thus the thermodynamic process is not reversible. The new dependence between pressure and temperature should include the load factor, K. Let us neglect the Hall effect and consider the speed constant. Making use of (93), (121), (124), and (127) the loading factor, K, becomes

$$K = \frac{E_y}{UB} = \frac{\zeta(U/J_{ny})\ (dh/dx)}{U(dp/dx)\ (1/J_{ny})} = \zeta \frac{dh}{dp} = \zeta C_p \frac{dT}{dp} \tag{130}$$

or finally

$$\frac{dT}{T} = K \frac{(\gamma - 1)\ dp}{\gamma\ p} \tag{131}$$

The solution to (130) is

$$\frac{T}{T_1} = (\frac{p}{p_1})^{K(\gamma-1)/\gamma} \tag{132}$$

where (T_1, p_1) correspond to the point 1 and (T, p) to an arbitrary point along the channel. The temperature drop along the channel when $K \neq 0$ is smaller than for the case of $K = 0$ because of the Joule losses, as shown by (129). It should again be remembered that the wall-friction viscosity losses are neglected. Also, thermal-wall losses have not been considered, and the process is assumed adiabatic. In the absence of electric power generation the adiabatic process would become isentropic.

5.3.5. Channel Design Guidelines

In general the initial thermodynamic data are pressure and temperature at the entry of the accelerating nozzle (p_0, T_0), the pressure at the exit of the diffuser (p_3), the Mach number at entry (M_1), and the loading factor K (Fig. 5.31). In the nozzle and diffuser the processes are considered isentropic, (129). Thus

$$\frac{T_1}{T_0} = \left(\frac{p_1}{p_0}\right)^{(\gamma-1)/\gamma} \; ; \quad \frac{T_3}{T_2} = \left(\frac{p_3}{p_2}\right)^{(\gamma-1)/\gamma} \qquad (133)$$

For MHD channel entry and exit ends (132) holds:

$$\frac{T_2}{T_1} = \left(\frac{p_2}{p_1}\right)^{K[(\gamma-1)/\gamma]} \qquad (134)$$

But according to (128),

$$M_2^2 = \frac{U^2}{\gamma R T_2} = \frac{U^2}{\gamma R T_1}\,\frac{T_1}{T_2} = M_1^2\left(\frac{p_1}{p_2}\right)^{K[(\gamma-1)/\gamma]} \qquad (135)$$

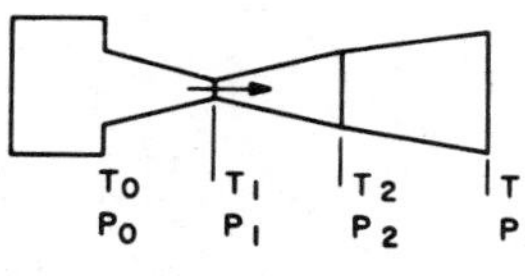

Fig. 5.31. Longitudinal view of the channel.

$$\frac{p_1}{p_0} = [1 + \frac{\gamma-1}{2} M_1^2]^{(\gamma-1)/\gamma} ;$$

$$\frac{p_2}{p_3} = [1 + \frac{1}{2}(\gamma - 1)M_2^2]^{(\gamma - 1)/\gamma)} \qquad (136)$$

Finally,

$$\frac{p_3}{p_0} = \frac{p_3}{p_2}\frac{p_2}{p_1}\frac{p_1}{p_0} = \frac{p_2}{p_1}$$

$$\cdot [\frac{1 + (1/2)(\gamma - 1)M_1^2(p_1/p_2)^{K(\gamma-1)/\gamma}}{1 + (1/2)(\gamma - 1)M_1^2}]^{\gamma/(\gamma - 1)} \qquad (137)$$

As M_1, p_3, p_0 are known the ratio (p_2/p_1) of the exit to entry pressures in the channel could be found by a numerical standard procedure, from (137). Further, from (132) it follows that

$$\frac{T_3}{T_0} = [(\frac{p_1}{p_2})^{(1-K)} \frac{p_3}{p_0}]^{(\gamma\frac{-1)}{\gamma}} \qquad (138)$$

And thus the temperature, T_3, at the diffuser exit is found. Considering the constant-speed case, from (122) the length of the channel could be calculated for the continuous electrode:

$$(1)_{U=ct} = \int \frac{dp(1 + \beta_n^2}{(1 - K)\sigma UB^2} = \frac{p_1(1 + \beta_n^2)}{(1 - K)\sigma UB^2}(1 - \frac{p_2}{p_1}) \qquad (139)$$

As p_1 can be calculated from (136), the channel length is directly obtainable from (139). Finally, from (133), T_1 and T_2 are found. In order to maintain the flow rate at a constant speed, the channel cross section should be increased toward the exit end. The ratio of cross sections is

$$\frac{A}{A_1} = \frac{\zeta_1}{\zeta} = \frac{p_1}{p}\frac{T}{T_1} = \left(\frac{p}{p_1}\right)^{K[(\gamma-1)/\gamma] - 1} \tag{140}$$

where A is the cross section at an arbitrary point along the channel. As p/p_1 varies linearly, in our case, along the channel length we have

$$\frac{p(x)}{p_1} = \left(1 - \frac{x}{l}\frac{p_1 - p_2}{p_1}\right) \tag{141}$$

Thus from (140) the variation of channel section along OX, A(x) could be calculated. Finally the adiabatic efficiency, η_{ad}, is defined as

$$\eta_{ad} = \frac{T_0 - T_3}{T_0 - T_3'} \tag{142}$$

T_3' is the temperature at the diffuser exit that would exist should the processes in the channel be isentropic:

$$\frac{T_3'}{T_0} = \left(\frac{p_3}{p_0}\right)^{(\gamma-1)/\gamma} \tag{143}$$

Finally, η_{ad} is given by

$$\eta_{ad} = \frac{1 - [(p_1/p_2)^{1-K}(p_3/p_0)]^{(\gamma-1)/\gamma}}{1 - (p_3/p_0)^{(\gamma-1)/\gamma}} \tag{144}$$

As expected the adiabatic efficiency approaches 100% when the loading factor is close to 1 and p_3/p_0 is sufficiently low. The thermodynamic efficiency, η_t, is used to characterize completely the MHD energy conversion:

$$\eta_t = \frac{IV - P_{aux}}{\Delta H} \tag{145}$$

where VI is the electric power output, P_{aux} is the auxiliary power for producing the magnetic field, electrode losses, seed recovery, and so on, and ΔH is the fuel enthalpy variation in the combustor. To find η_t all MHD plant aspects should be accounted for. This is beyond our scope. To illustrate the MHD channel design guidelines a numerical example is given.

5.3.6. A numerical example

Let us calculate an MHD channel with the following data: combustor exit temperature $T_0 = 3000$ K; combustor exit pressure $p_0 = 6$ atm; diffuser exit pressure $p_3 = 1.2$ atm; flux density $B = 4$ T; Hall parameter $\beta_n = 1.6$; entry geometry $b = 0.5$ m; $d = 2$ m; electrical conductivity $\sigma = 40.6\ (\Omega m)^{-1}$ at 1 atm; Mach number at entry $M_1 = 0.8$; loading factor $K = 0.5$; $\gamma = 1.2$.

The following unknowns are to be calculated: $T_1, p_1, p_2, T_2, \sigma_{av}$ (for the average pressure in the channel, $p_{av} = [(p_1 + p_2)/2]/U)$, u, J_{ny}, A_2, I, VI, T_3', η_{ad}. Making use of the expressions developed above we

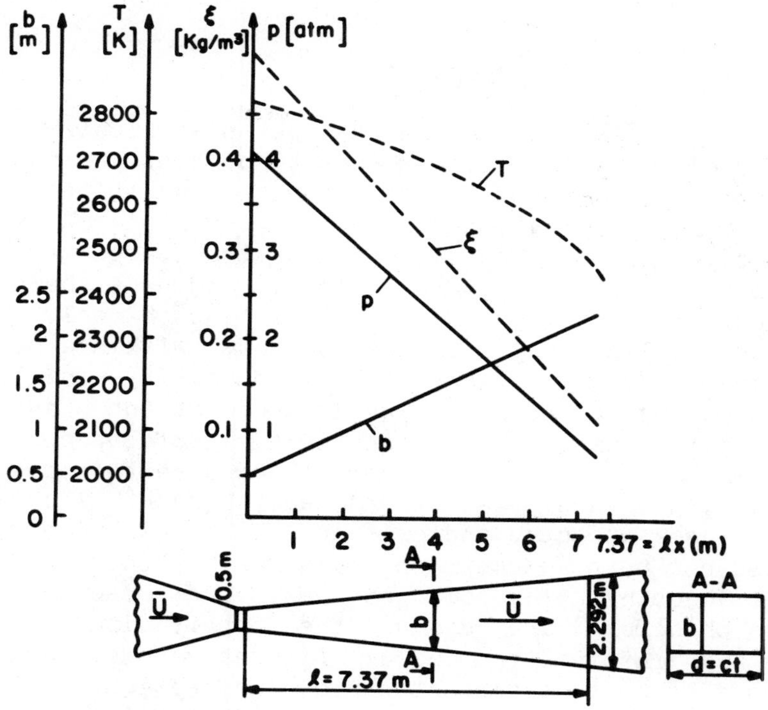

Fig. 5.32. Distribution along the channel length:
T = temperature, ρ = density, p = pressure,
b = channel height.

obtain: $T_1 = 2819.6$ K; $T_2 = 2455$ K; $\sigma_{av} = \sigma\sqrt{1/p_{av}} = 25.93\ (\Omega m)^{-1}$; $U = 787.7$ m/sec; $p_1 = 4.1352$ atm; $l = 7.37$ m; $J_{ny} = 11{,}474\ A/m^2$; $A_2 = 4.583\ m^2$; $p_2 = 0.7856$ atm; $I = 1.18 \cdot 10^5$ A; $V = 3150.8$ V; $P_e = I \cdot V = 371.88$ MW; $\eta_{ad} = 0.517$. The temperature, T, plasma density, ζ, pressure, P, and channel cross section along OX are plotted in Fig. 5.32.

5.3.7. Note on Theory Refinements

It has been assumed that the electrical conductivity of the plasma is the same throughout the channel. However, the temperature and pressure variations indicate that a variation of conductivity along OX exists. In the vicinity of the electrodes the temperature is notably lower than in the channel, and here the plasma conductivity is even smaller than in the channel. Thus, it is qualitatively indicated that within a refined design study the electrical conductivity should be included as a tensor. Furthermore, at entry and exit ends of the channel, the magnetic field of superconducting magnets attenuates to zero along a certain distance (Fig. 5.33), causing additional eddy current losses in the plasma at entry and exit ends. The electromagnetic forces per unit volume then would vary at entry and exit ends, causing flow disturbances. The consequences of these phenomena are called the longitudinal effect. Only by involved numerical (iterative) bidimensional methods--the electromagnetic flow interactions including the conductivity as a tensor, the viscosity and wall-friction losses, and so on--could an adequate mathematical treatment be given. This is beyond our scope. A pertinent engineering study of this complex problem, including theory, design, and experiments is presented in Ref. 19.

Finally, a typical open cycle MHD generator scheme, including, an electric power plant, is shown in Fig. 5.34. Closed cycle MHD generators are proposed for topping nuclear power plants.[17]

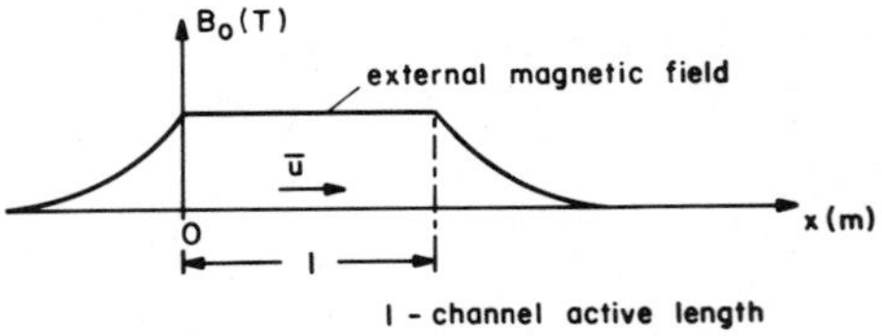

Fig. 5.33. Longitudinal profile of superconducting magnet flux density.

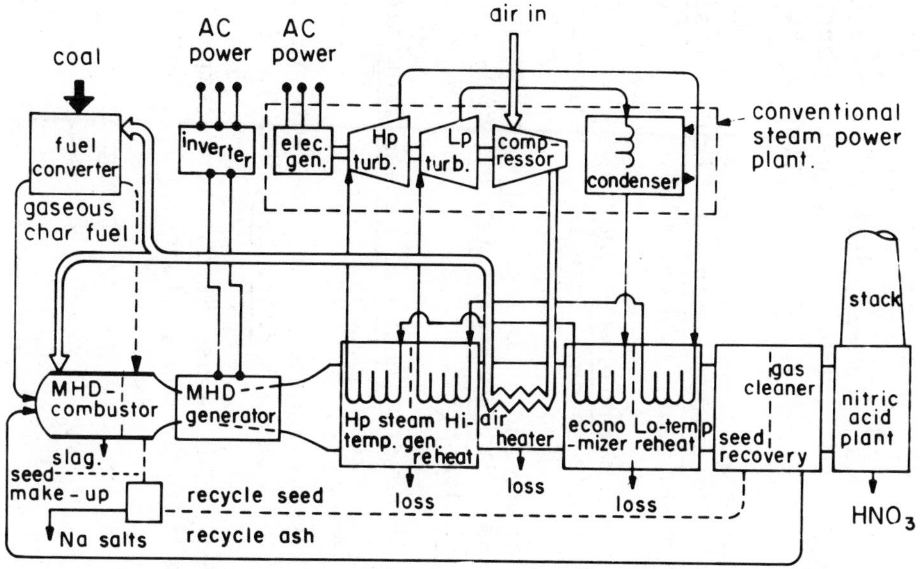

Fig. 5.34. Open cycle mhd generator scheme.

References

1. L. R. Blake, "Conduction and induction pumps for liquid metals," Rev. IEE, Vol. 104A, Nov. 13, 1957.

2. M. Kant and J. Robert, "Preliminary study of an MHD generator with liquid metal and traveling field" (in French), RGE, Vol. 76, No. 6, 1967.

3. N. M. Ohremenk, Basics of Theory and Design of Linear Induction Pumps for Liquid Metals (in Russian), Atomizdat, Moskow, 1968.

4. I. Voldek, Magnetohydrodynamic Machines for Liquid Metals (in Russian), Energhia, Leningrad, 1970.

5. S. A. Nasar and I. Boldea, Linear Motion Electric Machines, Wiley-Interscience, New York, 1976, chapters 4 and 8.

6. M. Poloujadoff, Theory of Linear Induction Motors, Clarendon Press, London, 1981.

7. H. Bolton, "Transverse edge effect in sheet-rotor induction motors," Proc. IEE, No. 5, 1969.

8. R. Hans and H. Wess, "Induction pumps for liquid metals" (in German), Siemens Rev., No. 4, 1977.

9. G. B. Kliman, "Large electromagnetic pumps," EME, Vol. 3, No. 2, 1979, pp. 129-142.

10. V. Fireteanu, "Contribution to the performance improvement of liquid metal electromagnetic pumping systems" (in Romanian), Ph.D. thesis, Polytechnic Institute of Bucharest, Romania, 1979.

12. A. Nicolaide, et al., "Study on the MHD pumping of mercury in the electrolysis installation for alcaline chlorides," Rev. Rom. Sc. Techn. Serie E.E., Vol. 18, Nov. 1, 1973.

13. E. Z. Anowich et al., "Creation of high temperatures high delivery induction pumps," Magnitnayahidrodinamica, Vol. 12, No. 2, 1976, pp. 71-78 (pp. 185-192 in English).

14. C. C. Yang and S. Kraus, "A large electromagnetic pump for high temperature LMFBR applications," Nuclear Eng. Design, Vol. 44, Nov. 3, 1977, pp. 383-395.

15. Y. A. Birzvalk, Basics of Theory and Design of Conduction MHD DC Pumps (in Russian), Zinatne, Riga, 1968.

16. L. A. Verte, The Magnetohydrodynamics in Metallurgy (in Russian), Metallurgy Publ., Moskow, 1975, Chapter II.

17. R. Rosa, Magnetohydrodynamic Energy Conversion, McGraw-Hill, New York, 1968.

18. G. J. Womack, MHD Power Generators, Engineering Aspects, Chapman and Hall Ltd., (London (England) 1969.

19. J. Raeder (Editor), MHD Power Generation. Selected Problems of Combustion MHD Generators.

20. I. Boldea, Direct Energy Conversion (in Romanian), The Polytechnical Institute of Timisoara Romania, Reprography, 1977.

21. A. De Montardy, in Magnetoplasmadynamic Power Generation (B.C. Lindley, Editor), IEE, London, pp. 22-27.

22. T. G. Cowling. Magnetohydrodynamics Adam Hilger, 1976.

23. C. Carter and J. B. Heywood, "Optimization studies on open gate MHD generators," Proc. IEEE, Vol. 56, 1968, pp. 1409-1419.

24. Z. M. Celinski, "Electrical parameters of the dc MHD generator with arbitrary connected electrodes," Adv. Energy Convers., Vol. 6, 1966, pp. 223-231.

25. H. Ogiwara. "Voltage distribution in the channel of segmented- electrode Faraday MHD generator," Elec. Eng. Jap., Vol. 89, No. 9, 1969, pp. 18-25.

26. L. L. Lengyel, "Two-dimensional current distributions in Faraday type MHD energy convertors operating in the non-equilibrium conduction mode," Energy Convers., Vol. 9, 1969, pp. 13-23.

CHAPTER 6

LINEAR INDUCTION MOTORS FOR TRANSPORTATION

6.1. Construction Guidelines

Linear induction motors (LIMs) have been widely used as propulsion systems for Maglevs (vehicles on magnetic cushion) and for new urban transportation systems on wheels. Of the various types of LIMs, a short on-board primary (double-sided or single-sided) and a passive secondary spread along the guideway constitute a solution acceptable in terms of performance and costs. The fixed-primary (active-guideway) LIM has a low efficiency and power factor[1,2], and is not treated here.

The short primary LIM, single-sided and double-sided (Fig. 6.1) consists of three main parts:

1. Primary magnetic core with open slots.

2. Three phase primary winding.

3. Secondary.

The primary core is made of thin longitudinal laminations and has, in general, open slots (Fig. 6.2) in order to reduce the slot leakage permeance, for a rather high mechanical airgap of 10 to 15 mm.

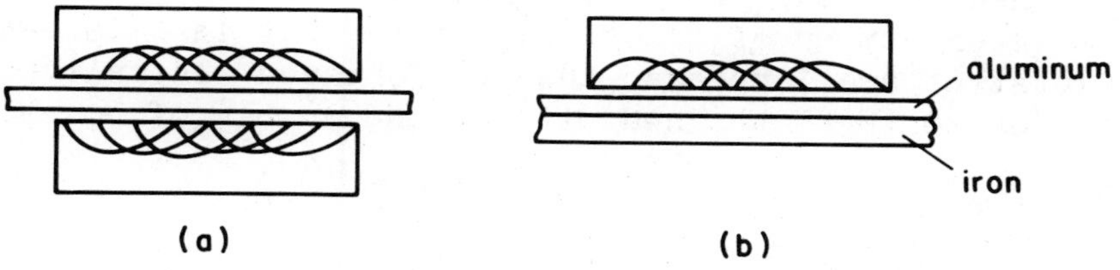

Fig. 6.1. Short-primary LIM with longitudinal flux. (a) Double-sided, (b) single-sided.

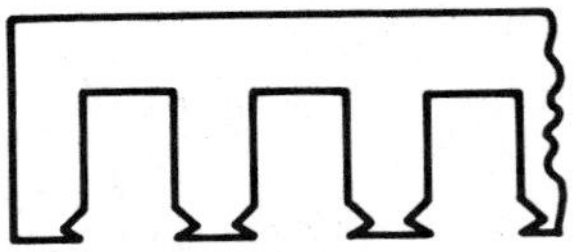

Fig. 6.2. LIM core with open slots.

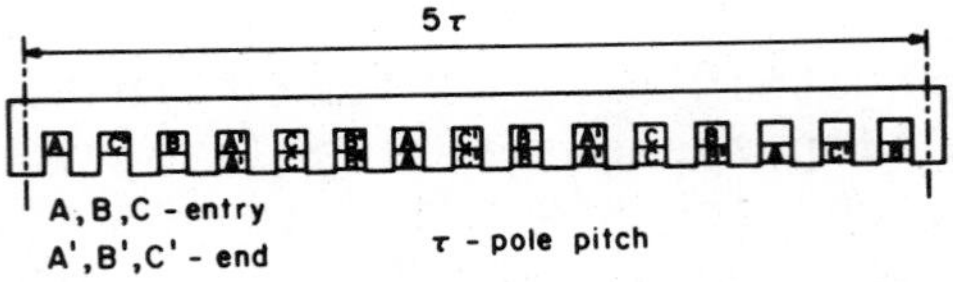

Fig. 6.3. Primary winding with odd total number of poles and half-filled end slots.

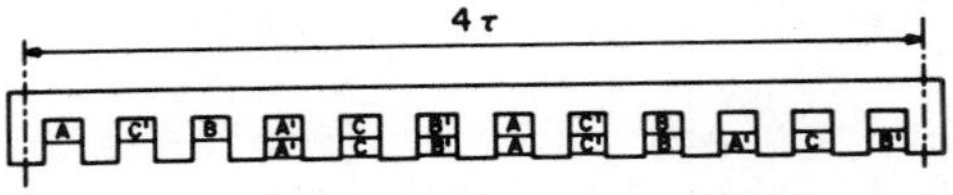

Fig. 6.4. Primary winding with even total number of poles and half-filled end slots.

Studies[3] have been carried out to obtain the optimal winding suitable for the primary longitudinally open magnetic circuit. Theoretical investigations followed by experimental tests on LIMs of up to 1800 kW and 110 m/sec have led to the conclusion that a double-layer fractional-pitch three-phase winding with a total odd number of poles, and half-filled end slots (Fig. 6.3) is economically superior to other choices. It should be mentioned that such a winding may also be built for a total even number of poles (Fig. 6.4), but in this case, at least at small speeds, the performance is slightly inferior to that obtained with an odd number of poles.

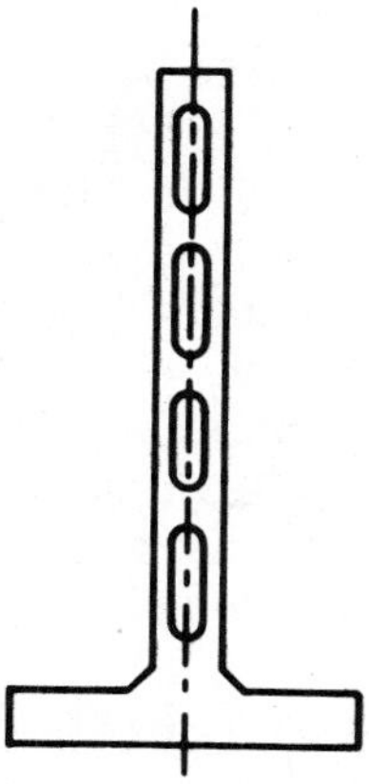

Fig. 6.5. Secondary of DSLIM.

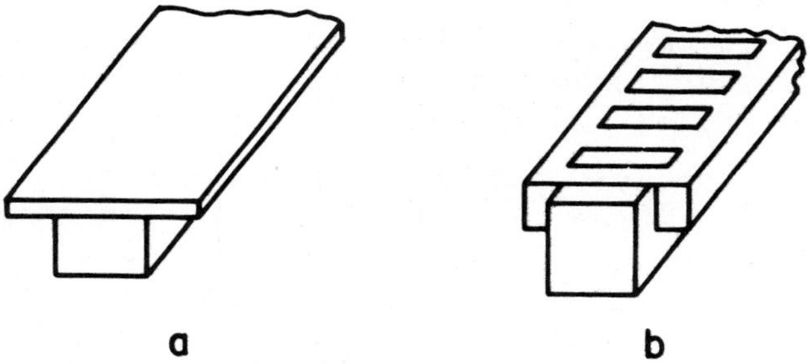

Fig. 6.6. Secondaries of SLIM. (a) Aluminum sheet on solid iron, (b) ladder secondary.

The secondary of a double-sided LIM (DSLIM) consists of an aluminum sheet with a simple or special cross section for increased mechanical strength (Fig. 6.5). For a single-sided LIM (SLIM) there are two main types of secondary (Fig. 6.6): aluminum sheet on solid iron, and ladder secondary (aluminum "cage" in laminated core). The secondary made of an aluminum sheet secured to a "solid" iron core made of only 2 to 3 laminations[4] is not expensive but yields a satisfactory performance.

The higher cost of the ladder secondary[5] is partly justified by a better performance, mainly because a 25 to 35% reduction in the "magnetic" airgap of LIM.

For a general mathematical treatment, we consider here an aluminum-sheet-on-iron secondary LIM in detail. The equivalence conditions of the ladder secondary with the sheet-on-iron secondary are also pointed out. Thus, in fact, the ladder secondary may be treated by the theory developed for the sheet-on-iron secondary LIM. Before proceeding with the analysis certain specific phenomena, occurring in LIMs applicable to Maglevs, are briefly and qualitatively summarized.

6.2. Specific Phenomena in High-Speed LIMs

6.2.1. End Effect

Specific phenomena introduced by the finite length of the primary core and winding along the direction of motion are called end effects. The end effect could be divided into static end effect and dynamic end effect.[6] The static end effect is caused by the open character of the primary magnetic circuit, implying an inevitable asymmetrical positioning of the three-phase coils with respect to the primary core entry and exit ends. Consequently the coupling inductances between phases differ from one another. However, for a number of poles $2p \geq 8$ the static end effect may be neglected for design purposes. In general, the LIMs for transportation fulfill this condition. So the static end effect is neglected here. The dynamic end effect is caused by the relative motion between the finite-length primary and the secondary of "infinite" length.

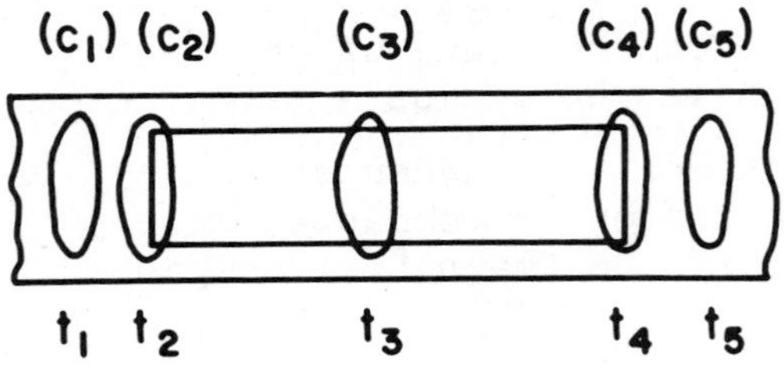

Fig. 6.7. Dynamic end effect (qualitative view).

To understand the dynamic end effect, we assume that the contour, C, of the secondary occupies, at conventional synchronism ($U = U_s$), the successive positions (C_1) through (C_5) at times t_1 through t_5, as shown in Fig. 6.7. No voltages are induced when the contour is in positions (C_1) and (C_5) because of the absence of any magnetic coupling with the primary magnetic field. The same situation occurs in position (C_3) but the reason here is that the flux in the contour does not vary because the speed, U, is equal to the synchronous primary-field speed, U_S. However, in positions (C_2) and (C_4) the flux in the contour varies and induced currents occur in the secondary. Moreover, when the contour (C) moves from (C_2) to (C_4) the entry end induced currents, (C_2), experience a rather slow attenuation. The exit end induced currents, (C_4), also attenuate, but faster, between positions (C_4) and (C_5). Consequently, even at synchronism, while neglecting the primary field harmonics, induced currents occur in the secondary causing additional forces and losses. These phenomena occur notably at speeds close to synchronism and constitute the dynamic end effect.

In general, for high-speed LIMs, the additional longitudinal force occurring in the low slip region has a braking character causing a notable reduction in efficiency and power factor. It is well understood that the end effect global influence decreases when the number of poles increases. The dynamic end effect is a maximum at synchronism but it deserves attention within a rather wide low-slip region: s = 0.2 to 0.0.

Because of the combined influence of static and dynamic end effects, the voltages induced in the three phases are not equal, even if the currents injected are symmetric. An additional performance deterioration is thus produced. Fortunately, even for high speeds, 100 to 120 m/sec, for optimally designed LIMs of Maglev, the rate of induced voltages asymmetry hardly goes beyond 2 to 3%, and thus, for engineering purposes, this asymmetry may be neglected.

The induced currents trajectories for synchronism ($U = U_s$) and for a subsynchronous speed ($U < U_s$) are qualitatively plotted in Fig. 6.8, showing the entry and exit end currents caused by end effect. It is well understood that the end-effect-induced currents will also distort the airgap flux density longitudinal distribution. With these qualitative remarks in mind, we now consider the transverse edge effect.

6.2.2. Transverse Edge Effect

The limited widths of the primary, 2a, and of the secondary, 2c, as well as the nature of the induced current trajectory, imply the existance of a longitudinal component, J_x, of current density, besides the transverse one, J_y (Fig. 6.8). The consequences of this situation are called transverse edge effects. Transverse edge effect decreases when the secondary width, 2c, increases with respect to 2a, up to $2c_{max} = 2a + 2\tau/\pi$ where τ is the pole pitch. A secondary wider than $2c_{max}$ would not result in a notably better performance since the induced current trajectory does not extend significantly beyond $2c_{max}$.

Because of longitudinal current density, J_x, in the active zone ($|y| <= a$), the current paths seem longer than in the case when J_x is absent, as if the equivalent secondary resistivity is increased by a factor of K_t. Moreover, a demagnetizing reaction field is created by J_x in the active zone ($|y| \leq a$). Consequently, the transverse edge effect leads to an increase in the secondary Joule losses and a reduction of power factor. Finally, when a lateral displacement of primary occurs ($c \neq b$), the interaction between J_x and B_g (Fig. 6.9a) produces a decentralizing lateral force, F_y. The lateral

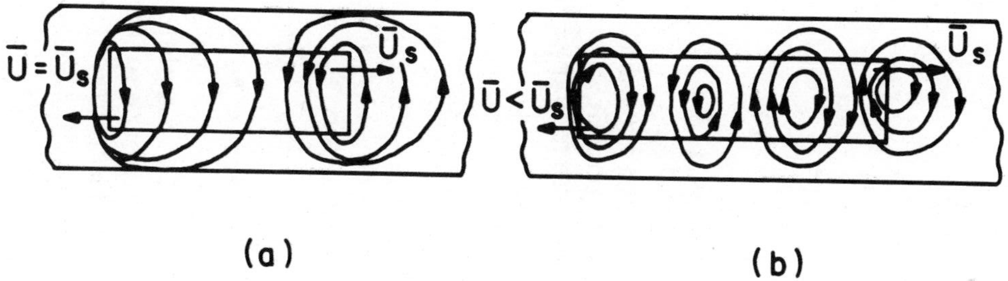

Fig. 6.8. Secondary-induced currents trajectory (qualitative). (a) $U = U_s$, (b) $U < U_s$.

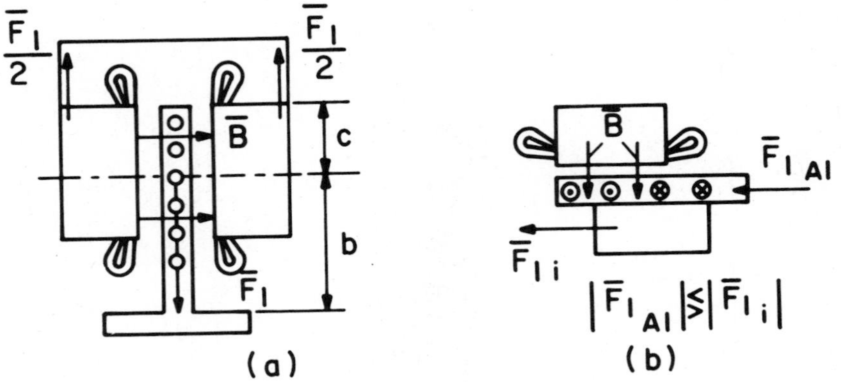

Fig. 6.9. Lateral decentralizing force.

force of a SLIM also exhibits an additional "centralizing" lateral force component, due to the alignment tendency of primary and secondary magnetic cores (Fig. 6.9b).

6.2.3. Normal Force

In contrast with the rotary machines the normal force in a LIM is generally not zero and has two components: an attraction force, F_{na}, and a repulsion force, F_{nr}, as shown in Fig. 6.10. The repulsion force is caused by the interaction between primary and secondary magnetomotive forces (mmfs) and the attraction force acts between the two magnetic cores of the LIM. In SLIMs for Maglev, the resultant normal force on the secondary is of attraction type for low slip frequencies and repulsion type for high slip frequencies.

For the DSLIM the repulsion normal force acting on the secondary sheet has a centralizing character, working like a spring. Besides, the repulsion normal force has a nonuniform longitudinal distribution, and thus a torque occurs tending to increase the airgap at entry end and to decrease it at the exit end. This phenomenon

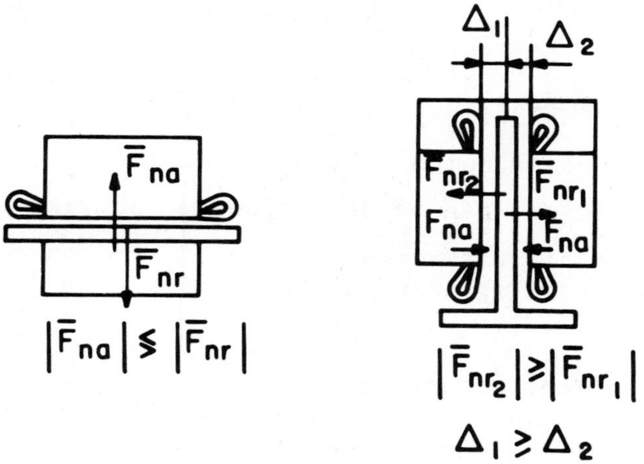

Fig. 6.10. Normal force. (a) For SLIM, (b) for DSLIM.

has been called the dolphin effect.[3] The normal force should always be accounted for in the design process of LIM frame structure.

6.2.4. Airgap Leakage

Mechanical and electric restraints lead to a secondary sheet effective thickness of 5 to 10 mm for a LIM rated at 2.5 MW and 100 m/sec. On the other hand the mechanical clearance between the secondary and the priamry is not less than 10 to 12 mm for speeds of 100 to 120 m/sec in order to prevent, with reasonable effort, motor-guideway collisions. Consequently the magnetic (total) airgap (magnetic core to magnetic core) is

Fig. 6.11. Airgap leakage.

between 15 and 17 mm for SLIM and 35 to 40 mm for DSLIM. Within this rather large magnetic airgap the flux density experiences a distortion in the sense that the "useful" normal airgap flux density, B_z, is accompanied by a longitudinal component, B_x (Fig. 6.11). The airgap leakage influence can be determined by solving the field equations within the airgap region.[7] In essence this effect causes an apparent increase of airgap represented by a ratio $K_1 \geqq 1$.

Skin effect occurs in the aluminum sheet, especially in the solid secondary core. A two-dimensional field analysis, numerical or analytical, can yield the current density distribution in the secondary. Finally, resistive and inductive correction coefficients may be derived in a similar manner as for rotary machines.

In reality the secondary frequency (or slip frequency) must be kept between 3 and 12 Hz, in order to keep the performance at acceptably high values. Under these conditions, for the aluminum sheet, the skin effect correction coefficients known from rotary machinery[8] may be used satisfactorily.

Theoretically, the diversity of specific phenomena, just briefly described, requires a three-dimensional field analysis. Two- and three-dimensional analytical,[9] as well as two-dimensional numerical, approaches[10] have been proposed for solving the field equations in LIMs. However, the computing time is generally too high for design purposes. Furthermore it has been proved, in theory and tests,[11] that very useful results can be obtained if the end effect is studied with a quasi-one-dimensional field approach after the transverse-edge, skin and airgap-leakage effects have been taken into account, using equivalent correction coefficients resulting from separate analyses of each of these phenomena. The correction coefficients should be determined for each value of primary mmf and slip frequency (given the geometrical dimensions).

This theoretical approach, called here the field technical theory, facilitates a rather straightforward identification of those factors that lead to an optimal LIM design, while drastically reducing the computing time. After a preliminary design has been accomplished, the two- and three-dimensional field theories are very useful for increasing the precision as well as for determination of the field and current patterns. However, only the field technical theory is developed in some detail in the present book. Thus to study the end effect, first the equivalent correction coefficients of transverse-edge airgap leakage and skin effects are evaluated.

6.3. Transverse-Edge Effect Correction Coefficients

Here only the SLIM is considered and the DSLIM is taken as a special case of the former. The solid iron core in the secondary of a SLIM causes additional eddy currents and magnetic core saturation. These aspects must be accounted for in any realistic analysis. For a given motor the degree of saturation depends upon primary mmf, F_{m1}, and secondary frequency, f_2. To account instantaneously for transverse edge effect and secondary iron core eddy currents and saturation, a multistage iterative approach[4] is used. Thus, first, the transverse edge effect resistive, K_{tr}, and inductive, K_{tm}, correction coefficients are determined from a field analysis that takes into account only this phenomenon,[12] assuming that the secondary iron core resistivity and permeability are infinitely high:

$$K_{tr} = \frac{K_x^2}{K_R} \frac{1 + (sG_{ep}K_R/K_x)^2}{1 + s^2G_{ep}^2} \geq 1 \qquad (1)$$

$$K_{tm} = \frac{K_R}{K_x} K_{tr} \leqq 1 \qquad (2)$$

where

$$K_R = 1 - \text{Re}[(1 - jsG_{ep}) \frac{\lambda}{\underline{\alpha}a_e} \tanh(\underline{\alpha}a_e)] \qquad (3)$$

$$K_x = 1 + \text{Re}[(sG_{ep} + j) \frac{sGep}{\underline{\alpha}a_e} \tanh(\underline{\alpha}a_e)] \qquad (4)$$

$$\underline{\lambda} = \frac{1}{1 + \sqrt{(1 + jsG_{ep})} \tanh \underline{\alpha}a_e \tanh (\pi/\tau)(c - a_e)} \qquad (5)$$

$$\underline{\alpha} = \frac{\pi}{\tau}\sqrt{1 + jsG_{ep}};$$

$$G_{ep} = \frac{2\mu_0 f_1 \sigma_{Al} \tau^2 d_{A1}}{\pi g_0 K_c (1 + K_p)}; \quad g_0 = g_m + d_{Al} \qquad (6)$$

f_1 = primary frequency; τ = primary pole pitch; g_m = mechanical clearance; d_{Al} = aluminum sheet thickness; σ_{Al} = sheet conductivity; s = slip; K_c = Carter coefficient; G_{ep} = goodness factor; k_p = an equivalent coefficient accounting for the secondary magnetic core saturation.

An approximate expression[4] of K_p is as follows:

$$K_p = \frac{\tau^2}{\pi^2} \frac{\mu_0}{\mu_i \delta_i g_0 k_c} \qquad (7)$$

where δ_i is the equivalent field penetration depth in the secondary magnetic core, and is given by

$$\delta_i = \text{Re} \frac{1}{[(\pi/\tau)^2 + 2j\mu_i \pi f_1 s \sigma_i / K_{ti}]^{1/2}} \tag{8}$$

In (8), K_{ti} is the correction coefficient accounting for the transverse edge effect corresponding to the eddy currents induced in secondary magnetic core, while μ_i is its equivalent permeability. The coefficient K_{ti} could be deduced from (1) through (6) if the iron eddy currents reaction is neglected ($G_{ep} \to 0$) and if the secondary iron width equals the primary core width ($c \to a_e$):

$$K_{ti} = \frac{1}{1 - (i\tau/\pi a_e) \tanh(\pi a_e/\tau i)} \tag{9}$$

where i is the number of "solid" longitudinal laminations (i = 1 to 3) of secondary core, and a_e is the equivalent half-width of the primary core:

$$a_e \simeq a + g_0/2 \tag{10}$$

In (10) the transverse fringing has been accounted for as approximated from conformal mapping results.

Now, given an initial value to the equivalent permeability, μ_i, (1) through (10) may be used to yield the correction coefficients, K_{tr} and K_{tm}, provided the geometrical dimensions and the slip frequency are known. However, to determine realistically the value of μ_i we first replace the real motor by an ideal motor (without transverse edge effect) having an ideal secondary core ($\sigma_i \to \infty$, $\mu_i \to \infty$), but equivalent otherwise to the real motor. The equivalent airgap, g_{e0}, and equivalent goodness factor, G_{e0}, of the ideal motor are finally

$$g_{e0} = \frac{g_0 K_c (1 + K_p)}{K_{tm}} \tag{11}$$

$$G_{e0} = \frac{2f_1 \mu_0 \tau^2}{\pi g_{e0}} \left(\frac{\sigma_{Al} d_{Al}}{K_{tr}} + \frac{\sigma_i \delta_i}{K_{ti}}\right) \tag{12}$$

In fact an equivalent secondary conductivity, σ_{e0}, may be identified in (12) such that

$$\sigma_{e0} = \frac{\sigma_{A2}}{K_{tr}} + \frac{\sigma_i \delta_i}{K_{ti} d_{A2}} \tag{13}$$

But still, g_{e0}, G_{e0} and σ_{e0} have been determined for the initial value of μ_i; the real value of μ_i strongly depends on primary mmf F_{m1}, and the secondary (slip) frequency.

The airgap flux density, B_g, is given by

$$B_g \simeq \frac{F_{m1} \mu_0}{g_{e0}(1 + jsG_{e0})} \tag{14}$$

with

$$F_{m1} \simeq \frac{3 \quad 2W_1 I_1 K_{d1}}{\pi p} \tag{15}$$

where 2p + 1 (or 2p) = total number of poles, K_{d1} = winding factor; W_1 = turns per phase; I_1 = phase current.

Assuming an exponential distribution of field penetration in the secondary core, the tangential flux density in iron on its upper surface (close to the aluminum sheet), B_{xi}, is

$$B_{xi} \simeq K_{1i} \frac{B_g}{\delta_i} \frac{\tau}{\pi} \tag{16}$$

The factor K_{1i} = 1.1 to 1.2, which accounts for the secondary core flux redistribution as caused by the end effect, in advanced saturation conditions[3] (B_{xi} = 2 to 2.2 T). Now the equivalent value of permeability, μ_e, corresponds to $0.9B_{xi}$ and is found from the actual magnetization curve. It may be argued that $0.9B_{xi}$ has been choosen without any theoretical justification. It is true, but the problem is far too complicated and the above value has yielded good practical results.[4]

Now, the value of μ_e is compared to that obtained in the previous computation cycle, μ_i, and if $(\mu_i - \mu_e)/\mu_i \geq$ 0.01 to 0.02, then the computation procedure is resumed with a new under-relaxed value, μ', given by

$$\mu' = \mu_e + 0.1(\mu_i - \mu_e) \tag{17}$$

The under-relaxation coefficient of 0.1, obtained after numerous attempts,[4] results in rather rapid convergence (only four to five iterations). Thus, for each value of primary mmf, F_{m1}, and secondary frequency, $f_2 = sf_1$, and given geometric data, the real LIM is replaced by an ideal one (without transverse edge effect and with an ideal secondary magnetic core) characterized by a different airgap, g_{e0}, and a different goodness factor, G_{e0} (in fact, a new secondary equivalent conductivity, σ_{e0}).

In the DSLIM the secondary core is absent and thus the equivalent airgap, g_{e0}, and the goodness factor, G_{e0}, are obtained directly from (10) and (11) with $K_p \to 0$ and $\sigma_i \to 0$.

6.4. Skin Effect and Airgap-Leakage Corrections

The resistive correction coefficient, K_{es}, accounting for aluminum sheet skin effect, for a SLIM, is obtained from the well-known expression used for rotary machines

$$K_{es} = \frac{2d_{Al}}{d_s} \frac{\sinh(2d_{A2}/d_s) + \sin(2d_{Al}/d_s)}{\cosh(2d_{Al}/d_s) - \cos(2d_{Al}/d_s)} \tag{18}$$

where d_s is the field penetration depth in the aluminum sheet:

$$d_s^{-1} = \left| \left[j\pi f_1 \mu_0 s \sigma_{Al} + \frac{1}{2} \left(\frac{\pi}{\tau}\right)^2 \right] \right|^{1/2} \tag{19}$$

The aluminum sheet equivalent conductivity, σ_{Al}, as influenced only by the skin effect, is

$$\sigma_{Al2} = \frac{\sigma_{Al1}}{K_{es}} \tag{20}$$

For the DSLIM, in (18) d_{Al} is replaced by $d_{Al}/2$ since the field penetrates the sheet from both sides.

The airgap-leakage correction, K_l, is obtained by solving the field equation in the airgap, in the absence of the secondary. (End and transverse edge effects are also neglected here.) Finally, K_l can be expressed as

$$K_1 = \frac{\tau}{\pi g_0 K_c} \sinh \left(\frac{\pi}{\tau} g_0 K_c\right) \geq 1 \qquad (21)$$

The equivalence condition used here consists in the conservation of normal airgap flux density (B_g) on the ideal secondary core surface (for SLIM). For DSLIM this means that in the middle of the airgap B_g is invariant.

As in reality the aluminum sheet thickness is rather a small part of total (magnetic) airgap, d_{Al}/g_0 = 0.25 to 0.33, and the above equivalent conditions assure practically the same propulsion force of ideal (without airgap leakage) and real (with airgap leakage) motor. Thus the airgap, as influenced only by the airgap leakage, g_{01}, is

$$g_{01} = g_0 K_1 \qquad (22)$$

Consequently, the skin effect and airgap leakage can be accounted for by using σ_{Al} and g_{01} in (1) through (10) instead of σ_{Al} and g_0. In what follows we determine the equivalence conditions through which the ladder secondary can be replaced (mathematically) by an equivalent sheet-on-iron secondary.

6.5. Ladder-Type Secondary

For urban systems having tracks 15 to 20 km long and high traffic densities it seems adequate to consider the ladder type. In this case the transverse edge effect does not occur. Consequently, to replace the ladder with an equivalent sheet, the resistance of the short-circuiting bars should be accounted for. First, the ladder secondary is replaced by a sheet having a thickness d_{Al} (Fig. 6.12):

$$d_{Al} = \frac{h_{s2} 3 z_2 b_{s2}}{\tau} \qquad (23)$$

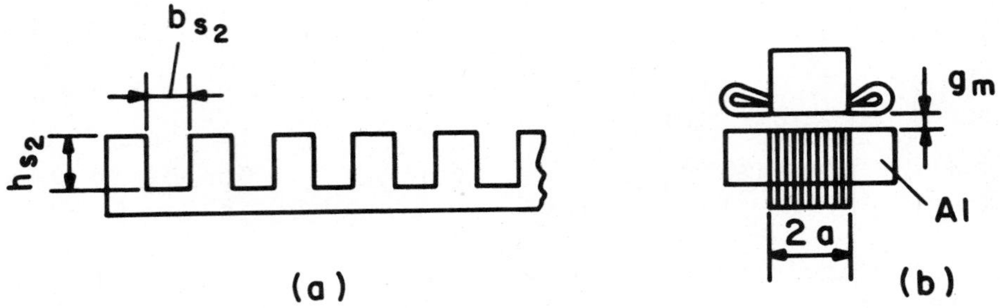

Fig. 6.12. Ladder secondary. (a) Longitudinal view, (b) cross section.

where $3z_2$ is the number of slots per pole. The short-circuiting bars equivalent width, for a thickness d_{Al}, becomes:

$$c - a = \frac{A_i}{d_{Al}} \tag{24}$$

The relation between the cross section, A_i, of the short-circuiting bars and that of the in-slot bars may be written as

$$\frac{h_{s2} b_{s2}}{A_i} = \frac{j_{bs}}{j_b} 2 \sin^2 \frac{\pi}{6z_2} \tag{25}$$

where j_{bs} and j_b are the design reference current densities chosen for the short-circuiting bars and for the in-slot bars, respectively. The magnetic airgap, g_0, is equal to the mechanical airgap, g_m, and the Carter coefficient should account also for secondary slotting. The secondary leakage reactance, X_2, should also be taken into account. Making use of the expressions for rotary machinery we get

$$X_2 = X_b + \frac{X_{bs}}{2\sin^2(\pi/6z_2)} \tag{26}$$

where X_b is the slot leakage reactance and X_{bs} is the short circuiting bar leakage reactance.

Similarly the equivalent resistance of one in-slot bar and one short-circuiting bar is

$$R_2 = R_b + \frac{R_{bs}}{2\sin^2(\pi/6z_2)}; \quad j_b = j_{bs} \tag{27}$$

where

$$X_b \simeq 2a\mu_0(\lambda_c + \lambda_{d2})\omega_1 \tag{28}$$

$$X_{bs} \simeq \frac{\mu_0\omega_1}{\pi}\,\frac{\tau}{3z_2}\,[0.25 + \ln(\frac{\tau}{3z_2r_e}); \quad r_e = \sqrt{\frac{A_i}{\pi}} \tag{29}$$

$$\lambda_c = \frac{hs_2}{3b_{s2}} + \frac{h_s}{b_s}; \quad \lambda_{d2} = \frac{5g_m/b_{s2}}{5 + 4g_m/b_{s2}} \tag{30}$$

λ_c = slot permeance; λ_{d2} = differential permeance. Now a secondary time constant, T_2, may be defined as

$$T_2 = \frac{X_2}{\omega_1 R_2} \tag{31}$$

Under these conditions the secondary equivalent sheet, whose dimensions result from (22) and (23), also has an inductive character, and consequently the equivalent goodness factor, G_{e0}, becomes now a complex number, $\underline{G}_{ec}$:

$$\underline{G}_{ec} \simeq \frac{G_{e0}}{1 + j(\omega_1 s T_2 / G_{e0})} \tag{32}$$

where

$$G_{e0} = \frac{2f_1 \mu_0 \tau^2}{\pi g_{e0}} \; \frac{\sigma_{Al} d_{Al1}}{K_{tr}} \tag{33}$$

The primary and secondary magnetic core saturation may be considered by a coefficient, K_s.

Thus the equivalent airgap becomes

$$g_{e0} = g_e K_c (1 + K_s) \tag{34}$$

Because of the absence of transverse edge effect in the active zone (under the primary core), the profile of the induced currents is rectangular (in the equivalent sheet that replaces the real ladder) and thus $K_{tm} = 1$ and K_{tr} is

$$K_{tr} = 1 + \frac{R_{bs}}{2R_b \sin^2(\pi/6z_2)} \tag{35}$$

where

$$R_b \simeq \rho_{Al} \frac{2a_e}{h_{s2} b_{s2}}; \quad R_{bs} = \rho_{Al} \frac{\tau}{3z_2 A_i} \tag{36}$$

The airgap leakage influence results from (21) with g_0 replaced by g_m. Finally, the skin effect correction coefficient is calculated from (18) where d_{A1} is replaced by h_{s2} (slot depth). Skin effect in a LIM is more important than in rotary machinery because the secondary frequency is, in general, higher. Thus it has been shown how the ladder secondary can be replaced by an equivalent sheet on ideal iron core secondary. The main purpose of this equivalence is the possibility of applying the field technical theory to the ladder secondary as well. However, it should also be kept in mind that during most of the vehicle acceleration time the end effect may be neglected, and thus the ladder-secondary LIM performance is computed using the procedure applied to the cage-type rotary induction motors.

6.6. The Field Technical Theory

The field technical theory is based on the following main assumptions:

1. The primary winding has a sinusoidal distribution and is fully characterized by its current sheet.

2. All the quantities vary sinusoidally in time.

3. Slotting, transverse edge, skin, and airgap-leakage effects are accounted for through corresponding correction coefficients as derived in previous sections.

4. According to assumption 3 the secondary-induced currents have only a transverse component (along OY).

5. Primary core permeability is infinite.

6. The primary core length is hypothetically considered infinite, but in order to account for its actual finite length, in front of and after the actual winding, an infinitely thin, purely reactive, sheet (k_4) is placed on the primary as shown in Fig. 6.13.

According to these assumptions the primary current sheet fundamental, $\underline{J}_m$, has, in complex form, the following form:

$$\underline{J}_m = J_m \exp^{[j(\omega_1 t - \frac{\pi x}{\tau})]} \tag{37}$$

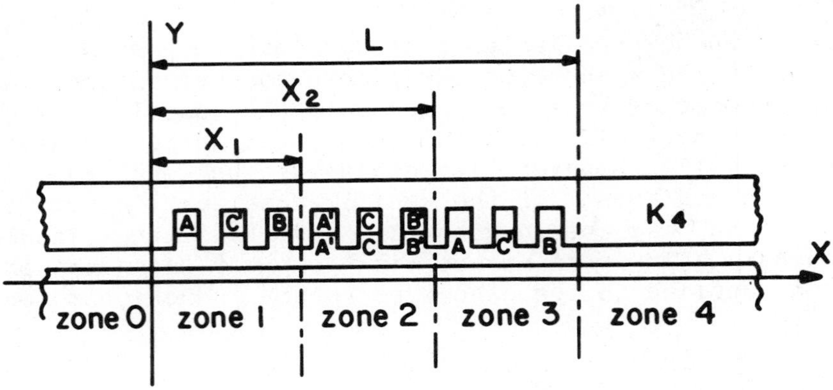

Fig. 6.13. Longitudinal mathematical model.

Now we can write the field equations.

6.6.1. Field Equations

Zone 2, (Fig. 6.13)

According to Ampere's law (along the contour abcd) we have

$$g_{e0} \frac{\partial H}{\partial x} = J_m e^{-j(\pi/\tau)x} + \underline{J}_2 d_{Al} \tag{38}$$

where $\underline{H}$ represents the airgap magnetic field and $\underline{J}_2$ the secondary current density. Sinusoidal time variation is understood.

Similarly the induction law yields

$$\frac{\partial \underline{J}_2}{\partial x} = j\omega_1 \mu_0 \sigma_{e0} \underline{H} + \mu_0 U \sigma_{e0} \frac{\partial H}{\partial x} \tag{39}$$

where U is the motor speed. Eliminating $\underline{J}_2$ from (38) and (39) we obtain

$$\frac{\partial^2 H}{\partial x^2} - \mu_0 \sigma_{e0} \frac{U d_{Al}}{g_{e0}} \frac{\partial H}{\partial x} - j\omega_1 \mu_0 \sigma_{e0} \frac{d_{Al}}{g_{e0}} \underline{H}$$

$$= - \frac{j\pi}{\tau g_{e0}} J_m e^{-j(\pi/\tau)x} \tag{40}$$

Using the goodness factor definition, (40) becomes

$$\frac{\partial^2 H}{\partial x^2} - G_{e0} \frac{\pi}{\tau}(1 - s)\frac{\partial H}{\partial x} - j\left(\frac{\pi}{\tau}\right)^2 G_{e0} \underline{H}$$

$$= - \frac{j\pi}{\tau g_{e0}} J_m e^{-j(\pi/\tau)x} \tag{41}$$

For the ladder secondary G_{e0} is replaced by $\underline{G}_{ec}$, (30), d_{Al} by d_{Al1}, (23), while g_{e0} results from (34). Similar equations are valid for zones 1 and 3 with J_m replaced by $J_m/2$. The solution to (41) is

$$\underline{H}_2 = \underline{A}_2 e^{\underline{\gamma}_1 (x-x_1)} + \underline{B}_2 e^{\underline{\gamma}_2 (x-x_2)} + \underline{B}_n e^{-j(\pi/\tau)x} \tag{42}$$

where

$$B_n = \frac{jJ_m}{g_{e0}(\pi/\tau)(1 + jsG_{e0})} \tag{43}$$

The third term in (42) is of conventional type, well known from rotary machines, and the first two terms account for the end effect. Similar expressions hold for zones 1 and 3; that is,

$$\underline{H}_1 = \underline{A}_1 e^{\underline{\gamma}_1 x} + \underline{B}_1 e^{\underline{\gamma}_2 (x-x_1)} + \frac{\underline{B}_n}{2} e^{-j(\pi/\tau)x} \tag{44}$$

$$\underline{H}_3 = \underline{A}_3 e^{\underline{\gamma}_1 (x-x_2)} + \underline{B}_3 e^{\underline{\gamma}_2 (x-L)} + \frac{\underline{B}_n}{2} e^{-j(\pi/\tau)x} \tag{45}$$

The coefficients $\underline{\gamma}_1$ and $\underline{\gamma}_2$ represent the roots of the characteristic equation of (41). Hence,

$$\underline{\gamma}_1 = \frac{a_1}{2}\left(\sqrt{\frac{b_1 + 1}{2}} + 1 + j\sqrt{\frac{b_1 - 1}{2}}\right) = \gamma_{1r} + j\gamma_i \quad (46)$$

$$\underline{\gamma}_2 = -\frac{a_1}{2}\left(\sqrt{\frac{b_1 + 1}{2}} - 1 + j\sqrt{\frac{b_1 - 1}{2}}\right) = \gamma_{2r} - j\gamma_i \quad (47)$$

where

$$a_1 = \frac{\pi}{\tau} G_{e0}(1 - s) \quad \text{and} \quad b_1 = \left[1 + \frac{16}{G_{e0}^2(1 - s)^4}\right]^{1/2} \quad (48)$$

For the ladder secondary the expressions of $\underline{\gamma}_1$ and $\underline{\gamma}_2$ differ from (45) through (47) because the real number G_{e0} is replaced by the complex number $\underline{G}_{ec}$. However the procedure of finding $\underline{\gamma}_1$ and $\underline{\gamma}_2$ is similar.

For entry (0) and exit zones (4), the current sheets $\underline{K}_0$ and $\underline{K}_4$, respectively, should be also accounted for as follows.

Entry and exit zones

Applying Ampere's and Faradays laws in these zones yields:

$$g_{e0} \frac{\partial \underline{H}_0}{\partial x} = \underline{K}_0 + \underline{J}_2 d_{Al} \quad (49)$$

$$\frac{\partial \underline{E}_0}{\partial x} = j\omega_1 \mu_0 \underline{H}_0 \quad (50)$$

$$\frac{\partial J_2}{\partial x} = j\omega_1 \sigma_{e0} \mu_0 \underline{H}_0 + \mu_0 U \sigma_{e0} \frac{\partial \underline{H}_0}{\partial x} \tag{51}$$

$$\underline{K}_0 = j \sigma_r \underline{E}_0 \tag{52}$$

where σ_r is the specific magnetic susceptance of the purely reactive sheet. Finally the field equation for these zones becomes

$$\frac{\partial^2 \underline{H}_0}{\partial x^2} - G_{e0} \frac{\pi}{\tau} (1 - s) \frac{\partial \underline{H}_0}{\partial x}$$

$$- [jG_{e0} (\frac{\pi}{\tau})^2 - \frac{\mu_0 \sigma_r \omega_1}{g_{e0}}] \underline{H}_0 = 0 \tag{53}$$

The solutions of this equation are

$$\underline{H}_0 = \underline{A}_0 e^{\underline{\gamma}_0 x}; \quad \text{for } x \leq 0 \tag{54}$$

$$\underline{H}_4 = \underline{A}_4 e^{\underline{\gamma}_4 (x-L)}; \quad \text{for } x \geq L \tag{55}$$

where $\underline{\gamma}_0$ and $\underline{\gamma}_4$ are given by

$$\underline{\gamma}_0 = \frac{a_1}{2} [\sqrt{\frac{b_2 + b_e}{2}} - 1 + j \sqrt{\frac{b_2 - b_e}{2}}]$$

$$= \gamma_{0r} + j\gamma_{0i} \tag{56}$$

$$\underline{\gamma}_4 = -\frac{a_1}{2}\left[\sqrt{\frac{b_2 + b_e}{2}} - 1 + j\sqrt{\frac{b_2 - b_e}{2}}\,\right]$$

$$= \gamma_{4r} - j\gamma_{0i} \tag{57}$$

and with:

$$\mu_0\sigma_r\omega_1 = \frac{1}{2g_{e0}}; \quad b_e = 1 + \frac{2}{(g_{e0}a_1)^2};$$

$$b_2 = b_e^2 + \frac{16}{G_{e0}^2(1 - s)^4} \tag{58}$$

The constants $\underline{A}_0$, $\underline{A}_1$, $\underline{A}_2$, $\underline{A}_3$, $\underline{A}_4$, $\underline{B}_1$, $\underline{B}_2$, $\underline{B}_3$ are determined from the boundary conditions. The resultant field and induced current density are considered continuous for $x = 0, x_1, x_2, L$. The specific magnetic susceptance, σ_r, as defined by (58), results as an approximate expression from conformal mapping.[3]

From (40) we now deduce the current densities

$$\underline{J}_0 = \frac{g_{e0}}{d_{Al}}\underline{\gamma}_0\underline{A}_0 e^{\underline{\gamma}_0 x}; \quad \text{for } x \leq 0 \tag{59}$$

$$\underline{J}_1 = \frac{g_{e0}}{d_{Al}}\,[\underline{\gamma}_1\underline{A}_1 e^{\underline{\gamma}_1 x}$$

$$+ \underline{\gamma}_2\underline{B}_1 e^{\underline{\gamma}_2(x-x_1)} + j\,\frac{sG_{e0}J_m e^{-j(\pi/\tau)x}}{2g_{e0}(1 + jsG_{e0})}\,] \tag{60}$$

$$0 \leqq x \leqq x_1$$

$$\underline{J}_2 = \frac{g_{e0}}{d_{Al}} [\underline{\Upsilon}_1 \underline{A}_2 e^{\underline{\Upsilon}_1 (x-x_1)}$$

$$+ \underline{\Upsilon}_2 \underline{B}_2 e^{\underline{\Upsilon}_2 (x-x_2)} + j \frac{sG_{e0} J_m e^{-j(\pi/\tau)x}}{g_{e0}(1 + jsG_{e0})}] \qquad (61)$$

$$x_1 \leqq x < x_2$$

$$\underline{J}_3 = \frac{g_{e0}}{d_{Al}} [\underline{\Upsilon}_1 \underline{A}_3 e^{\underline{\Upsilon}_1 (x-x_2)}$$

$$+ \underline{\Upsilon}_2 \underline{B}_3 e^{\underline{\Upsilon}_2 (x-L)} + \frac{jsG_{e0} J_m e^{-j(\pi/\tau)x}}{2g_{e0}(1 + jsG_{e0})}] \qquad (62)$$

$$x_2 \leqq x \leqq L$$

$$\underline{J}_4 = \frac{g_{e0}}{d_{Al}} \underline{\Upsilon}_4 \underline{A}_4 e^{\underline{\Upsilon}_4 (x-L)} \quad ; \quad x \geqq L \qquad (63)$$

In mathematical form the boundary conditions are:

$$(\underline{H}_0)_{x=0} = (\underline{H}_1)_{x=0}; \quad (\underline{J}_0)_{x=0} = (\underline{J}_1)_{x=0} \qquad (64)$$

$$(\underline{H}_1)_{x=x_1} = (\underline{H}_2)_{x=x_1}; \quad (\underline{J}_1)_{x=x_1} = (\underline{J}_2)_{x=x_1} \tag{65}$$

$$(\underline{H}_2)_{x=x_2} = (\underline{H}_3)_{x=x_2}; \quad (\underline{J}_2)_{x=x_2} = (\underline{J}_3)_{x=x_2} \tag{66}$$

$$(\underline{H}_3)_{x=L} = (\underline{H}_4)_{x=L}; \quad (\underline{J}_3)_{x=L} = (\underline{J}_4)_{x=L} \tag{67}$$

A system of eight equations and eight unknowns is thus obtained. This system with coefficients in complex forms may be solved, either by successive elimination or by computer standard subroutines (the system matrix is rather sparse).

6.6.2. Primary Core Flux Density

Let h_c be the primary core thickness. The primary core flux density can then be written as

$$\underline{B}_{ci}(x) = \frac{\mu_0}{h_c} \int_0^x \underline{H}_i(x)dx \tag{68}$$

where $\underline{H}_i(x)$ takes the expressions $\underline{H}_1(x)$, $\underline{H}_2(x)$ and $\underline{H}_3(x)$ for x in the interval (0 to x_1), (x_1 to x_2), and (x_2 to L).

The end effect causes a notable distortion of the core flux density distribution in comparison with the case of rotary machines. A knowledge of the primary core flux density distribution is useful for LIM primary core design.

6.6.3. Propulsion Force

The propulsion force can be calculated by integrating the force densities:

$$F_x = - a_e \mu_0 d_{Al} \text{ Re } [\int_0^L (\underline{J}_i(x)\underline{H}_i(x)) + \int_L^\infty (\underline{K}_4^*\underline{H}_4)dx] \quad (69)$$

The longitudinal force corresponding to x ≦ 0 (in front of motor entry) has been neglected since the magnetic field in the zone is practically zero. The first term in (69) represents the active zone contribution (0 ≦ x ≦ L), and the second term accounts for the reluctance-type braking force occurring "behind" the primary core (x ≧ L).

6.6.4. Normal Force

The normal force acting on the primary has two components: an attraction force, F_{na}, and a repulsion force, F_{nr}:

$$F_{na} = \frac{a_e \mu_0}{2} \int_0^\infty |\underline{H}_i(x)|^2 \, dx \quad (70)$$

$$F_{nr} = a_e d_{Al} \mu_0 \text{Real } [\int_0^L J_{mi} e^{-j(\pi/\tau)x} \underline{J}_i^* \, dx] \quad (71)$$

where

$$J_{mi} = \frac{1}{2} J_m \text{ for } 0 \leqq x \leqq x_1$$

$$\text{and } x_2 \leqq x \leqq L$$

$$= J_m \text{ for } x_1 \leqq x \leqq x_2 \quad (72)$$

The resultant normal force, F_n, acting on the primary (or of one side of it, for DSLIM) becomes

$$F_n = F_{na} - F_{nr} \quad (73)$$

The preceding analysis enables the determination of $\underline{H}_i(x)$, $\underline{J}_i(x)$, propulsion, F_x, and normal, F_n, forces showing most of end effect consequences. To prepare for performance calculation the secondary powers are now calculated.

6.6.5. Secondary Performance

First, we calculate the secondary losses, P_2, and then the reactive power, Q_2, corresponding to magnetic energy stored in the airgap and secondary:

$$P_2 = \frac{a_e d_{Al}}{\sigma_{e0}} \int_{-\infty}^{\infty} |\underline{J}_i|^2 \, dx \tag{74}$$

$$Q_2 = 2F_{na} g_{e0} \omega_1 \tag{75}$$

The theoretical expressions presented so far suffice for LIM performance computation, either in the motor or in the generator mode of operation. Moreover, the theory developed here is valid for three different primary windings, that is for an even number of poles with filled or half-filled end slots, and for an odd number of poles with half-filled end slots. The expressions developed above are now exploited to demonstrate the end effect main consequences. Though most of expressions exhibit a remarkable degree of generality, to get a feeling of magnitudes, a numerical example of practical interest is considered.

6.6.6. End Effect Waves.

The first two terms in the flux density expression, (42), are known as end-effect waves.[3] To evaluate quantitatively their importance it is useful to explore the relative depth of penetration of the end-effect waves, $(\tau^{-1}\gamma_{1r}^{-1})$, $(\tau^{-1}\gamma_{2r}^{-1})$, $(\tau^{-1}\gamma_{0r}^{-1})$, $(\tau^{-1}\gamma_{4r}^{-1})$, in the airgap,

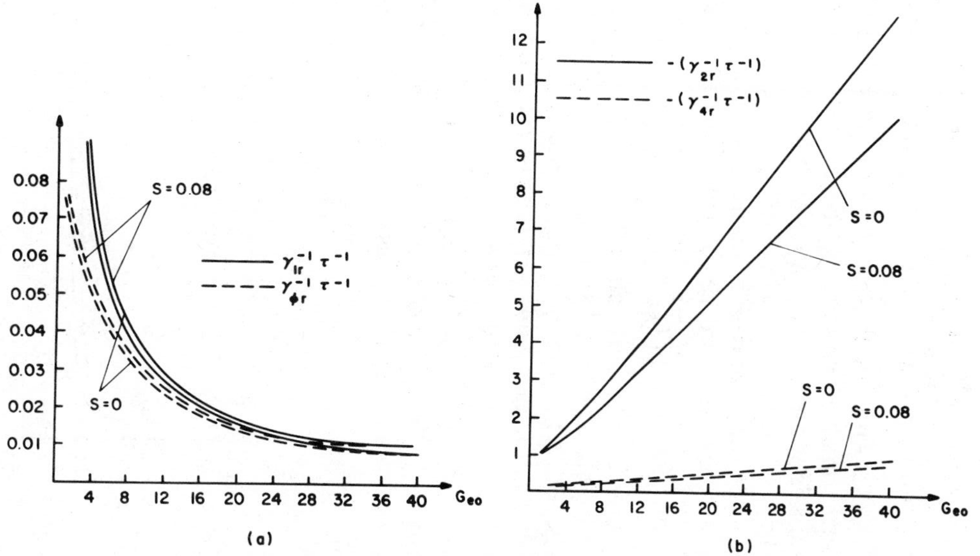

Fig. 6.14. Relative penetration depth of end-effect waves. (a) Entry wave, (b) exit wave.

against the goodness factor, G_{e0}, for given slip (Fig. 6.14). Also of notable significance is the ratio between the end-effect waves pole pitch and the primary winding pole pitch as plotted in Fig. 6.15. The end-effect waves corresponding to the coefficients $\underline{\gamma}_1$ and $\underline{\gamma}_0$ are called entry end waves and those related to $\underline{\gamma}_2$ and $\underline{\gamma}_4$ are exit end waves. The entry end waves are inverse, while the exit end ones are direct waves with respect to the conventional traveling wave expressed by the third term of expression (42). The entry end waves attenuate practically to zero within a few tenths of a pole pitch while the exit end waves attenuate slowly. However, a faster attenuation takes place behind the primary core ($x \geqq L$, see that $|\gamma_{4r}^{-1}| \leqq |\gamma_{2r}^{-1}|$ in Fig. 6.14). Finally, the pole pitch of end-effect waves is always greater than the pole pitch of conventional traveling wave (Fig. 6.15). As a first-order approximation it is reasonable to assume that if the exit end wave depth of penetration in airgap $|1/\gamma_{2r}|$, for synchronism, goes beyond 5 to 8% of motor length, L, the end effect should be accounted for. Therefore for a given goodness factor, G_{e0}, a lower limit of motor length, L_1, could be defined as

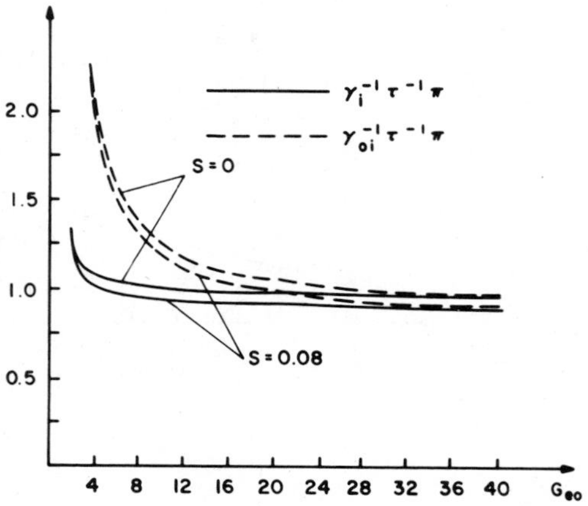

Fig. 6.15. Relative pole pitch of end-effect waves.

$$L_1 = \frac{12 \text{ to } 20}{\gamma_{2r}} \tag{76}$$

Now, if the motor length, L, is greater than L_1, the end effect may be neglected; otherwise the end effect should be accounted for.

To obtain reasonably high performance, G_{e0} is selected, in general, within the interval G_{e0} = 10 to 30. On the other hand, to negotiate curves of small (medium) radius, the motor length should not be over 2.0 to 4.0 m, that is generally $L < L_1$. Consequently, for Maglev purposes, the end effect should always be considered in the low slip region. It is significant that the end-effect waves characteristics do depend only upon equivalent goodness factor, G_{e0}, and slip, s. They depend upon speed only to the extent to which the speed is a term in the expression for goodness factor.

6.7. Specific Phenomena--Numerical Example

Let us consider a high-speed SLIM with the following initial data: rated frequency f_1 = 210 Hz; pole pitch τ = 0.25 m; total number of poles 2p + 1 = 13; primary core width 2a = 0.24 m; secondary sheet width 2c = 0.4 m; aluminum sheet thickness $d_{Al} = 4 \cdot 10^{-3}$ m; mechanical airgap $g_m = 12 \cdot 10^{-2}$ m; aluminum electrical conductivity $\sigma_{Al} = 2.16 \times 10^7$ (Ω0.5'$^{-1}$m^{-1}); iron electrical conductivity $\sigma_i = 3.52 \times 10^6$ (Ω^{-1}m^{-1}); number of turns per phase W_1 = 72; chording ratio Y/τ = 10/12; slots per pole and phase q_1 = 4; primary current I_1 = 800 A; primary current sheet fundamental $J_1 = 1.5 \times 10^5$ ampturns/m.

The secondary core is considered to be manufactured from one, two, or three "solid" longitudinal laminations.

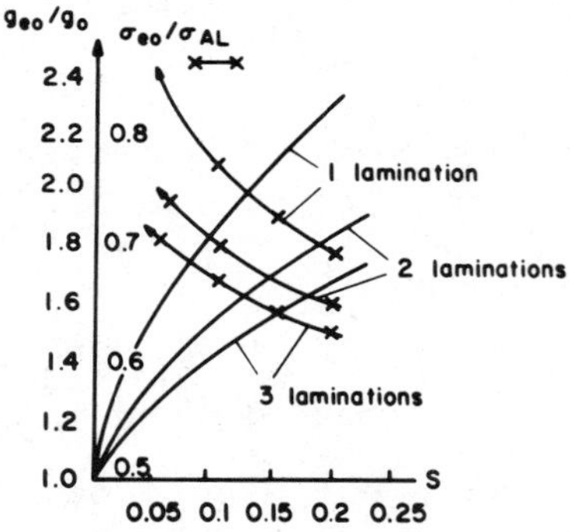

Fig. 6.16. Equivalent relative airgap (g_{e0}/g_0) and conductivity (σ_{e0}/σ_{Al}).

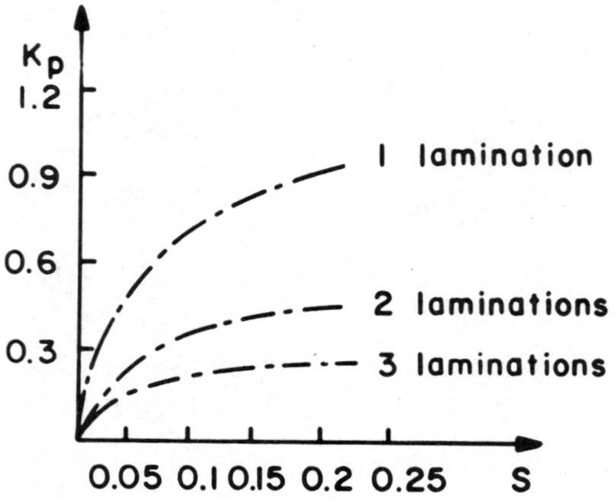

Fig. 6.17. Magnetic reluctance of secondary core divided by airgap reluctance, K_p.

6.7.1. Transverse Edge Effect and Secondary Iron Influence

To determine the transverse edge effect as well as the secondary iron-core induced-currents and saturation influence on the equivalent airgap, g_{e0}, and conductivity σ_{e0}, we use (1) through (18) for values of slip within the interval s = 0.045 to 0.2, that is, for speeds of 100 to 84 m/sec. The final results are plotted in Figs. 6.16, 6.17, and 6.18.

The influence of transverse edge effect and secondary iron eddy currents and saturation upon the equivalent airgap g_{e0} is evident from Fig. 6.16. The depth of eddy currents (or field) penetration in secondary core strongly decreases when the secondary frequency increases. The same is valid for G_{e0}. Thus, it is necessary to keep the secondary frequency below 15 Hz in order to obtain a satisfactory performance. A reasonably good performance is obtained by making the secondary core of three "solid" longitudinal laminations.

6.7.2. Propulsion and Normal Forces

Based on the results obtained in the previous section and making use of (42) through (75) we can determine

1. The propulsion force, F_x.

2. Normal force, F_n.

3. Secondary efficiency:

$$\eta_2 = \frac{F_x U}{F_x U + P_2} \tag{77}$$

4. Secondary power factor:

$$\cos \phi_2 = \frac{1}{[1 + (Q_2 \eta_2 / F_x U)^2]^{1/2}} \tag{78}$$

The final data thus obtained are plotted in Figs. 6.19 and 6.20.

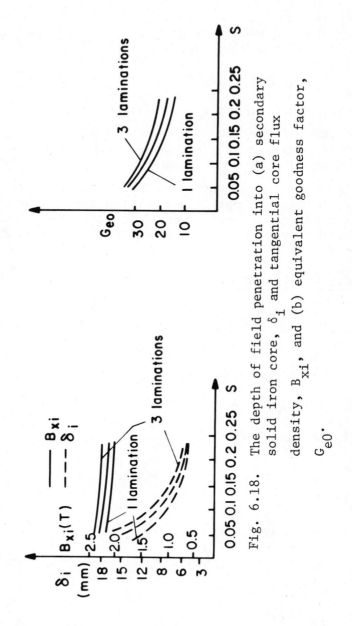

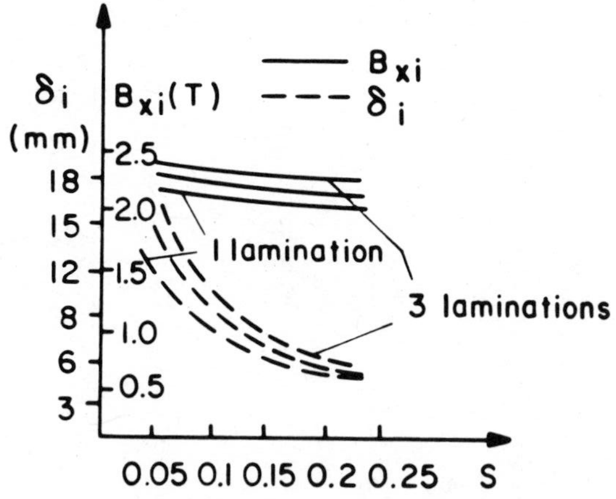

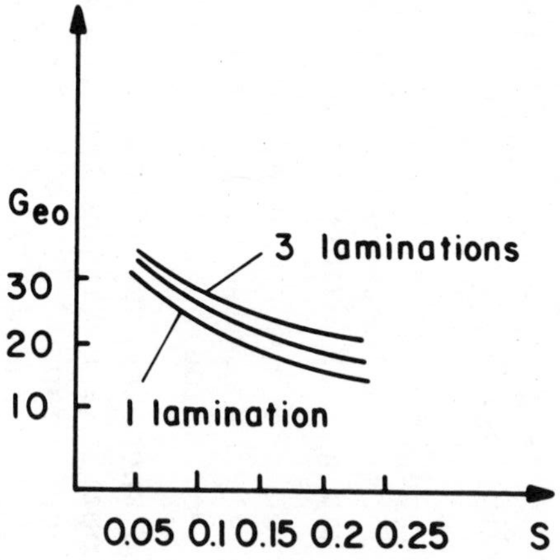

Fig. 6.18. The depth of field penetration into (a) secondary solid iron core, δ_i and tangential core flux density, B_{xi}, and (b) equivalent goodness factor, G_{e0}.

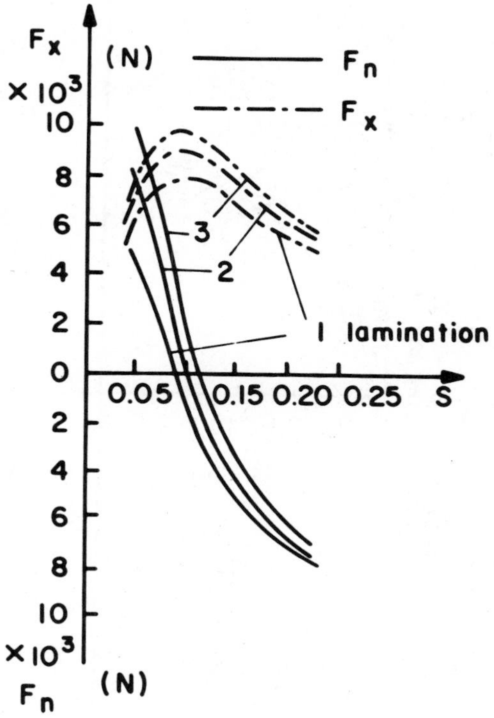

Fig. 6.19. Propulsion, F_x, and normal, F_n, forces.

It should be noted that the normal force, F_n, has an attraction character for small secondary frequencies and changes quickly to repulsion for higher secondary frequencies. Thus, for the range of secondary frequencies 2 to 15 Hz used for LIMs of Maglev, the normal force is attractive and varies rapidly with secondary frequency. In the design of the levitation system of Maglev, the SLIM normal force influence should also be considered. The beneficial effect of a three-lamination secondary core is evident only in a notable increase of the propulsion force (Fig. 6.19). However, the relative small value of equivalent airgap, g_{e0} (Fig. 6.16), indicates that the power factor as seen from primary will also increase. Making the secondary core of more than three laminations would negligibly improve the performance while appreciably increasing the secondary costs.

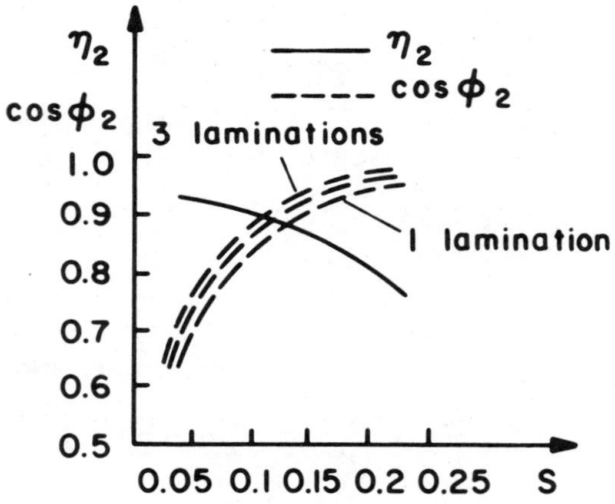

Fig. 6.20. Secondary efficiency η_2 and power factor, $\cos\phi_2$.

Thus three laminations in secondary core seems a realistic cost-energy compromise.

6.7.3. Longitudinal Distribution of Some Electromagnetic Quantities

The longitudinal distribution of airgap flux density, $|\mu_0 \underline{H}_i(x)|$, the primary core flux density, $|\underline{B}_{ci}(x)|$, the secondary current density, $\underline{J}_2(x)$, and the secondary losses ($P_2(x)$) are all calculated for two different values of slip (s = 0.05 and 0.10), and plotted in Fig. 6.21. These distributions are far from being uniform and thus the end effect acts strongly, causing these distortions. The current density increases notably at entry end, exercising a powerful demagnetizing effect on the airgap flux density. The rather "inductive" nature of the additional end-effect-induced currents at entry end cause a small propulsion (or even braking) force produced by the first poles of motor (Fig. 6.22). The normal force longitudinal distribution (Fig. 6.23) indicates (especially for s = 0.10) a net repulsion force obtained at entry end in contrast with the strong attraction force at exit end. This

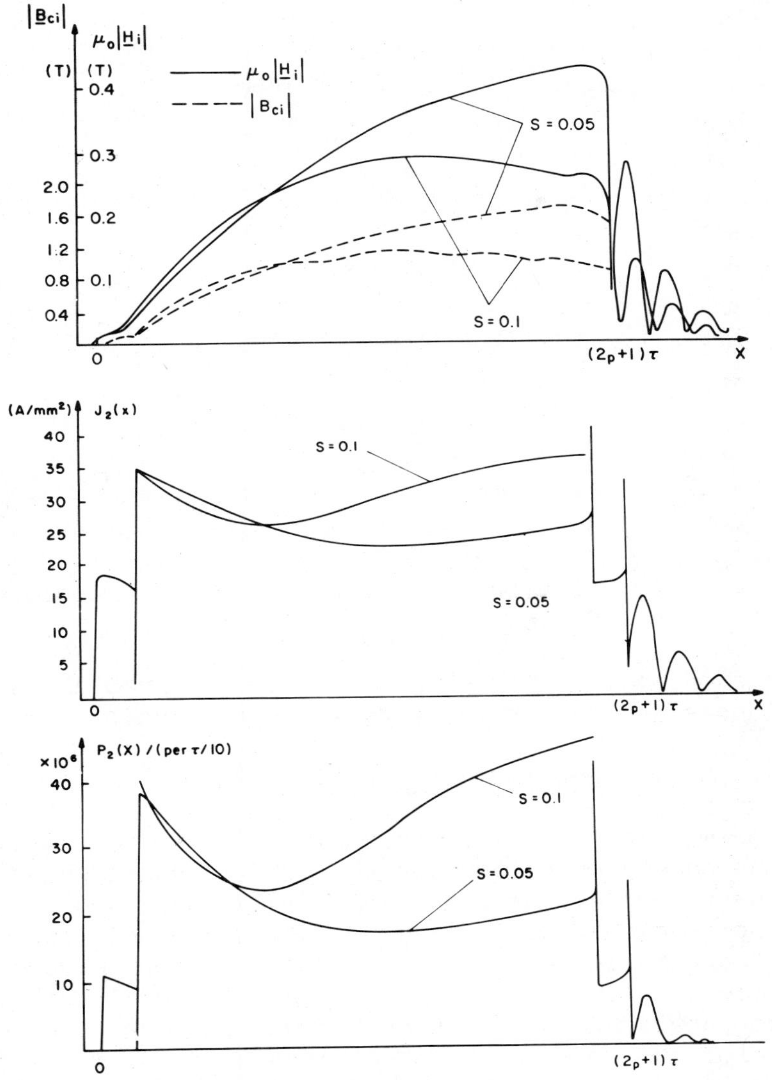

Fig. 6.21. Longitudinal distributions. (a) Air gap, $|\mu_0 H_i(x)|$, and primary core $|B_{ci}, (x)|$, flux densities, (b) secondary current density, $|J_2(x)|$, (c) specific Joule losses, $P_2(x)$.

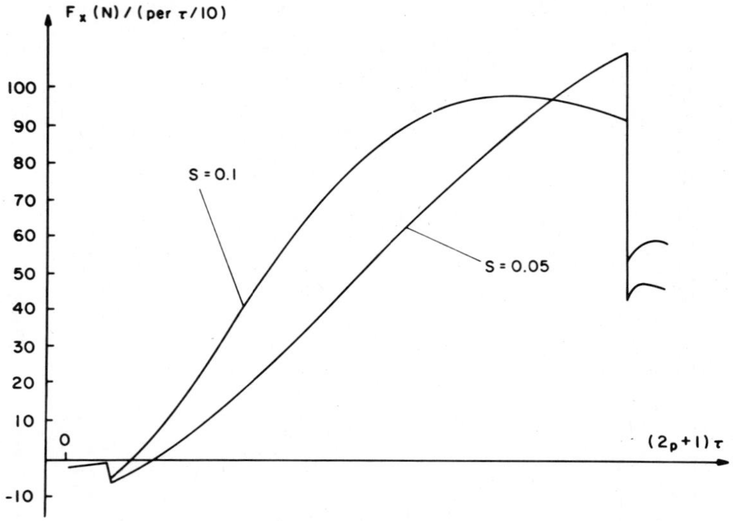

Fig. 6.22. Longitudinal distribution of propulsion force.

phenomenon causes the dolphin effect and should also be considered in the LIM frame design and dynamics.

6.7.4. Normal Repulsion Force in DSLIM[12-22]

As mentioned above, the DSLIM can also be analyzed by the theory developed for the SLIM by considering the secondary core as ideal and infinitely thin and placed in the middle of DSLIM airgap. We thus obtain two Smotors, each having half of the total airgap. However, the field technical theory does not allow directly for the computation of repulsion force on the secondary sheet of DSLIM when the sheet is placed asymmetrically in the airgap. When the secondary sheet is symmetric, $\Delta_1 = \Delta_2$ in Fig. 6.24 and the expressions used for SLIM are still valid.

To extend the field technical theory we now calculate the repulsion forces, F_{nr1} and F_{nr2}, acting on the secondary sheet as a result of the interaction between

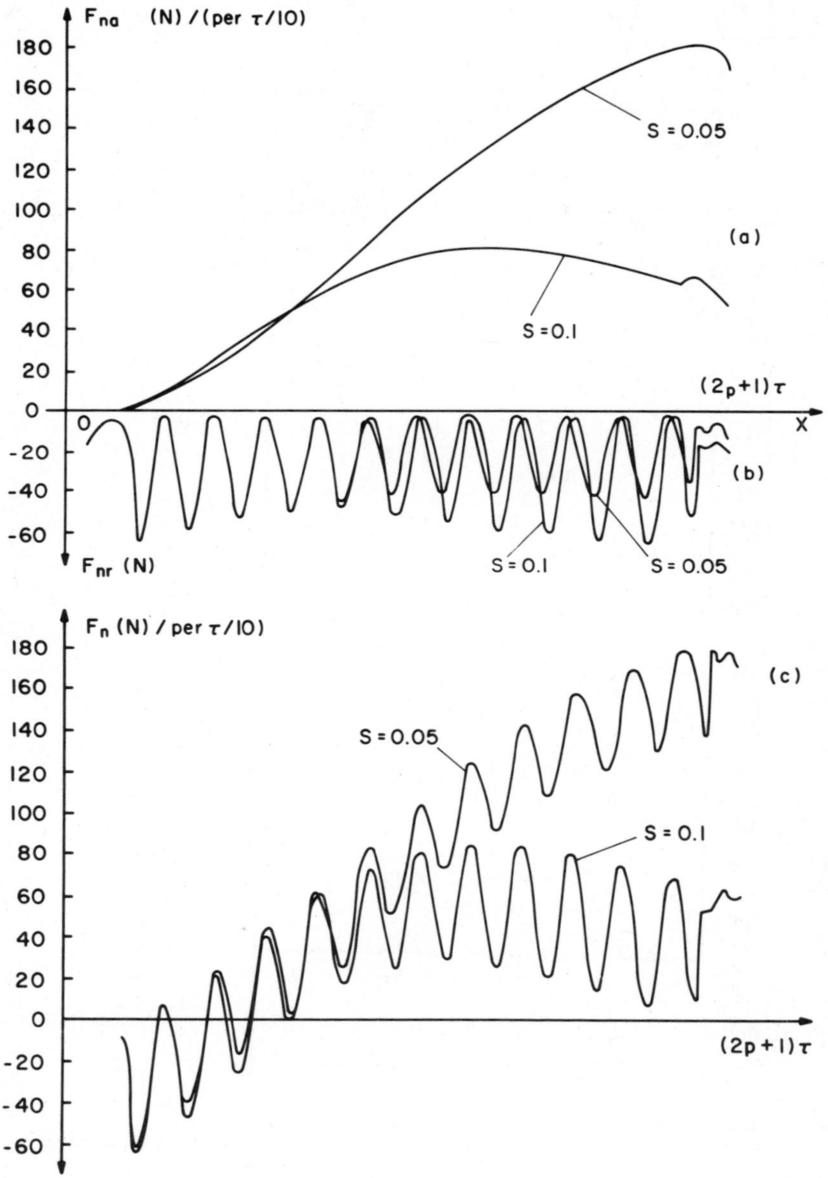

Fig. 6.23. (a) Normal attraction force component, (b) normal repulsive force component, (c) resultant normal force.

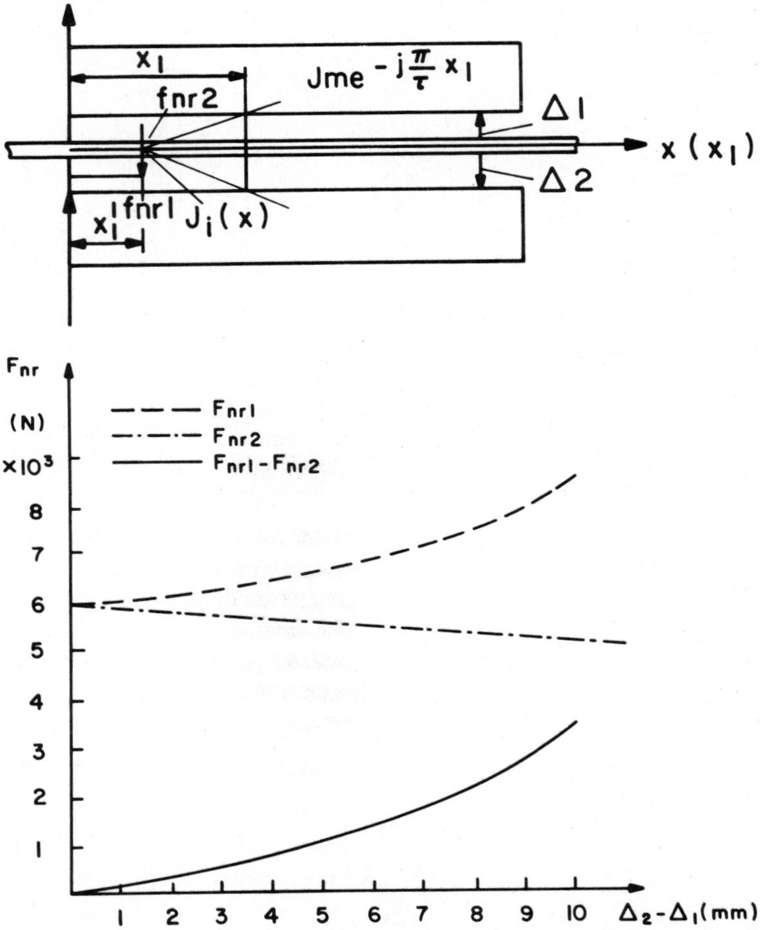

Fig. 6.24. Repulsion force of DSLIM. (a) Mathematical model, (b) numerical example.

the two primary current sheets and the secondary-induced currents (Fig. 6.24):

$$F_{nr1} = \frac{\mu_0 a_e d_{Al}}{4\pi}$$

$$\text{Re}\left[\int_0^L \int_0^L J_{me} e^{-j(\pi/\tau)x_1'} \underline{J}_i^*(x) \frac{\Delta_1 \, dx \, dx_1'}{\Delta_1^2 + (x - x_1')^2}\right] \quad (79)$$

$$F_{nr2} = \frac{\mu_0 a_e d_{Al}}{4\pi}$$

$$\text{Re}\left[\int_0^L \int_0^L J_{me} e^{-j(\pi/\tau)x_i'} \underline{J}_i^*(x) \frac{\Delta_2 \, dx \, dx_1'}{\Delta_2^2 + (x - x_1')^2}\right] \quad (80)$$

where

$$J_{me} = J_m \quad \text{for } x_1 \leqq x \leqq x_2;$$

$$J_{me} = J_{m/2} \quad \text{for } 0 \leqq x \leqq x_1 \text{ and } x_2 \leqq x \leqq L \quad (81)$$

The integrals (79) and (80) can be evaluated numerically.

Now if $\Delta_1 < \Delta_2$, then $F_{nr1} > F_{nr2}$, and consequently the net repulsion force tends to centralize the sheet, acting like a spring constant, C_s, such that

$$C_s = \frac{2(F_{nr1} - F_{nr2})}{\Delta_1 - \Delta_2} \quad (82)$$

The equivalent spring constant, C_s, depends on geometric dimensions, primary mmf, and secondary frequency. The mechanical lateral guidance system of DSLIM should be designed keeping this phenomenon in mind.

As an additional consequence of transverse edge effect, in a lateral asymmetric position of secondary ($c \neq b$), a lateral pure decentralizing force occurs in a DSLIM. For a SLIM this force is counteracted by the alignment tendency of primary and secondary magnetic cores. However, to determine this force a two-dimensional theory is necessary if the end effect is neglected;[3] or a three-dimensional analysis if the end effect is simultaneously accounted for.[9] It is beyond our scope here to treat this problem in detail.

In this section we have explored quantitatively the main aspects of specific phenomena occurring in a LIM. An extensive study of generator operation could also be carried out by using the technical field theory results directly when slip is negative. The generator mode can be used for regenerative Maglev braking, as shown later in this chapter. Finally, it is worth mentioning that new theories of end effect are still being proposed[23,24] and the subject is still open to future improvements. The field technical theory allows, however, for a rapid determination of the factors that determine the end effect and performance. Thus, it has been clearly shown that the relative airgap longitudinal penetration of end-effect waves depends only on slip and equivalent goodness factor, whereas the global influence of end effect on LIM performance is a function of only four parameters: slip, equivalent goodness factor, number of poles, and the ratio between equivalent airgap, g_{e0}, and the pole pitch. We are now ready to define an optimum goodness factor.

6.8. Optimum Goodness Factor

The existence of an optimum goodness factor may be visualized because both the conventional performance (in the absence of end effect) and the deteriorating influence of end effect increase when the equivalent goodness factor, G_{e0}, increases. However, the precise determination of an optimum goodness factor, while accounting both for motor weight (costs) and performance, is a very difficult task. In this sthe taskuation, it is to define a simplified optimum goodness factor corresponding to the case when the motor develops a zero total propulsion force at synchronism[14]:

$$(F_x)_{s=0} = 0 \tag{83}$$

To simplify the propulsion force expression, (69), we consider the case of an even number of poles with filled end slots ($x_1 = 0$ and $x_2 = L = 2p\tau$), while neglecting the reluctance-type braking force occurring at exit end ($\sigma_r = 0$):

$$(F_x)_{s=0}$$

$$= \frac{a_e \mu_0 \tau}{g_{e0}} J_m^2 \operatorname{Re}\left[\frac{-j(\underline{\gamma}_1\tau)(e^{(\tau\underline{\gamma}_2 - \pi j)2p} - 1)}{((\underline{\gamma}_2\tau) - (\underline{\gamma}_1\tau))(\frac{\tau\underline{\gamma}_2}{\pi} - j)}\right] = 0 \tag{84}$$

The numerical solution of (84) leads to the optimum goodness factor, (G_0), as a function of the number of poles (2p), as shown in Fig. 6.25. For odd number of poles G_0 is found by interpolation from Fig. 6.25.

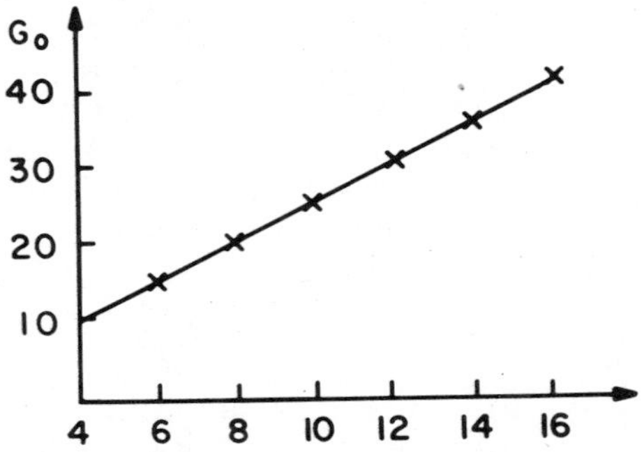

Fig. 6.25. Optimum goodness factor.

The equivalent goodness factor, G_{e0}, of a SLIM with sheet-on-iron secondary depends on the primary mmf, the secondary frequency, and the geometric dimensions. Consequently, the optimum goodness factor should be reached at rated speed and primary current.

The design of a high power DSLIM[14] for the optimum goodness factor $G_0 = 30$ and $2p = 12$, at $s = 0.07$, $U = 110$ m/sec, led to good performance, that is, to secondary efficiency $\eta_2 = 0.89$ and power factor $\cos \phi_2 = 0.82$. In general the rated slip varies from 0.04 to 0.1 in order to limit the end-effect deteriorating consequences. Though simple enough, the optimum goodness criterion leads to good LIM performance. It is, however, recognized that this criterion is not complete and refinements may be made searching for better performance around G_0. Whereas this method yields a reasonable reduction of end effect by adequate parameters selection, many efforts have been made to compensate the end effect.

6.9. Note on End-Effect Compensation Schemes

From the early beginning of LIM study it was hoped that some schemes[1] could be developed for compensating the end effect, thus achieving performance comparable to that of corresponding rotary machines. It was, however, later shown[15] that for a LIM designed with an optimum goodness factor, G_0, the compensation schemes would bring rather small overall improvements (under 5%) and thus could hardly be justified in practice. In fact, compensating schemes are just "moving the end-effect losses in the primary," in the sense that compensating windings require notable quantities of copper and iron, since rather high compensating mmfs are necessary.[15] For the same overall quantity of primary iron and copper, improvements of global efficiency and power factor occurring in a compensated LIM are small compared to those of an uncompensated LIM. More details on compensation schemes are available in Reference 15.

6.10. Design Guidelines

The LIM literature is rich in analytical work, but few significant attempts have been made to develop a thorough design procedure. A design procedure is, however, necessary to build a LIM. Based on the field technical theory we now present the main guidelines of LIM design. As is well known, even rotary machine design today, more than 50 years after their fundamental theory was developed, is a procedure depending upon data from past practical experience. The existence of end effect and other specific phenomena in LIMs makes their design a difficult task. Moreover, for transportation applications, it should be kept in mind that, whereas for urban applications (under 35 m/sec) the propulsion force at cruising speed is less than 25% of that needed for vehicle acceleration, for high-speed applications (100 to 120 m/sec) the propulsion force required at cruising speeds is about half of that needed for Maglev vehicle starting (for a typical 1 m/sec^2 acceleration). This notable difference between medium- and high-speed propulsion forces occurs because aerodynamic drag rises rapidly with speed. Hence, the design strategy should be considered separately for each of the two basic applications. First, the urban LIM Maglev is treated.[16]

6.10.1. Design of a Medium-Speed Maglev

The design procedure is divided into four major steps:

1. Preliminary data and calculations.

2. Principle dimensions and parameters.

3. Secondary thermal design.

4. Numerical results.

These four steps are illustrated for a SLIM example.

Preliminary data and calculations

The LIM initial data are: peak starting thrust F_{es0} = 12,000 N; rated thrust for cruising speed F_{xn} = 3000 N; rated speed U_n = 34 m/sec; pole pitch τ = 0.25 m; mechanical airgap $g_m = 1.10^{-2}$ m; starting primary phase current I_{1s} = 400 to 500 A.

Also, from past experience, the following constraints have been found:

1. The average air-gap flux density B_{gm} = 0.25 to 0.35 T for a SLIM with sheet-on-iron secondary, and B_{gm} = 0.35 to 0.55 T for a ladder-secondary SLIM and for DSLIM.

2. The primary current sheet fundamental, J_m, lies in the interval $J_m = 0.8$ to $1.3\ 10^5$ ampturns/m. These values correspond to one side of a DSLIM primary. The smaller values correspond to a ladder-secondary LIM.

3. The effective thickness of the aluminum sheet of the secondary $d_{Al} = 4$ to $6\ 10^{-3}$ m for SLIM, and $d_{Al} = 6$ to $10\ 10^{-3}$ m for DSLIM.

4. The pole pitch of primary winding is $\tau = 0.20$ to 0.28 m for SLIM (to limit the secondary core depth and saturation) and $\tau = 0.2$ to 0.35 m for DSLIM (to limit the length of the end connections of primary coils).

5. The primary frequency corresponds to Maglev cruising speed and $f_1 \leqq 75$ Hz for urban applications, while $f_1 \leqq 250$ Hz for high-speed Maglevs.

6. The primary stack width $2a = 0.75$ to $1.25\ \tau$. The lower limit is meant to reduce the end-connections length, and the upper limit aims to keep the secondary costs within reasonable limits.

At the start, the end effect may be neglected, and thus, the thrust, F_{xs}, is

$$F_{xs} \simeq 2a_e\left(p - \frac{1}{4}\right) \frac{\tau^2 J_m^2}{g_{e0}\pi} \frac{\mu_0 s G_{e0}}{(1 + s^2 G_{e0}^2)} \tag{85}$$

The factor $(p - 1/4)$ in (85) roughly accounts for the half-filled end poles. To yield the maximum starting thrust ($s = 1$) it is evident from (85) that

$$(G_{e0})_{s=1} = 1 \tag{86}$$

Based on the above mentioned constraints the starting frequency is about $f_{1s} = 4$ to 8 Hz; (1.5 to 3 Hz for a ladder secondary).

For given values of f_{1s}, the transverse edge, skin, airgap-leakage, and secondary-core saturation effects are calculated using (1) through (6). Thus, we finally determine the dependence of G_{e0} on f_{1s}. The lower the f_{1s}, the smaller the secondary losses, but the resultant normal (attractive) force is higher.

Now, if the SLIM is placed on a Maglev such that its normal force helps the levitation system, it is desirable to reach the condition $(G_{e0})_{s=1} = 1$ for as small a starting frequency as possible. This is, in practice, achievable only for ladder-secondary LIMs.

Ladder secondary

Making use of the results of Section 6.5, the propulsion force, F_{xse}, for small speeds becomes

$$F_{xse} \simeq 2a_e\left(p - \frac{1}{4}\right) \frac{\tau^2 \mu_0}{g_{e0}\pi} J_m^2 \frac{sG_{e0}}{1 + s^2(G_{e0} + 2f_{1s}\pi T_2)^2} \tag{87}$$

Here $(f_{1s})_{s=1} = 1.5$ to 3 Hz, and $f_2 = 1.5$ to 3.0 Hz for small speeds (0 to 15 m/sec).

For a given starting thrust, the smaller the f_{1s}, the stronger, and more expensive, the secondary ladder, but with the bonus of lower secondary losses. The choice of f_{1s} is, in fact, an involved optimum problem where the traffic density and all the costs (including the energy used by the system) should be considered.

In general, in the vicinity of stop stations, stronger (deeper) ladders should be used, and for cruising speed the dimensions should be adjusted such that the optimum goodness factor is achieved (Section 6.8). The ladder depth (and heating) within the stop stations is notably influenced by the traffic density. As a general rule the ratio between the secondary slot width, b_{s2}, and slot pitch, τ_{s2}, is around 0.6 ($b_{s2}/\tau_{s2} = 0.6$), and the slot depth $h_{s2} = 1.25b_{s2}$ to $3b_{s2}$. The selection of secondary slot pitch (or slots per pole) follows the rules known from rotary induction machines. From (87), the maximum starting force occurs when

$$G_{e0}(f_{1s}) + 2\pi f_{1s} T_2 = 1.0 \tag{88}$$

The starting frequency is thus obtained.

Now, in (85) and (87) we introduce the actual value of the starting force F_{xs0}. By choosing the primary current sheet, J_m, within the now-known limits, (85) provides the necessary condition to find p. By a slight adjustment of J_m and a, p is made an integer.

If $2p + 1 > 11$, two or more motors are used such that $2p + 1 = 7$ to 11. This limitation is caused by the fact that the motor length should not go beyond 2.0 to 3 m. The motor maximum length depends on the track curves imposed on an urban system.

Principal dimensions and parameters

Making use of the above data and of field technical theory, after accounting again for the transverse edge, skin, secondary-core saturation and eddy currents (for aluminum-on-iron secondary only), we determine the rated primary current sheet, J_{mn}, required at cruising speed, for the given rated slip: $s_n = 0.06$ to 0.1. By knowing the acceleration, cruising, and braking schedules between typical stop stations we can calculate the actual dependence of required propulsion force on the time between those stop stations. Then, assuming $f_2 = 4$ to 6 Hz, the variation of J_m in time between two stop stations is determined. A "thermal" squared average primary current sheet may thus be defined:

$$J_{mth} = \left[\frac{1}{t_s} \int_0^{t_s} J_m^2(t)\, dt \right]^{1/2} \tag{89}$$

where t_s is the time of travel between two typical stop stations.

Primary slots

If the primary winding has Class H insulation, even without forced cooling, the design-reference primary-current density is $J_{c0} = 4.5$ to 5.5×10^6 A/m^2. The overall heat transmission surface will be increased such that for the worst situation the temperature of the coils will be under 180 C.

Three or four slots per pole per phase are used ($q_1 = 3$ to 4), and the primary tooth flux density (1.7 to 1.75 T). Because of the high value of starting thrust, the maximum value of airgap flux density, B_{gmax}, occurs at start (s = 1):

$$B_{gmax} = (B_g)_{\substack{s=1\\G_{e0}=1}}$$

$$= \left|\frac{j\mu_0 J_{ms}\tau}{g_{e0}(1 + jsG_{e0})\pi}\right|_{\substack{s=1\\G_{e0}=1}} = \frac{\mu_0 J_{ms}}{\sqrt{2}g_{e0}}\frac{\tau}{\pi} \tag{90}$$

The ratio between the primary slot width, b_{s1}, and the pole pitch, τ (Fig. 6.26), is such that

$$\frac{b_{s1} + 0.003}{\tau} = \left(1 - \frac{B_{gmax}}{B_{tm}}\right)\frac{1}{q_1} \tag{91}$$

In (91), 0.003 accounts for the presence of the wedge. The primary slot useful height, h_{s1}, can now be determined:

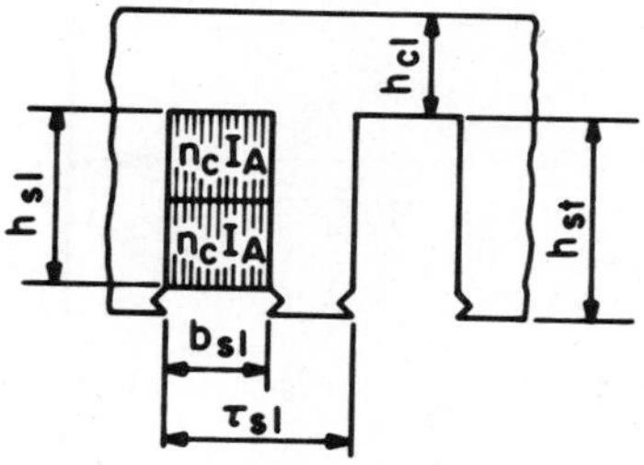

Fig. 6.26. Primary slot.

$$h_{s1} = \frac{2\eta_c I_{1e}}{j_{c0} K_f} \tag{92}$$

where I_{1e} = the "thermal" equivalent current corresponding to J_{mth}; η_c = turns per primary coil; K_f = slot-fill factor (K_f = 0.7 to 0.80 for rectangular cross-sectional conductors). On the other hand, from the definition of the current sheet

$$\eta_c I_{1e} = \frac{J_{mth}}{6\sqrt{2} q_1 K_{d1}}; \quad W_1 = 2pq_1\eta_c \tag{93}$$

where K_{d1} is the winding factor:

$$K_{d1} = \frac{\sin(\pi/6)}{q_1 \sin(\frac{\pi}{6q_1})} \sin(\frac{\pi}{2}\frac{Y}{\tau}) \tag{94}$$

For even number of poles $W_1 = (2p - 1)q_1\eta_c$. The total slot height, h_{st}, is

$$h_{st} = h_{s1} + (0.003 \text{ to } 0.004) \text{ m} \tag{95}$$

To complete the electric circuit design, W_1 should now be calculated. This is done using the initial values of J_{ms} and I_{1s}:

$$W_1 = \frac{J_{ms} p\tau}{3\sqrt{2} I_{1s} K_{d1}} \tag{96}$$

The value of W_1 should then be adjusted to conditions (93).

Finally, the primary phase voltage, V_1, should be determined for the entire speed range implied by the "travel schedule." We remember that by now we know the primary current variation between two typical stops as a function of secondary frequency for a given travel schedule. However, to find V_1 we must first determine the primary phase resistance, R_1, and leakage inductance, L_1:

$$R_1 = \rho_{c0} \frac{l_c W_1 j_{c0}}{I_{1th}};$$

$$l_c \simeq 4a_e + 2K_{s1}\tau; \quad K_{s1} \simeq 1.3 \text{ to } 1.35 \tag{97}$$

$$L_1 = \frac{2\mu_0}{p} \left[\left(\frac{\lambda_{s1} + \lambda_{d1}}{q_1} \right) 2a + \lambda_e K_{s1} \tau \right] W_1^2 \tag{98}$$

where λ_{s1}, λ_{d1}, and λ_e have the expressions

$$\lambda_{s1} = \frac{h_{s1} + 3(h_{st} - h_{s1})}{12 b_{s1}} (1 + 3\beta_1);$$

$$\lambda_{d1} = \frac{5 g_{e0}/b_{s1}}{5 + 4 g_{e0}/b_{s1}} \tag{99}$$

$$\lambda_e = 0.3\,(3\beta_1 - 1); \quad \beta_1 = \frac{Y}{\tau}$$

In (98) it would have been possible to account for the half-filled end-slots actual permeances. However, for $2p + 1 > 7$, (98) introduces negligible errors.

Now, the power balance yields the primary phase voltage, V_1

$$V_1 = \frac{1}{3I_1}\,[(3R_1I_1^2 + F_xU + P_2 + P_i)^2$$

$$+ (Q_2 + 6\pi f_1 L_1 I_1^2)^2]^{1/2} \qquad (100)$$

where P_i represents the primary core losses, and could be determined for any speed, U, current, I_1, and secondary frequency, f_2, by first finding the airgap flux density. Then standard procedures used for rotary machines could be used to find P_i.

It must be recognized that, if the actual longitudinal distribution of primary core flux density (as distorted by the end effect) is to be taken into account, the precise computation of iron losses, P_i, becomes a tedious task. The primary core losses are, in general, less than 8 to 10 times smaller than the primary winding losses. Thus an approximation of P_i is reasonable. From (78) and (75) P_2 and Q_2 are directly evaluated. Finally, the global efficiency, and power factor are given by

$$\eta_1 = \frac{F_xU}{F_xU + P_2 + 3R_1I_1^2 + P_i} \qquad (101)$$

$$\cos \phi_1 = \frac{1}{[\, 1 + [\, (Q_2 + 6\pi f_1 L_{1\sigma} I_1^2)\eta_1 / F_x U \,]^2 \,]^{1/2}} \qquad (102)$$

We now have all the expressions necessary to calculate the LIM performance accounting for all specific phenomena.

Secondary thermal design

The thermal design of the primary is quite similar to that commonly used for rotary induction machines. There are, however, two main differences. First, the primary is swept by the ambient air at a speed dependent on vehicle speed. Secondly, the rather high value of the airgap prevents any heat exchange between the primary and the secondary.

For a high-speed LIM of Maglev the primary should be forced-cooled with air, water, or oil. For urban purposes natural cooling seems preferable.

The secondary cooling raises special problems. First, between the stop stations, where the speed is close to cruising speed, the secondary does not have sufficient time to reach high temperature levels. On the other hand, in the vicinity of stop stations the secondary could easily be overheated. The thermal design of the secondary is based on the determination of the minimum lead time between two successive Maglevs that can start moving from the same place such that the secondary temperature be kept between 200 and 240 C. An exact computation procedure for the secondary heating/cooling in a stop station for diverse traffic conditions is very difficult and is strongly dependent on the peculiarities of the urban transportation system studied. However, a rather simple "sheltering" approach could also lead to very useful practical results.[17]

The worst situation occurs at the stop stations when the starting force has its maximum value, F_{xs}. The average acceleration, a_{ms}, at starting is

$$a_{ms} \simeq \frac{F_{xs}}{M} \tag{103}$$

where M is the vehicle mass. The thermal energy, W_p, stored in one meter length of secondary, while a vehicle passes over it, is

$$W_p = \frac{n}{L} \int_0^{t_1} P_2(t)dt; \quad t_1 \simeq \sqrt{\frac{2L}{a_{ms}}} \tag{104}$$

where n is the number of longitudinally successive motors of a vehicle and L is the motor length.

In (104) the most unfavorable situation is encountered; that is, no cooling takes place between the first and last motor passage over the same secondary point. The secondary losses at start are, approximately,

$$(P_2)_{s=1} = F_{xs} 2f_{1s}\tau \tag{105}$$

The heat transmission along longitudinal direction is also neglected. Moreover, the Joule losses $P_2(t)$ are considered constant in time and equal to $(P_2)_{s=1}$, whereas in reality they decrease when the vehicle starts moving.

Let T be the minimum interval between two successive transport units, and let m' be the number of transport units that had continuously circulated since their start. In this case the secondary temperature, $\theta(t)$, is

$$\theta(t) = \theta(m'T + \varepsilon)\exp[-\frac{\alpha L_p}{c_p \gamma S_t}(t - m'T)] \tag{106}$$

for

$$m'T + \varepsilon \leqq t \leqq (m' + 1)T - \varepsilon \tag{107}$$

When the secondary temperature stabilizes,

$$\theta'[(m' + 1)T - \varepsilon] = \theta(m'T - \varepsilon) \tag{108}$$

and thus

$$\theta(m'T - \varepsilon) = - \frac{W_p}{c_s \gamma S_t} + \theta(m'T + \varepsilon) \tag{109}$$

From (106) through (109) the value of T results as

$$T = \frac{c_s S_t \gamma}{L_p} \ln \left[\frac{\theta(m'T + \varepsilon)}{\theta(m'T - \varepsilon)}\right] \tag{110}$$

where S_t = cross section of secondary; c_s = equivalent specific heat of secondary; L_p = perimeter of the cross section along which the heat transfer takes place; α = heat transfer coefficient in W/m^2C; γ = equivalent specific weight of secondary.

The overheating temperature, $\theta(m'T + \varepsilon)$, is selected: $\theta(m'T + \varepsilon)$ = 200 to 240 C and, with this value, $\theta(m'T - \varepsilon)$ is obtained from (109). Finally, (110) yields T.

The heat transfer is more efficient for a ladder secondary. Also, in this case the Joule losses are smaller on account of smaller starting frequency. Consequently, for a ladder secondary the time, T, could be reduced by 1.8 to 2.5 times as compared to the case of sheet-on-iron secondary. The traffic capacity is thus notably increased. Finally, we should not rule out completely the possibility that near the stop stations the secondary may be forced cooled. The heat thus obtained could be profitably used. Of course, the traffic capacity could thus be further increased.

Numerical results

Following the design procedure just described, for the numerical data given in this section, the following final results have been obtained: number of turns per phase $W_1 = 90$; starting current $I_{1s} = 420$ A; current for 20 m/sec $I_{1n} = 180$ A; number of poles $2p + 1 = 11$; primary stack width $2a = 0.27$ m; starting frequency $f_{1s} = 6$ Hz; phase voltage at start $U_{1s} = 225$ V; phase voltage for 20 m/sec $U_{1n} = 200$ V; efficiency for 20 m/sec $\eta_1 = 0.8$; power factor for 20 m/sec $\cos \phi_1 = 0.55$.

If such an urban transportation unit is made of three articulated vehicles of 56 tons (including 400 passengers and 4 LIMs) the minimum head time is T = 90 sec, meaning 16,000 passengers per hour. For the peak traffic hours two or three such units could be coupled together, and by using a ladder secondary, at least in the vicinity of stop stations, the transport capacity increases to a maximum of 48000 passengers per hour. We now present some of the just-designed LIM performance characteristics. First, the motor regime was explored (Fig. 6.28) by imposing the current-frequency and frequency-speed dependencies (Fig. 6.27). The generator

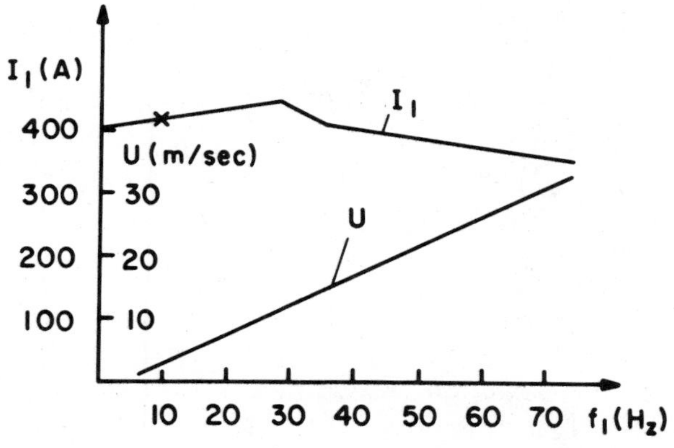

Fig. 6.27. Imposed current and speed dependence of frequency.

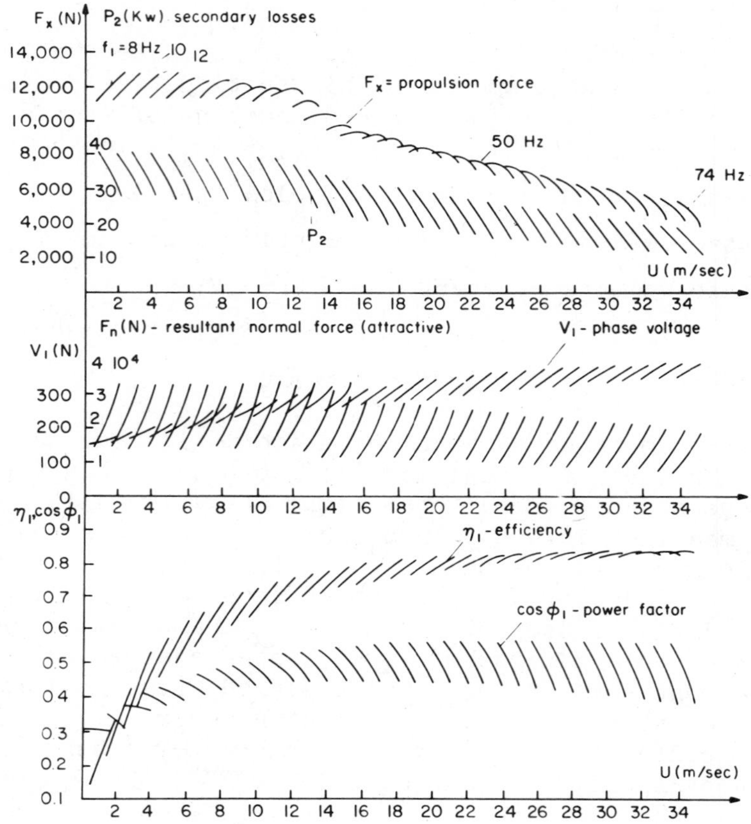

Fig. 6.28. Motor regime performance.

regime was also studied and the results are plotted in Fig. 6.29.

A discussion of results

In both regimes, motor and generator, the efficiency is rather high, up to 0.86 for maximum speed. The power factor is, however, low, that is up to 0.6. The use of a ladder secondary would have notably increased the power factor for small speeds. But at high speeds the power factor would have reached only 0.65 to 0.68. The resultant normal force (Fig. 6.27)

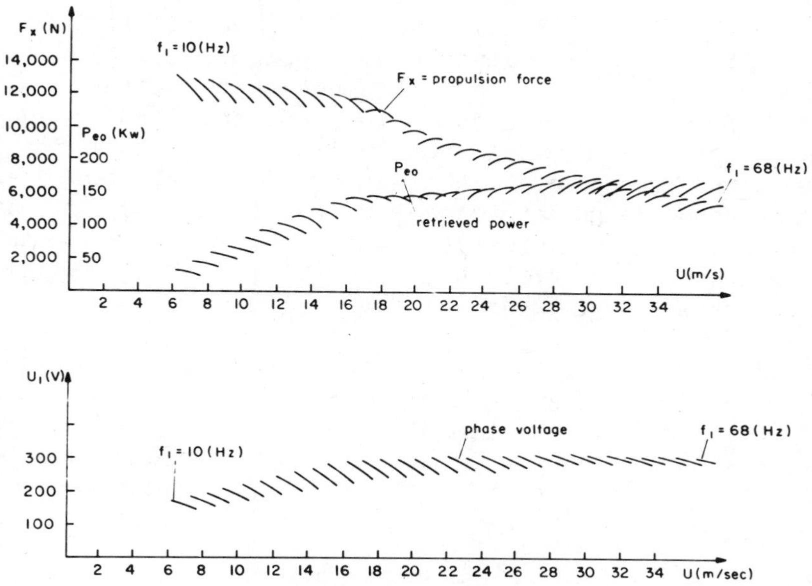

Fig. 6.29. Generator regime performance.

has an attraction character and is generally 3 to 4 times greater than the propulsion force. The use of a ladder secondary would yield an even higher normal (attraction) force. Thus, at least during vehicle starting (roughly up to half the cruising speed) the SLIM can provide up to 50% of levitation force, easing the job of levitation system.

Propulsion force time variations (to obtain the travel schedule) are slow for the levitation control systems and can easily be handled by them. We could also consider an automatic closed-loop control system of primary current and secondary frequency such it would always have the same normal (attraction) force, but this would lead, for speeds higher than half of cruising speed, to very low power factors of LIM. This is so because the propulsion force requirements decrease gradually when approaching the cruising speed, thus implying lower secondary frequency in order to keep the normal force constant.

For high speeds very low values of secondary frequencies (1.5 to 3 Hz) allow a high end effect leading to low power factor values and, consequently, low propulsion efficiency. Thus it seems that maintaining the normal force at a constant high value is hardly justified in practice, though not entirely ruled out. On the other hand from Fig. 6.29 it is evident that noticeable active power quantities are retrieved during generator regime, a strong indication that regenerative braking should be extensively used, together with a mechanical braking system on skids.

We now conclude that the theoretical approach developed in this chapter provides the necessary data for a rather complete LIM design and extensive performance study. However, not all LIM problems have been studied in detail and thus there is still much room for improvement.

6.10.2. Interurban High-Speed Vehicles

For high-speed LIMs the Maglev propulsion force required for constant speed (zero acceleration) increases rapidly with speed because of increase in the aerodynamic drag. A typical variation of propulsion

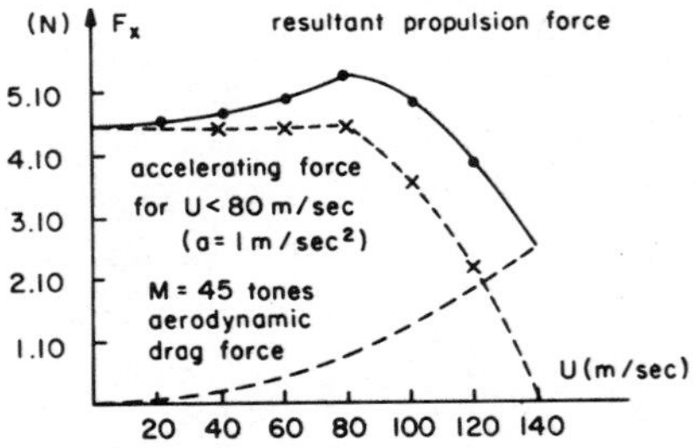

Fig. 6.30. Typical propulsion force profile for a 45-tonne Maglev.

force with speed required for a 45-tonne vehicle, when a constant acceleration of 1 m/sec^2 is maintained from standstill up to 80 m/sec, is plotted in Fig. 6.30. Therefore, the main objective of designing a high speed LIM is optimal running at cruising speeds. The design expressions are in fact those used for urban LIM but in this case the pole pitch, rated frequency, and secondary sheet thickness are selected so as to obtain the optimum goodness factor (Fig. 6.25) for the cruising speed. This is so because the end effect is now very important. Again the starting frequency is chosen to yield a maximum starting propulsion force. Finally, it should be mentioned that the entire design procedure is also similar to that developed for urban LIM.

6.11. Static Power Converters for LIM

From the above remarks it is clear that a high performance can be obtained only by supplying the LIM from a variable-voltage and variable-frequency static power converter. A number of converter schemes[19] have been proposed to meet the LIM requirements of high reactive power (rather low power factor). Without entering into great detail we mention here only two main types of converters suitable for LIMs.

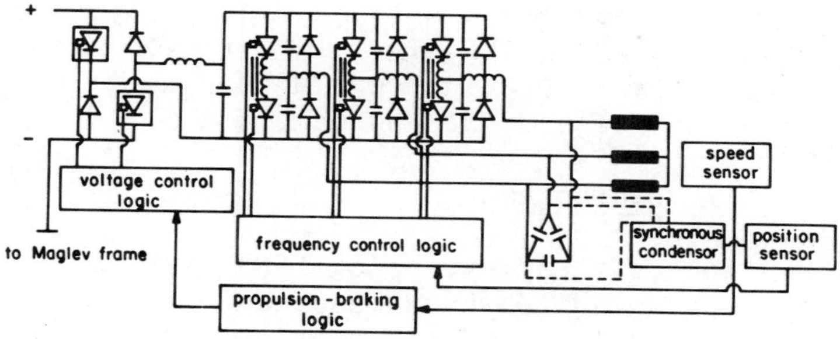

Fig. 6.31. DC-AC static power converter for LIM (urban).

6.11.1. DC/AC Converters

The dc power supply of a vehicle is generally superior to the ac solution, in urban applications. Thus the power line located along the track may be fed from power substations consisting of transformers, diode-rectifiers, and filters at voltages of 0.5 to 1.5 kV, for urban purposes. In this case the static power converter located on board consists of an inverter providing a variable frequency (0 to 75 Hz) and variable voltage. In principle the scheme of such an inverter is shown in Fig. 6.31.

Static capacitors are commonly used to supply all the reactive power required by LIM (up to zero power factor angle) and an additional quantity to secure a corresponding leading angle required for the turn-off process of thyristors. Static capacitors may be replaced by a high speed, low noise, synchronous condenser (4500 rev/min at 75 Hz). The weight of such a synchronous condenser is smaller than that of equivalent static capacitors. Moreover the reactive power control for optimum performance may be obtained by controlling the field current of the synchronous condenser. However, some means for thyristors turn-off during the starting period (up to 2 to 3 m/sec) should be provided, since the induced voltage is not sufficient to do this job below 2 to 3 m/sec. All in all, though the synchronous condenser requires a special control system, it is a very "attractive" solution. The retrieved energy during regenerative braking could be used by then-accelerating vehicles or stored on board.

6.11.2. AC/AC Converters[25,26,28-39]

For high-speed Maglevs using LIMs, the main objectives are reducing power substation costs and vehicle weight, and providing for regenerative braking. In an ac power system this can be easily done. The power line will be three-phased with line voltages up to 8 kV. In this case a typical ac-ac static power converter[20] (Fig. 6.32) is composed from a phase-delay rectifier and an

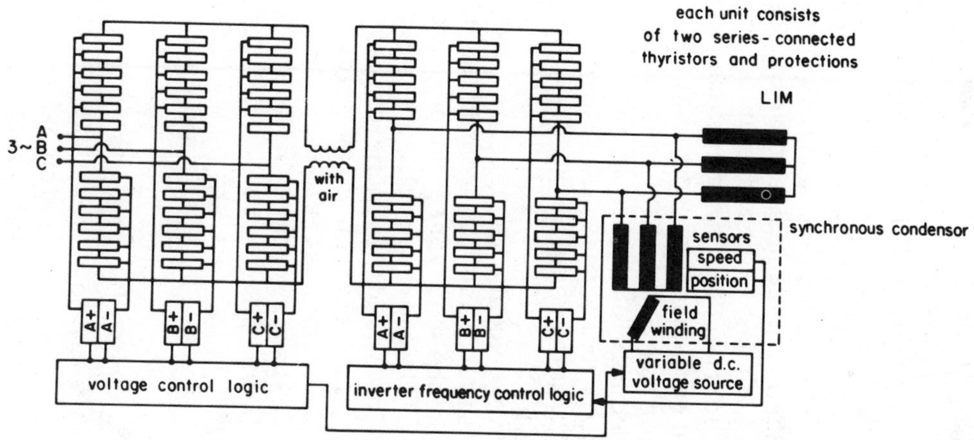

Fig. 6.32. AC-AC static power converter.

inverter that can run both ways. The high-speed synchronous, low-noise capacitor[20] now represents the "ideal" low weight controllable reactive power source, by which the thyristors are commutated for speeds above 10% of cruising speed. For starting, after each group of firing pulses that lead to a kind of LIM "step motion," the rectifier is switched to the inverter mode long enough for the adequate thyristors to be "line turned-off." Then two adequate thyristors are fired and the process goes on until the speed (frequency) achieved is sufficient, so that the synchronous-condenser-induced voltage is high enough for commutating the inverter-thyristors. During the starting, due to line commutation, the LIM propulsion force experiences rather high time pulsations, but the high inertia of the vehicle prevents significant jerks.

Regenerative braking is obtained by switching the inverter to rectifier and the rectifier to inverter mode and pumping back the recovered power directly into the ac power grid. Maglev propulsion could, of course, be fully automated to yield minimum energy per passenger and kilometer, for a given commercial speed (or travel time) between given stop stations. The power line could also run at voltages above 8 kV (27 kV one phase for example), but, in this case, a power step-down transformer (of light weight) should be provided on-board.

There are also other static power converter schemes[27] that could meet the LIM requirements. Finally, the high power transfer to Maglev as required by LIM is another important problem that must be dealt with, but this is beyond our scope here.

References

1. E. R. Laithwaite, Induction Machines for Special Purposes, George Newness, London, 1966.

2. S. Yamamura, The Theory of Linear Induction Motors, John Wiley, 1972, pp. 125-136.

3. S. A. Nasar and I. Boldea, Linear Motion Electric Machines, Wiley-Interscience, 1976.

4. E. R. Laithwaite (Editor), Transport Without Wheels, Westview Press, Boulder, Colorado, 1977.

5. P. K. Budig, A.C. Linear Motors, (in German), VEB Verlag, Berlin, 1978, p. 76.

6. M. Poloujadoff, The Theory of Linear Induction Machines, Clarendon Press, Oxford, 1980.

7. I. Boldea, Vehicles on Magnetic Cushion, (in Romanian), Romanian Acad. Publish. House, Bucharest, 1981, chapter 2.

8. M. Poloujadoff "Linear induction machines," Parts I and II, IEEE Spectrum, 1971, pp. 72-86.

9. H. Bolton, "Transverse edge effect in sheet rotor induction motors," Proc. IEE, Vol. 116, 1969, pp. 725-739.

10. Y. X. K. Chun, "Transverse edge effect in linear induction motors taking account of the fringing flux," EME, Vol. 7, 1982.

11. H. May, H. Mosebach and H. Weh, "Three dimensional effects in linear motors," International Conference on Electric Machines, L 3/4, Brussels, Belgium, 1979.

12. I. Boldea and S. A. Nasar, "Improved performance from high-speed single sided linear induction motor: A theoretical study," Int. Quat. Elec. Mach. Electromech., Vol. 2, No. 2, 1978, pp. 155-166.

13. K. Oberretl, "Single-sided linear motor with ladder secondary" (in German), Arch. fur Elektrotechn., Bds. 56, 1976, pp. 305-319.

14. L. Del Cid, Jr., "Methods of analysis of linear induction motors," Ph.D. dissertation, University of Kentucky, 1973.

15. I. Boldea and S. A. Nasar, "Quasi-one-dimensional theory of linear induction motors with half-filled end slots," Proc. IEE, Vol. 122, 1975, pp. 61-66.

16. E. M. Freeman and C. Papageorgiou, "Spatial Fourier transforms: A new view of end effects in linear induction motors," Proc. IEE, 1978.

17. S. Yamamura, M. Ito, and H. Masuda, "Three dimensional analysis of double-sided linear induction motor with composite secondary," Elec. Eng. Jap., Vol. 99, No. 2, 1979, pp. 100-104.

18. I. Boldea and M. Babescu, "A multilayer approach to the analysis of single sided linear induction motors," Proc. IEE, Vol. 125, No. 4, 1978, pp. 283-288.

19. S. Nonaka and K. Konyama, "Feasibility of hollow aluminum reaction rail for high speed linear induction motors," Elec. Eng. Jap., Vol. 99, No. 4, 1979, pp. 89-98.

20. S. Nonaka, N. Fujii, and N. Shinada, "Experimental study on effect of hollow aluminum reaction rail for high speed LIM," Elec. Eng. Jap., Vol. 99, No. 4, 1979, pp. 80-88.

21. T. A. Lipo and T. A. Nondahl, "Pole-by-pole d-q model of a linear induction machine," IEEE, PES Winter Meeting, New York, 1978.

22. K. O. Sharples, I. Van Bueren, and S. M Mahendra, "On the magnetic flux distribution and forces in linear induction motors," International Conference on Electric Machines, L 25, Brussels, Belgium, 1979.

23. W. Deleroi, "Airgap field, induced voltage and thrust in short stator linear motors" (in German), Arch. Elek., Vol. 62, 1980, pp. 233-242.

24. S. Nakamura, Y. Takeuchi and M. Takahoshi, "Experimental results of the Slinear induction motor," IEEE Trans., Vol MAG-15, 1979, pp. 1434-1436.

25. K. Yoshida, K. Harada, and S. Nonaka, "Analysis of short-primary LIM's with odd poles taking into account ferromagnetic end effect," Elec. Eng. Jap. Vol. 99, No. 1, 1979, pp. 43-50.

26. R.B. Powell, "Linear induction motor electrical performance test," Final report, Pb. 261856, prepared for U.S. Department of Transportation, June 1976, Washington, DC 20590.

27. I. Boldea and S. A. Nasar, "Optimum goodness criterion for linear-induction-motor design," Proc. IEE., Vol. 123, No. 1, 1976, pp. 89-92.

28. S. A. Nasar, I. Boldea, and N. Laguna, "Performance of linear induction motors with dual windings," IEE Conference on Linear Electric Machines, London, Oct. 1974, pp. 191-196.

29. S. Nonaka and M. Matsuzaki, "Analysis of performances of various primary windings of high speed linear induction motors," Elec. Eng. Jap., Vol. 99, No. 3, 1979, pp. 103-109.

30. A. Lang, "Propulsion systems for magnetically suspended vehicles," International Conference on Electric Machines, L 3/2, Brussels, Belgium, 1979.

31. M. Rentmeister, "Comparison between asynchronous and synchronous linear motors of short stator construction," International Conference on Electric Machines, L 3/5, Brussels, Belgium, 1979.

32. R. M. Katz and T. R. Eastham, "Single sided linear induction motors with cage and solid-steel reaction rails for integrated magnetic suspension and propulsion of guided ground transport," IEEE-IA 1980, Annual Meeting.

33. E. Nicolescu, "The heating of LIM secondary" (in Romanian), Electrotehnica, Vol. 22, No. 5-6, 1974, pp. 162-166.

34. G. O. D'Sena and J. E. Leney, "Linear induction motor research vehicle speed upgrading tests," Report FRA-ORDRD-74-1.P8.224878, June 1973, prepared for Federal Railroad Administration, Washington, DC 20590.

35. A. Wiart, "LIM propulsion system with static power converter" (in French), RGE, Vol. 84, No. 2, 1975, pp. 112-120.

36. A. K. Walace, J. M. Parker and G. E. Dawson, "Slip control for LIM propelled transit vehicles," IEEE Trans., Vol. MAG-16, No. 5, 1980, pp. 710-712.

37. H. Kolm, P. Mongeau, and F. Williams, "Electromagnetic launchers," IEEE Trans., Vol. MAG-16, No. 5, 1980, pp. 719-721.

38. M. Guarino, Jr., "Integrated linear electric motor propulsion systems for high speed transportation," International Symposium on Linear Electric Motors, Lyon, France, May 1974.

39. "Intermediate capacity transit systems," booklet issued by UTDC-2, St. Clair Ave., W. Toronto, Canada 194 V 1 L 7, 1981.

CHAPTER 7

LINEAR INDUCTOR MOTORS

7.1. Introduction

The linear inductor motor (LIM) has been proposed[1,2,6] for the propulsion of future medium- and high-speed vehicles. The main assets of this motor are:

1. High energy conversion efficiency.

2. Passive-low cost-guideway.

When used as an integrated propulsion-levitation means, the linear inductor motor shows higher specific power/weight performance in comparison with LIMs plus levitation magnets. In the following we present some construction guidelines, some expressions representing longitudinal effects and inverter-fed performance under quasi-stationary conditions, some numerical examples, and test results. Finally we discuss the state equations of the motor during levitation control.

7.2. Construction and Principle of Operation

In a LIM both the field winding and the armature three-phase winding are placed on the primary core as shown in Fig. 7.1a. Longitudinal[1,2,4] (Fig. 7.1a) or transverse[6,11] (Fig. 7.1b) laminations may be used to make the primary core. The secondary is made either of laminated (or solid iron) segments (Fig. 7.2a) or of a continuous variable reluctance structure (Fig. 7.2b). When the secondary is made of segments it is necessary to use twisted coils in the armature winding (Fig. 7.3a) in order to add the induced voltages under the two twin primary cores. Straight coils (Fig. 7.3b) in the armature winding are to be used when the variable reluctance structures of secondary parts are shifted a pole pitch

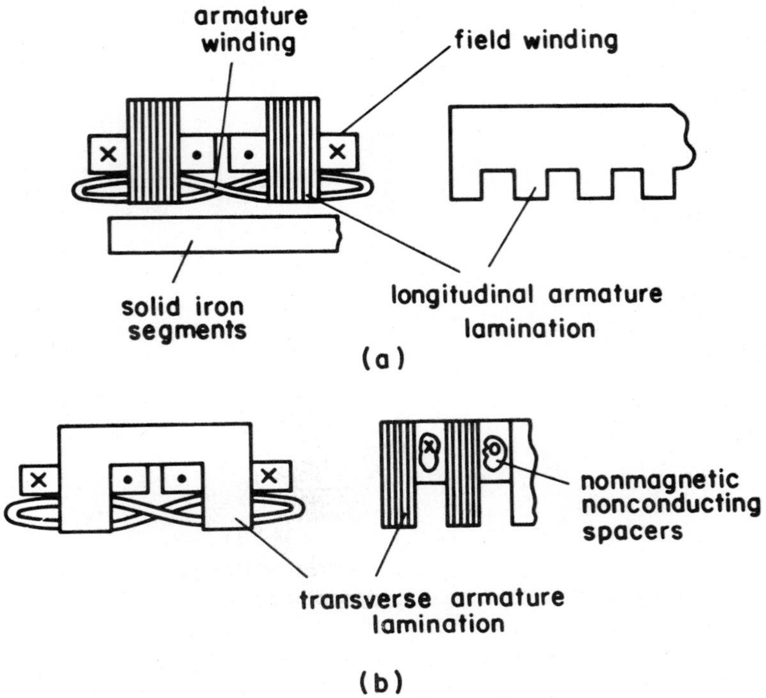

Fig. 7.1. Layout of LIM. (a) Longitudinal lamination, (b) transverse lamination.

with respect to each other. The configuration using transverse laminations and twisted coils in armature winding with straight solid-iron segments in the secondary is considered the best because the secondary (track) weight and cost are low, secondary construction is simple, primary core weight is low, and a better cooling of the winding is possible. These features are deemed to offset the more complicated (twisted) coils in the armature winding.

The field winding is dc fed. Its airgap field exhibits a pulsating character along the primary length (Fig. 7.4). The leakage between segments apparently increases the segment length, l_i, to l_e, thus lowering

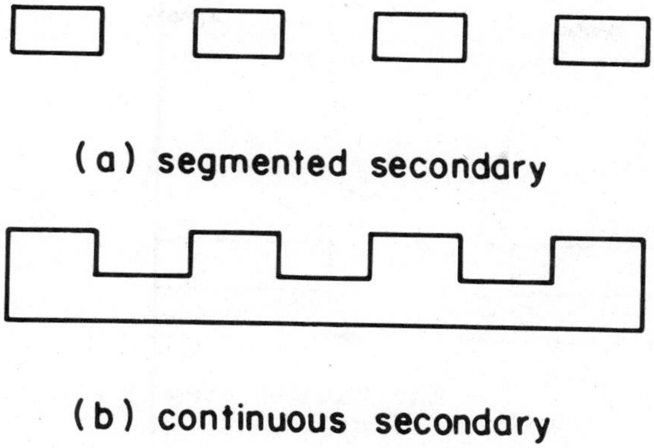

Fig. 7.2. Secondary construction variants. (a) Segmented, (b) continuous.

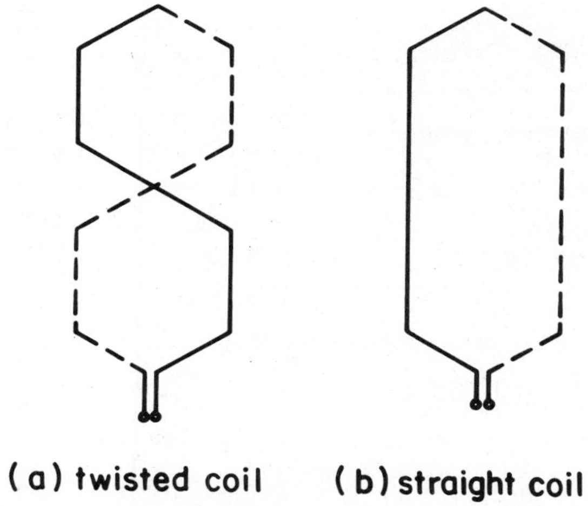

Fig. 7.3. Armature coils. (a) Twisted, (b) straight.

the fundamental of airgap field. However, this leakage flux contributes to the attraction force between the secondary segments and the primary core. The optimum ratio between the secondary segments of length l_i (along the direction of motion) and the pole pitch τ is a

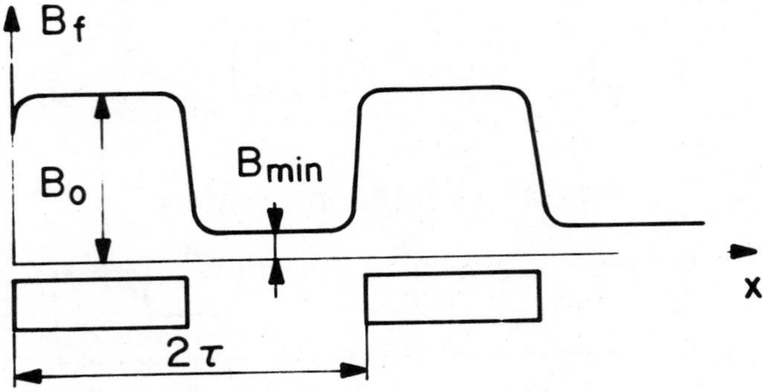

Fig. 7.4. Excitation airgap field.

serious problem. When used only as a propulsion means the optimization criterion should be: maximum ratio between the fundamental, B_1, and average, B_0, values of airgap field.

Values of $l_i/(2\tau)$ between 0.35 and 0.4 have been found most suitable for this case. However, when used as an integrated propulsion-levitation system the longer the segment the higher the attraction (or levitation) force. But when $(l/2)\tau > 0.5$ the fundamental of the airgap magnetic field decreases, lowering the thrust considerably. A practical compromise would be $l_i/\tau = 1 - g_0/\tau$. Thus when considering the airgap leakage, the equivalent length of segments, l_i, becomes τ. To reduce the leakage, the ratio between the airgap, g_0, and the segments thickness, d_0, should be less than 0.25. In this case the ratio between the minimum, B_i, and the maximum, B_0, values of the airgap flux density produced by the field winding, computable by conformal mapping, is $B_{min}/B_0 \leqq 0.15$ to 0.24.

7.2.1. Armature Winding

To avoid fluctuations in the attraction force between the primary and the secondary, due to continuous change of total secondary segments surface placed simultaneously above the primary, the number of poles of the primary should be an even number ($2p_1$). Following the same reasoning, the total length of the primary core should be as close as practically possible to $2p\tau$. Some chording in the armature coils is used to fulfill this requirement and the chording pitch is one slot pitch and not more. A typical, easy to build, two-layer winding suitable for the purpose is shown in Fig. 7.5. The winding resembles two twin three-phase windings, common in linear induction motors. A major drawback of the linear inductor motor consists of its poor use of copper in the armature winding because of long end connections in the coils. The length of end connections may be limited by keeping the pole pitch τ below 0.2 m.

In the transverse laminations case the paths of fields of both the excitation and armature winding are mainly trapped in the transverse plane. A transverse

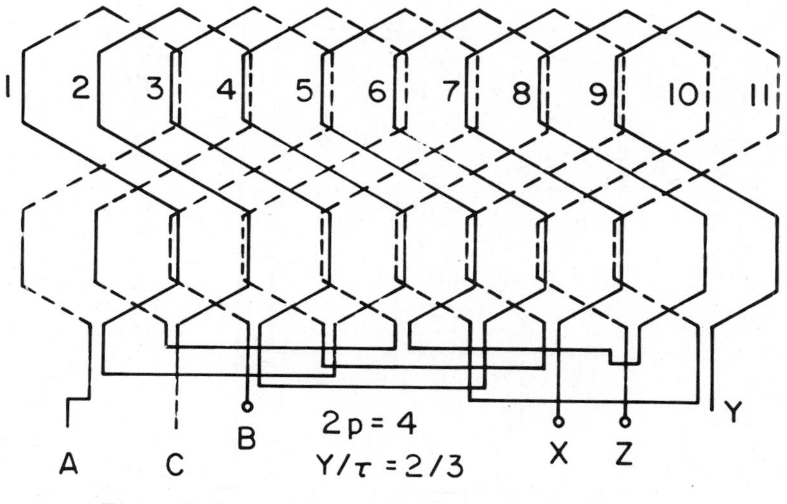

Fig. 7.5. Typical armature winding.

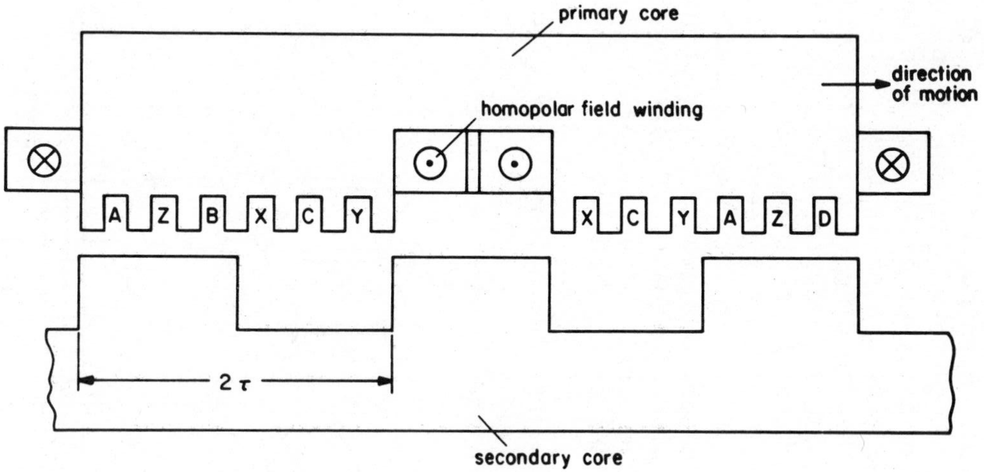

Fig. 7.6. Longitudinal flux LIM.

flux motor is thus obtained (rotary motors are in general of longitudinal flux type). It may be argued that a fully longitudinal flux LIM can also be built (Fig. 7.6). But to keep the primary and secondary core thicknesses within reasonable limits only two poles per core shoe should be utilized and a continuous secondary laminated structure is necessary. The higher weights and costs of the primary and the secondary cores do not seem to be offset by the better usage of copper in the armature windings due to considerably shorter end connections. In rotary inductor machines, however, this solution could prove superior to any other because the rotor and stator costs are rather balanced. But in our case the costs of the secondary prevail.

Let us now represent the reaction field for the transverse field motor (Fig. 7.7). The fundamental of reaction field, B_{r1}, may be written as

$$B_{r1} \simeq \frac{3\sqrt{2}W_1 I_1 K_{w1}\mu_0}{2p_1 g_e} \sin\left(\frac{\pi x}{\tau} + \delta\right) \qquad (1)$$

The excitation field is

$$B_0 = \frac{\mu_0 W_f I_f}{2g_e} \qquad (2)$$

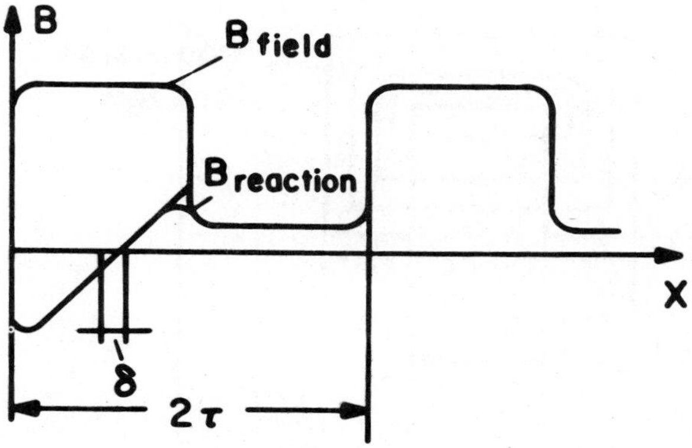

Fig. 7.7. Transverse reaction field.

where g_e = equivalent airgap; W_1 = turns/phase; I_1 = current/phase, δ = internal angle (phase lag between excitation-induced voltage and the phase current).

7.2.2. Transverse Leakage

Both the excitation and armature windings produce leakage fields in the window of transverse laminations as shown in Fig. 7.8. The maximum value, B_{tn}, of this field occurs for $x = \tau/2$ (at a segment corner) and is given by

$$B_{tn} = \frac{\mu_0}{L_1}\left(W_f I_f + \frac{W_1}{p} I_1 \sqrt{2} \cos \delta_0\right) \quad (3)$$

Thus the maximum flux density, B_{tmax}, obtainable in the primary teeth occurs also for $x = \tau/2$ (Fig. 7.7), and

$$B_{tmax} = B_{tn} + \frac{\tau_s}{b_s}\left[B_{r1}\left(\frac{\tau}{2}\right) + B_0\right] \quad (4)$$

where τ_s = slot pitch and b_s = slot width. The tooth maximum flux density, B_{tmax}, is a significant design

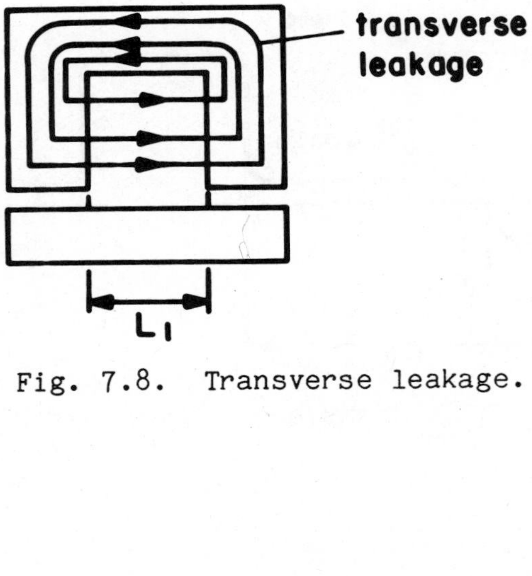

Fig. 7.8. Transverse leakage.

limitation that must be checked in order to avoid high saturation of the primary teeth. As should be foreseen, lowering the window width, L_1 (Fig. 7.8), causes a lowering of armature coils end connections but also an increase of transverse leakage and excitation winding leakage inductance, and hence a higher time constant of the excitation winding. A new problem of optimum (or compromise) arises here. However, considering that a 30° inclination angle for armature coils end connections is mandatory for construction purposes, we assume that a practical choice for L_1 is:

$$L_1 = 0.02 + \frac{\tau\sqrt{3}}{2} \qquad (5)$$

7.3. Lumped Parameters

So far we have neglected both saturation and eddy currents induced in the secondary solid segments because of the motion of the excitation field--the so-called longitudinal end effect. Thus the lumped parameters are similar to those of rotary synchronous homopolar motors with some notable exceptions. Since $l_e = \tau$ the motor has no saliency, but has a uniform double airgap. The magnetizing reactance, $X_{dm} = X_{qm}$, is

$$X_{dm} = \frac{12\mu_0}{\pi} f_1 \, [(W_1 K_{w1})^2] \, \frac{a_e \tau}{p g_e} \qquad (6)$$

When $l_e \neq \tau$ the expression for X_{dm} and X_{qm} ($X_{dm} \neq X_{qm}$) may be taken from the rotary machine literature. Also taking into account the even number of poles and the type of armature winding (Fig. 7.5), the primary leakage reactance, $X_{1\sigma}$, is

$$X_{1\sigma} = 4\mu_0 \pi f_1 \, [q(4p - 3)(\lambda_s + \lambda_d)4a$$

$$+ \lambda_f \frac{4\tau\beta q^2}{\sqrt{3}_d})4a + \lambda_f \frac{4\tau\beta q^2}{\sqrt{3}} (p - 1)] \, W_c^2 \qquad (7)$$

where λ_s and λ_d = the slot and differential specific permanence, β = the chording factor; q = the slot/pole/phase; and W_c = the turns/coil:

$$\beta = 1 \text{ or } 3q - \frac{1}{3q}; \quad W_1 = (2p - 1)qW_c \tag{8}$$

$$\lambda_s = \frac{h_s}{24b_s}(1 + 3\beta);$$

$$\lambda_d = \frac{5g_e/b_s}{5 + 4g_e/b_s}; \quad \lambda_f = 0.3(3\beta - 1) \tag{9}$$

The phase resistance, R_1, is

$$R_1 \simeq \rho_{co}(8a + 0.08 + \frac{8\tau}{\sqrt{3}}) \frac{W_1}{I_{1n}} J_{co} \tag{10}$$

where ρ_{co} = copper resistivity, J_{co} = copper design current density, and I_{1n} = rated phase current.

7.4. Induced Voltage

Only the fundamental of airgap magnetic field produced by the excitation winding, B_{1g}, is considered:

$$B_{1g} \simeq \frac{2}{\tau} \int_0^{\tau/2} (B_0 - B_{min}) \cos \frac{\pi x}{\tau} dx$$

$$= \frac{2}{\pi}(1 - \frac{B_{min}}{B_0}) B_0 \tag{11}$$

Finally, the voltage V_{e0} (RMS) induced by the excitation winding in an armature phase is

$$V_{e0} \simeq \pi \sqrt{2}\, f_1 W_1 K_{w1} [B_{1g} (\frac{2}{\pi}) \tau 4a] \tag{12}$$

where $f_1 = U/2\tau$ and U = speed.

7.5. The Longitudinal End Effect

The field winding has a finite length and moves at a constant speed, U, with respect to the solid-iron secondary segments. Each segment entering or leaving the airgap zone experiences a flux variation and, as a result, induced currents occur in it, thereby producing a drag force. These induced currents attenuate along the primary length, and even in high-speed motors, at the exit end of the motor they are practically zero. Thus, the induced currents in the entering and exiting segments are a rough approximation of the phenomenon called longitudinal end effect.

When considering only the entering and exiting segments the following assumptions are made:

1. The currents induced in the secondary segments have only a Z component and the longitudinal components (along OX) are accounted for by the well-known transverse effect correction coefficient:

$$K_t$$

$$= \frac{1}{1 - \dfrac{\tanh(a_e \pi/\tau)}{(\dfrac{\pi a_e}{\tau})\{1 + \tanh(a_e \pi/tau) \tan[((\pi/\tau)(c - a_e)]\}}} \tag{13}$$

with

$$c \simeq \frac{L_1}{2 + a}; \quad l_e = \tau \tag{14}$$

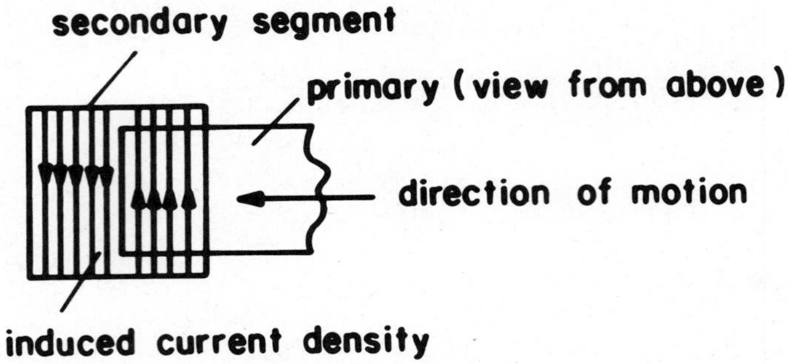

Fig. 7.9. Theoretical induced currents.

2. The magnetic field is zero outside the primary core.

3. The induced current density is uniform along the length of segments. (Fig. 7.9).

4. The iron permeability is constant but may be calculated by iteration.

5. The leakage flux of induced currents is neglected. Thus designating by I the current induced in a segment and by R the equivalent resistance of a segment, the total flux satisfies the equation

$$-\frac{d\psi_t}{dx} U = RI; \quad \text{with } R = \frac{2K_t a_x}{\sigma_i d_i} \left(\frac{1}{x} + \frac{1}{\tau} - x\right) \tag{15}$$

where

$$\psi_t = \psi_f + \psi_a \tag{16}$$

ψ_f = excitation field and ψ_a = reaction field. Also

$$\frac{d\psi_f}{dx} = 2B_0 a_e; \quad \frac{d\psi_a}{dx} = \frac{2a_e I \mu_0}{g_e}; \quad B_0 \simeq \frac{\mu_0 W_f I_f}{2g_e} \tag{17}$$

From (15) through (17) we obtain the current, I, expression:

$$I = \frac{-UB_0}{(\mu_0 U/g_e) + [K_t \tau/d_i \sigma_i(\tau - x)x]} \tag{18}$$

The current is maximum in (18) when $x = \tau/2$ and pulsates in time, being zero both for $x = 0$ and $x = \tau$; that is, just before the entrance and after complete entrance in the airgap zone. The phenomena at the exit end should be similar to those at the entry end.

Considering the time average current of (18) the total drag force per motor, F_{dax}, is given by

$$F_{dax} = \frac{4RI^2}{U}; \quad I_{av} = \frac{I_{max}}{2} \tag{19}$$

$$F_{daxav} = \frac{4}{U\tau} \int_0^\tau RI^2\, dx \tag{20}$$

$$F_{dax} = \frac{8k_t a_e}{\sigma_i d_i}$$

$$\int_0^\tau \frac{UB_0^2\, dx}{(\tau - x)\{(\mu_0 U/g_e) + [2K_t a_e/d_i \sigma_i(\tau - x)x]\}} \tag{21}$$

A rough approximation would be to consider the length of field penetration d_i equal to the thickness of the segment d_0. However $d_i \lessgtr d_0$ and thus the drag force would be overestimated. Because the reaction field is smaller than the excitation field the equivalent value of permeability, μ, is considered that corresponding to the tangential field at the segment surface B_{x0}:

$$B_{x0} = \frac{2a_e B_0}{d_i} \tag{22}$$

Up to this point, d_i is not known and for initializing the calculations d_i would be assigned the initial value of d_0. Using the magnetization curve, stored in the computer memory, the corresponding H_{x0} and an initial μ are found. Due to the rather large airgap involved the core magnetic reluctance may be neglected.

<u>Penetration depth, d_i</u>

Applying the Ampere and induction laws for a contour of a segment entering the airgap we obtain:

$$g_e \frac{\partial H_r}{\partial x} = J_z d_i \tag{23}$$

$$\frac{\partial J_z}{\partial x} = \sigma_i \mu u \frac{B_0}{\mu_0} + H_r) \tag{24}$$

where H_r is the reaction field. According to (23) and (24) the current density is no longer uniform along OX in spite of the above assumptions. Also, the longitudinal current density components (along OX) cannot be neglected in (23) and (24). For a sinusoidal variation along OZ

$$H_r(x,\ z) = H_r(x) \cos \frac{\pi z}{2c} \tag{25}$$

And from (23) through (25) it follows that

$$\frac{\partial^2 H_r(x)}{\partial x^2} - \frac{u d_i \sigma_i \mu_0}{g_e} \frac{\partial H_r(x)}{\partial x} - \left(\frac{\pi}{2c}\right)^2 H_r(x) = 0 \tag{26}$$

The variation of $\underline{H}_r$ along OX is sensed by the segment as a function of time. Thus an equivalent frequency ω_e may be defined as

$$\omega_e = U|\gamma_e| \tag{27}$$

where $\gamma_e i$ is the negative solution of the characteristic equation of (26):

$$\gamma_e = \frac{\mu_0 d_i \sigma_i u}{2g_e} - \sqrt{(\frac{\mu_0 d_i \sigma_i u}{2g_e}) + (\frac{\pi}{2c})^2} \tag{28}$$

Finally, the equivalent depth of penetration, d_i, is

$$d_i \simeq \sqrt{\frac{2}{\mu\omega_e\sigma_i}} \tag{29}$$

Iteratively from (27) through (29) the value of d_i is calculated provided μ is known in (29). A separate iteration cycle is necessary to obtain convergence for μ.

The above procedure is a simplified approach and should be used for preliminary design purposes. For a refined analysis a numerical two- or three-dimensional approach should be used. It has been demonstrated, however, that even for speeds up to 100 m/sec the field winding drag force (with solid-iron secondary segments) is less than 10% of rated thrust of the motor.[46] Since such a motor should be supplied from a variable-frequency, variable-voltage power supply the performance in such a case should be investigated. Such an investigation follows.

7.6. Inverter-Fed Performance

The linear inductor motor armature winding is considered to be fed from a controlled rectifier current

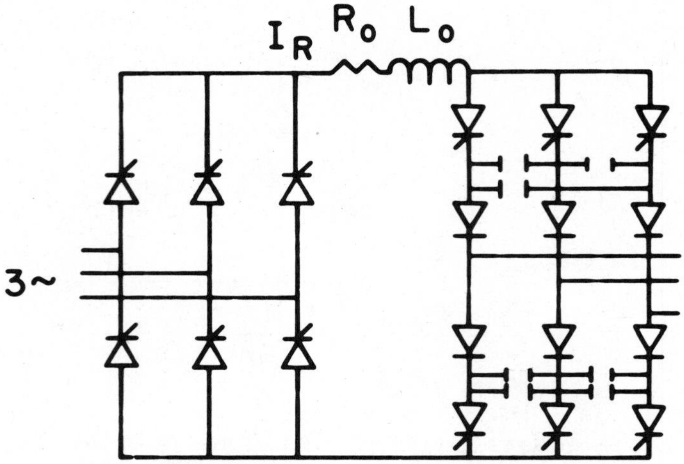

Fig. 7.10. Liner inductor motor fed from an inverter.

inverter with induced voltage commutation (Fig. 7.10). The following assumptions are made:

1. The dc link current, I_r, is time-invariant and the current inverter is a zero impedance instantaneous switching device.

2. The motor is in the steady state and the wye-connected armature winding is sinusoidally distributed along the motor length.

3. The machine shows in fact no saliency since the pole and interpole lengths are equal to each other in order to provide high, nonpulsating levitation force.

4. The field-current-induced voltage is sinusoidal in time.

5. The airgap is constant and the field winding is fed from an ideal dc voltage source.

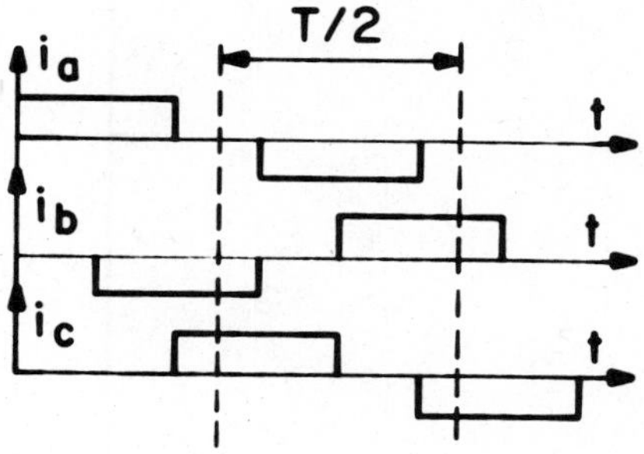

Fig. 7.11. Motor phase currents.

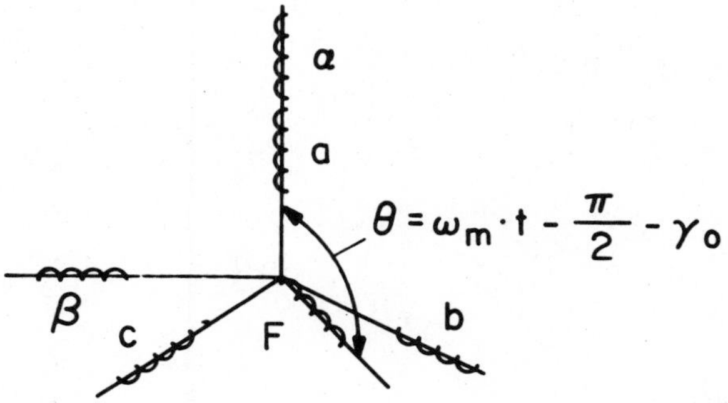

Fig. 7.12. Major equivalent circuit.

6. All motor parameters are constant and the longitudinal effect[3] is neglected.

7. The solid-iron secondary has a negligible damping effect.[2]

According to these assumptions the motor phase currents are as shown in Fig. 7.11. The motor is schematically shown in Fig. 7.12 exhibiting only the armature and field windings.

7.6.1. Motor Equations

First, the three-phase armature winding (Fig. 7.12) is transformed into an equivalent two-phase α_1, β_1 winding. The power angle is

$$\theta = \omega_m t - \frac{\pi}{2} - \gamma_0 \tag{30}$$

for motor mode, and

$$\theta_g = \omega_m t + \frac{\pi}{2} + \gamma_0 \tag{31}$$

for regenerative braking. When $\gamma_0 > 0$ the machine is overexcited, whereas for $\gamma_0 < 0$ it is underexcited as known from conventional synchronous machine theory. The α_1, β_1 currents are[12]:

$$i_{\alpha 1} = \sqrt{\frac{2}{3}}\,[i_a - \frac{1}{2}(i_b + i_c)\,]$$

$$i_{\beta 1} = \sqrt{\frac{2}{3}}(i_b - i_c) \tag{32}$$

The ideal distribution of $i_{\alpha 1}$ and $i_{\beta 1}$, shown in Fig. 7.13, was obtained by applying (32) to the ideal motor phase currents of Fig. 7.11.

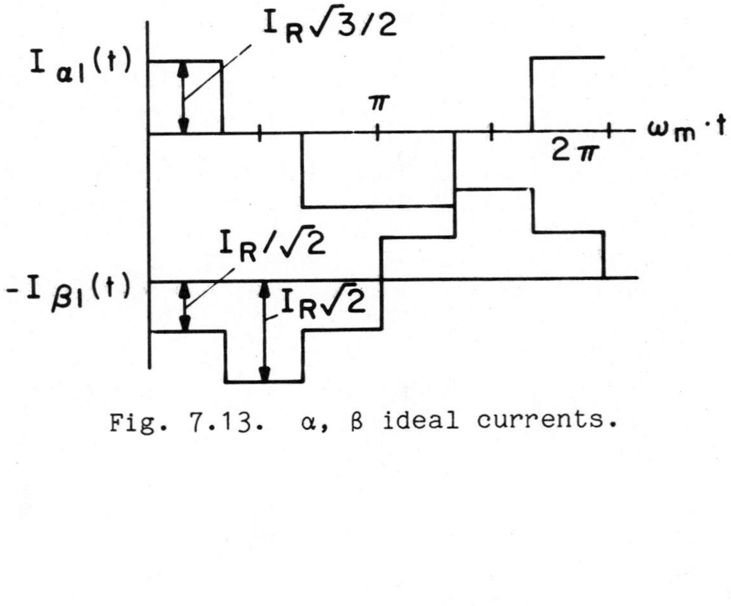

Fig. 7.13. α, β ideal currents.

The machine equations in α_1, β_1 coordinates,[12] in the absence of damper windings, become

$$\begin{bmatrix} V_{\alpha 1} \\ V_{\beta 1} \\ V_F \end{bmatrix} \begin{bmatrix} R_1 + pL_1 & 0 & M'p\cos\Theta \\ 0 & R_1 + pL_1 & M'p\sin\Theta \\ M'p\cos\Theta & M'p\sin\Theta & R_F + pL_F \end{bmatrix} \begin{bmatrix} i_{\alpha 1} \\ i_{\beta 1} \\ i_F \end{bmatrix} \quad (33)$$

where

$$M' = \sqrt{\frac{3}{2}} \cdot \frac{V_{e0}\sqrt{2}}{I_F} \cdot \frac{1}{2\pi f_1} \quad (34)$$

The thrust, F_x, may be obtained by considering the synchronous torque, T_s, of an imaginary equivalent rotary machine:

$$T_s = \omega_m M' I_F(t)(i_{\alpha 1}\sin\theta - i_{\beta 1}\cos\theta) \quad (35)$$

The electromagnetic propulsion power, P_s, of the linear counterpart is given by

$$P_x = F_x u \quad (36)$$

From (34) through (36) we get

$$F_x = \frac{\pi}{\tau} M' i_F(t)(i_{\alpha 1}\sin\theta - i_{\beta 1}\cos\theta) \quad (37)$$

The phase voltage, V_a, of the three-phase motor is

$$V_a = \sqrt{\frac{2}{3}}\left(\frac{V_0}{\sqrt{2}} + V_{\alpha 1}\right) = \sqrt{\frac{2}{3}} \cdot V_{\alpha 1} \quad (38)$$

By making use of the first equation of (33) V_a becomes

$$V_a = \sqrt{\frac{2}{3}}[(R_1 + pL_1)\, i_{\alpha 1} + M'pi_F \cos\theta] \qquad (39)$$

where R_1 = the phase resistance, R_F = field winding resistance, M' = mutual inductances between field and armature of two-phase model ($M' = \sqrt{(3/2)}M$); M = the mutual field-armature inductance for the actual three-phase machine; L_F = field winding self-inductance; $L_{1\sigma}$ = armature phase inductance of two phase model:

$$L_1 = L_{1\sigma} + \frac{3}{2} L_1' \qquad (40)$$

$L_{1\sigma} = X_{1\sigma}/2\pi f_1$ = leakage inductance; and $L_1' = X_{dm}/2\pi f_1$, phase self-inductance part corresponding to the useful field for the real three-phase machine.

7.6.2. Field Current Waveform

In (33) the armature α_1, β_1 currents (Fig. 7.13) are time-invariant over an interval of $\pi/3$ radians. Thus, the third equation of (33) becomes

$$V_F = i_{\alpha 1}M'p \cos\theta + i_{\beta 1}M'p \sin\theta + R_F i_F + L_F p i_F \qquad (41)$$

Within the interval $0 < \omega_m t < \pi/3$, $i_{\alpha 1} = I_R\sqrt{3/2}$ and $i_{\beta 1} = I_R/\sqrt{2}$ as shown in Fig. 7.13. And, the solution for i_F in (41) is

$$i_F(t) = \frac{V_F}{R_F} - i_{FM} \cos(\theta + \frac{\pi}{3} - \phi_F) \qquad (42)$$

where

$$i_{Fm} = \frac{I_R M' \omega_m{}^2}{R_F^2 + \omega_m^2 L_F^2}; \quad \phi_F = \tan^{-1}\left(\frac{\omega_m L_F}{R_F}\right) \tag{43}$$

and

$$\theta = \omega_m t - \gamma_0 - \frac{\pi}{2}$$

Hence, the field current exhibits an additional sinusoidal component induced through the motion of the armature currents. The rapid variation of armature currents at the commutation instants produces some additional field current but this is not considered here.

7.6.3. Thrust Pulsations

For the first interval $0 < \omega_m t < \pi/3$ the thrust $F_x(t)$ (37) is

$$F_x(t) = -\frac{\pi}{\tau} M i_F'(t)\left(I_R \sqrt{\frac{3}{2}} \sin\theta - \frac{I_R}{\sqrt{2}} \cos\theta\right)$$

$$= -\frac{\pi}{\tau} M i_F'(t) I_R \sqrt{2} \sin\left(\theta - \frac{\pi}{6}\right) \tag{44}$$

It may be shown that the thrust time variation is identical to that of (44), in all the other five intervals of a period. A look at (44) reveals that the thrust pulsates at two frequencies. The average thrust, F_{xav}, is

$$F_{xav} = \frac{3\omega_m}{\pi} \int_0^{\pi/3\omega_m} F_x(t)\, dt \tag{45}$$

Making use of (42), (44), and (45) F_{xav} becomes

$$F_{xav} = \frac{3\ 2}{\tau}\frac{V_F}{R_F} M'i_R \cos\gamma_0$$

$$- \frac{M^2 I_R^2 \omega_m}{\tau\sqrt{R_F^2 + \omega_m^2 L_F^2}} \qquad (46)$$

$$[\pi \cos\phi_F + \frac{3\sqrt{3}}{2}\cos(2\gamma_0 + \phi_F)]$$

The second term in the average thrust (46) is due to the time-varying component of field current. Its character is that of a drag force if

$$\cos\phi_F + \frac{3\sqrt{3}}{2}\cos(2\gamma_0 - \phi_F) > 0 \qquad (47)$$

Whereas for high speeds the high frequencies of thrust pulsations [6 and 12 times $f_m (f_m = \omega_m/2\pi)$] cannot be followed by a vehicle high inertia, at small speeds they could produce propulsion instabilities and (or) mechanical resonance conditions.

7.6.4. Normal Force Pulsations

The normal force has an attraction character and it is produced by the field and armature windings. Assuming a sinusoidal distribution of armature mmf the resultant airgap flux density, B(x, t), becomes

$$B(x, t) = B_0(t)$$

$$+ \frac{4}{\pi}\ \frac{W_1}{(2p_1 - 1)}\sqrt{\frac{3}{2}}\ \frac{1}{K_s g_0}\ \frac{\sin(\pi/6)}{q\sin(\pi/6q)}$$

$$[\,-i_{\alpha 1}(t)\sin\left(\omega_m t + \frac{\pi}{\tau}x - \gamma_0\right)$$

$$+\,i_{\beta 1}(t)\cos\left(\omega_m t + \frac{\pi}{\tau}x - \gamma_0\right)];$$

$$0 \leqq x \leqq \tau \tag{48}$$

where

$$B_0(t) = \frac{\mu_0 W_F i_F(t)}{2g_0 K_s} \tag{49}$$

and is produced only by the field current, and where $W_1\sqrt{3/2}$ = number of turns per phase for the α, β model; p_1 = number of pole pairs; q = slots per pole and phase; g_0 = airgap; K_s = saturation correction factor; W_F = field turns; $i_{\alpha 1}(t)$ and $i_{\beta 1}(t)$ are given in Fig. 7.13.

The normal force per motor is

$$F_n(t) = \frac{4ap_1}{2\mu_0}\int_{-\tau/2}^{\tau/2} B^2(x, t)\,dx \tag{50}$$

where 2a is the width of one of the two primary stacks. The average normal force F_{nav} corresponding to the first interval of the period 0 to $\pi/3\omega_m$ may be obtained from the integral

$$F_{nav} = \frac{3\omega_m}{\pi}\int_0^{\pi/3\omega_m} F_n(t)\,dt \tag{51}$$

7.6.5. Phase Voltage Waveform

The expression for the phase voltage may be developed from (39) by use of $i_F(t)$ as given by (42). Hence

$$V_a(t) = \sqrt{\frac{2}{3}} \{ R_1 i_{\alpha 1}(t) + L_1 \frac{di\alpha_1(t)}{dt} + M'\omega_m[i_F(t) \cos(\omega_m t - \gamma_0) + i_{Fm} \sin(\omega_m t - \gamma_0) \cos(\omega_m t - \gamma_0 + \frac{\pi}{3} - \phi_F)]\} \quad (52)$$

The voltage peaks that would occur for instantaneous commutation should be theoretically infinite because

$$\left(\frac{di_{\alpha 1}}{dt}\right)_{t=0,3\pi/\omega_m} \rightarrow \infty \quad (53)$$

In reality this current derivative is finite and may be approximated by accounting for the commutation phenomenon through an equivalent commutation resistance, R_c, as given by[11]

$$R_c \simeq \frac{3\omega_m}{\pi} L_1 \quad (54)$$

Thus the voltage peaks, V_k, occurring at the commutation instants are

$$(V_K)_{t=0,\pi/3\omega_m} \simeq \frac{3}{\pi} \omega_m L_1 I_R \sqrt{\frac{3}{2}} \quad (55)$$

The phase voltage equation, (52), shows also the presence of a constant and a double frequency component due to the time-varying field current component.

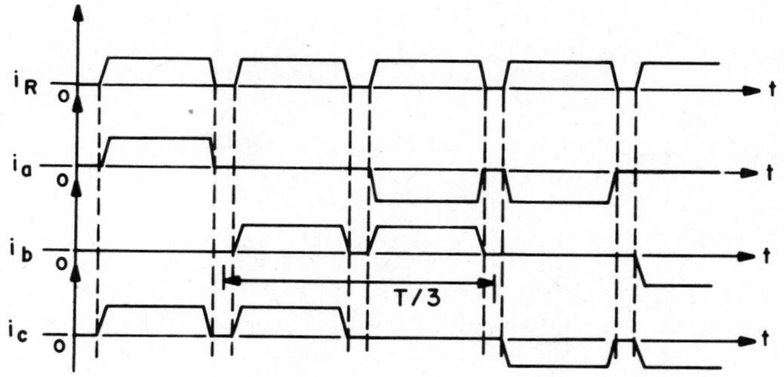

Fig. 7.14. Ideal phase currents during starting.

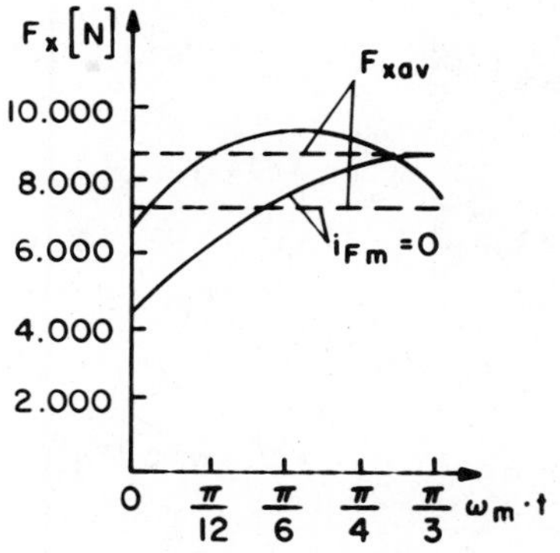

Fig. 7.15. Thrust pulsation in time.

7.6.6. Starting Procedure by Line Commutation

For speeds smaller than 10% of the rated speed, the induced voltage is too small to provide "natural" commutation. In the absence of the rather costly switching capacitors some kind of line commutation should be provided for these conditions occurring during vehicle starting. In a standard line commutation procedure the phase ideal currents are as shown in Fig. 7.14. During a time interval $\Delta t_1 = \alpha_1$ over ω_m ideally there is a discontinuity in the motor current. The resulting $i_{\alpha 1}(t)$ and $i_{\beta 1}(t)$ currents could be obtained from (32) and Fig. 7.15. Frequencies up to 250 Hz are envisaged for high speed vehicles. Even for slow thyristors, t_{of}

= 300 μsec turn-off time; this would correspond to α_1 = 5.4°. To allow for the overlapping to occur in any real commutation process, α_1 should be increased to α_{1m} = 8 to 10°. Thus the conducting angle is α = 50 to 52°, instead of 60° as it was for the ideal running conditions. Short current and thrust breaks occur also during such a starting process. Now as the starting process is slow in comparison with the electromagnetic phenomena, the thrust, normal force, phase voltage waveform could be calculated with the same equations as above, allowing for the current discontinuities.

The average thrust, F_{xav}, and normal force, F_{nav}, are now

$$F_{xav} = \frac{3\omega_m}{\pi} \int_{\alpha_1/2\omega_m}^{(\alpha_1/2+\alpha)/\omega_m} F_x(t)\, dt \tag{56a}$$

$$F_{nav} = \frac{3\omega_m}{\pi} \int_{\alpha_1/2\omega_m}^{(\alpha_1/2+\alpha)/\omega_m} F_n(t)\, dt \tag{56b}$$

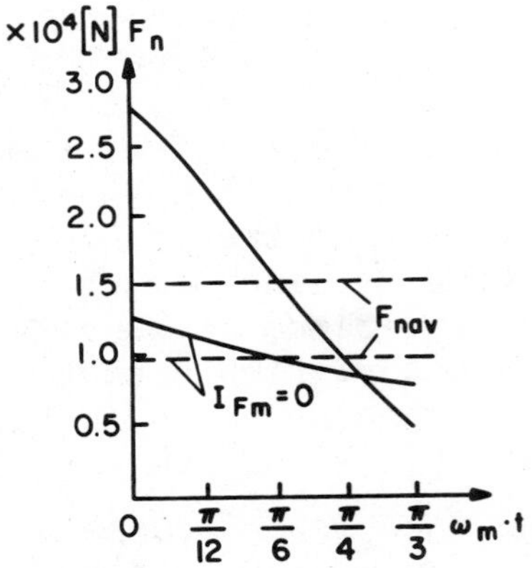

Fig. 7.16. Attraction force pulsation in time.

7.6.7. A Numerical Example of Practical Interest

We now consider a linear inductor motor of practical interest having the following data: $M_F = 0.011$ H, $L_F = 0.064$ H, $R_F = 0.557\ \Omega$, $V_F = 55.7$ V, $I_R = 200$ A, $\omega_m = 20$ rad/sec, $\tau = 0.2$ m, $p_1 = 4$ pole pairs, $W_F = 140$ turns, $W_1 K_w = 110$ turns, $g_0 = 10^{-2}$ m, $K_s = 1.25$, $\gamma_0 = \pi/6$. Substituting these in (43), (44), and (48) through (50) we have computed the time variation of thrust (Fig. 7.15) and normal force (Fig. 7.16) for steady-state running conditions. The special case when the component of field current is zero is also pointed out in Figs. 7.15 and 7.16. From these figures we draw two main conclusions:

1. The thrust and normal force pulsate significantly because of the rectangular form of armature currents.

2. The presence of motion-induced ac field current accentuates this phenomenon. This pulsation could add to the difficulties in stabilizing levitation and propulsion functions.

7.6.8. Some Test Data

In view of the pulsations in the thrust and normal force observed in theory (Figs. 7.15 and 7.16) a suggestion has been made: Introduce capacitors in parallel with the motor, partly to provide fail-safe commutation conditions of inverter thyristors and to filter out the thrust and normal force pulsations. In order to illustrate this idea, we present some test results obtained from a rectifier current-inverter LIM laboratory model. The laboratory motor model has an arch-type primary and a rotating secondary.

An inductive-type position transducer fires the inverter thyristors. Data pertaining to the motor are: pole pitch $\tau = 0.084$ m; stack length $2a = 0.055$ m; airgap $g_0 = 0.006$ m; segment length $l_i = \tau$; number of poles

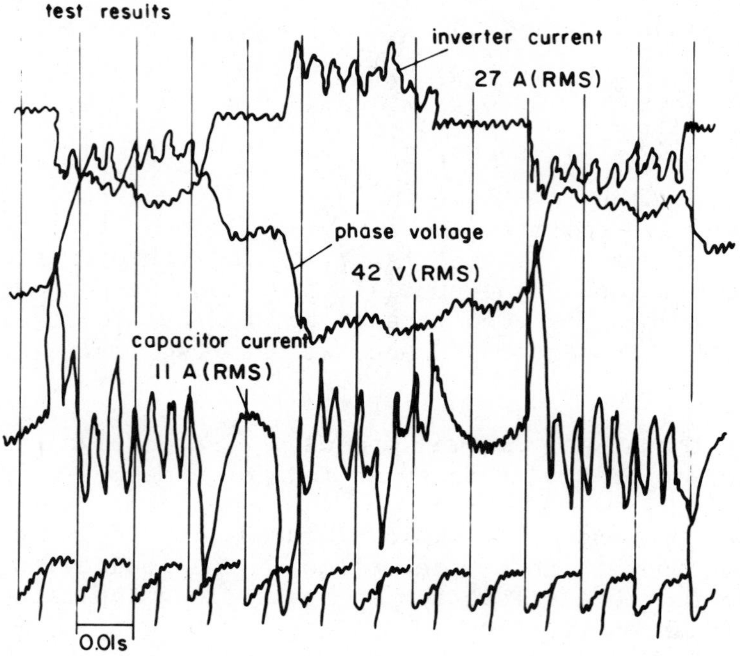

Fig. 7.17. Inverter current, capacitor current, phase voltage waveforms during steady-state motoring.

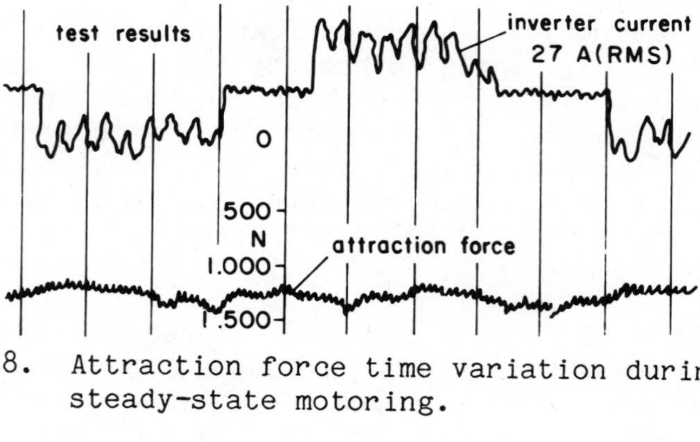

Fig. 7.18. Attraction force time variation during steady-state motoring.

$2p + 1 = 7$; number of slots/pole/phase $q = 2$; turns/phase $W_1 = 456$. Recordings of the inverter phase (before the capacitors) current, phase voltage and capacitor current for steady-state motoring are as shown in Fig. 7.17. Also the normal (attraction force) is shown in Fig. 7.18.

It should be noted that the filtering of dc-link current is incomplete while the normal force pulsations are mild due to the beneficial influence of capacitors at the motor mains. The pulsations would be even smaller in a motor with an even number of poles.

The generator braking is illustrated in Figs. 7.19 and 7.20. In Fig. 7.19 the ac supply voltage and currents are recorded during the switching from motoring to generator braking. An important, but short in duration, peak current occurs during switching. The genera-

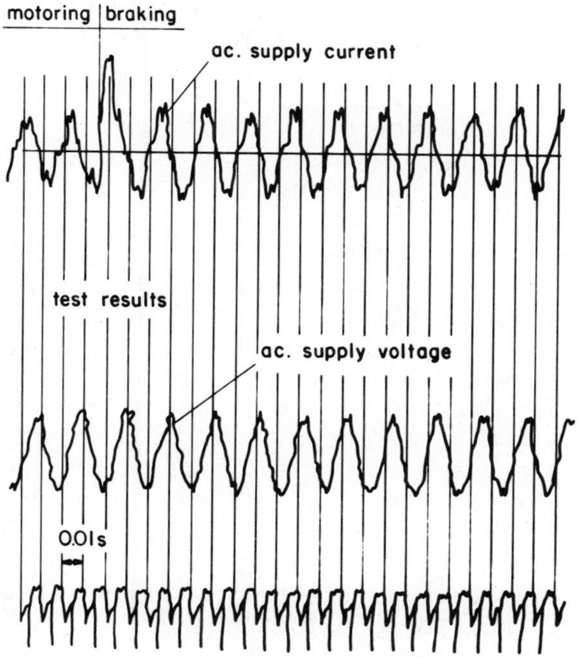

Fig. 7.19. AC supply voltage and currents during motor-generator braking switching.

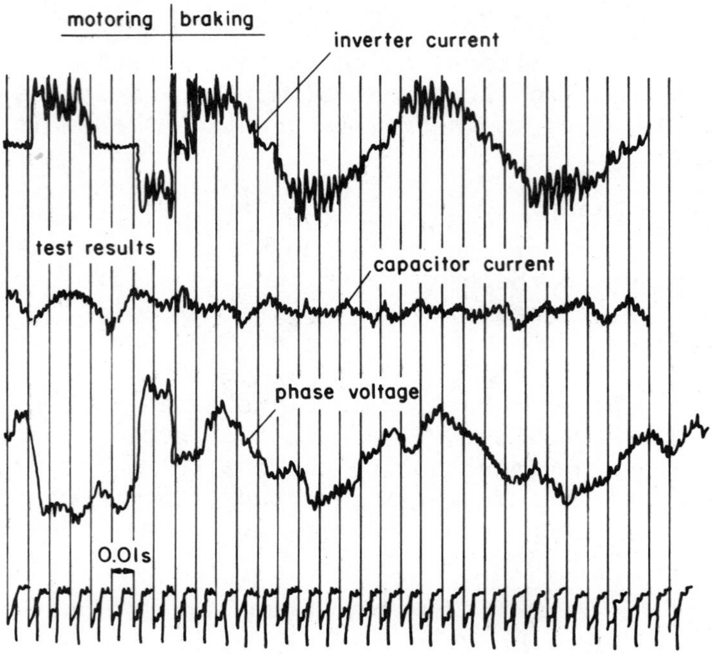

Fig. 7.20. Inverter current, capacitor current, and motor phase voltage during motor-generator braking switching.

tor braking is obtained by turning off the position transducer information and turning the inverter into a rectifier and the rectifier into an inverter.

The inverter and capacitor currents and the motor phase voltage have also been recorded during motor-generator switching as shown in Fig. 7.20. No significant pulsation in the field current or speed (due to thrust pulsations) have been observed during the measurements, indicating that in reality the capacitors are filtering the pulsations. This is an important result since the levitation control is simplified in the absence of normal force pulsations due to the inverter. Also the starting procedure with line-commutation (inverter brake) has been implemented and tested both with and without capacitors. We present the results

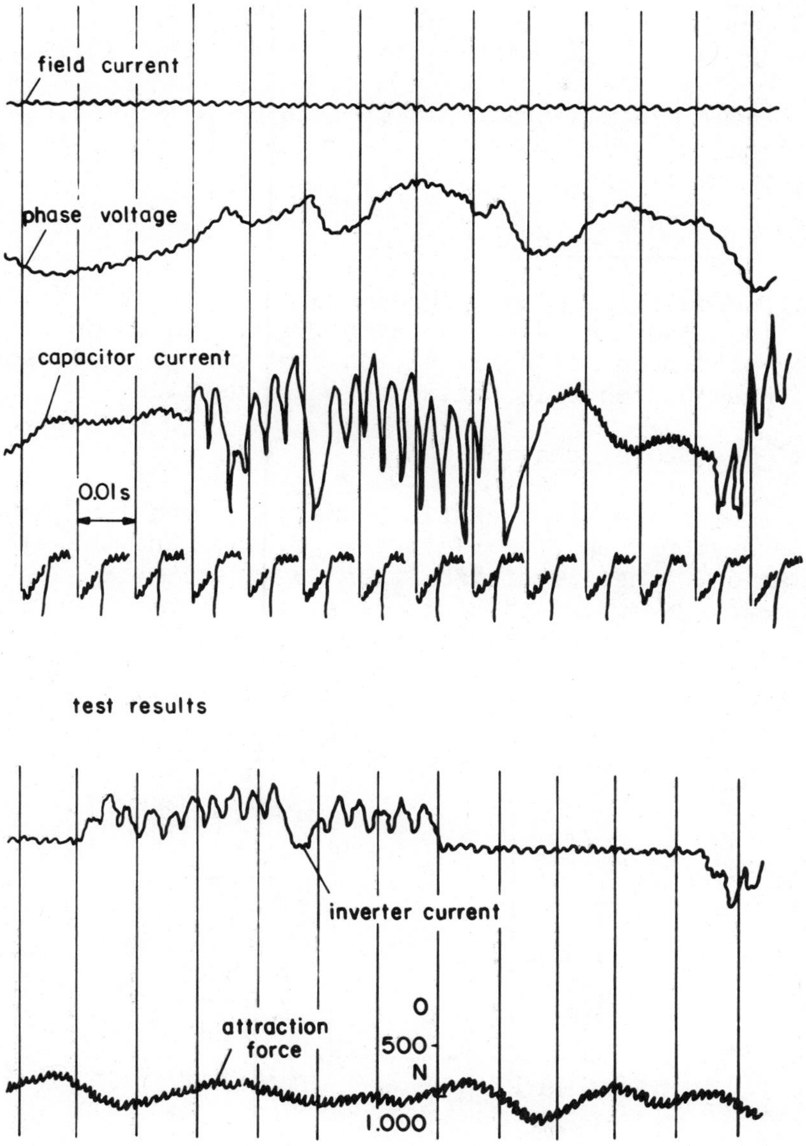

Fig. 7.21. Line commutation starting performance.

with capacitors in Fig. 7.21. Since the speed is small, the induced voltage is also small and the phase voltage is far from being sinusoidal. A peak in the capacitor current is observed during the current discontinuity. The normal force pulsations are more important than those occurring during the steady state (Fig. 7.18).

These experimental results must be considered when designing a LIM for propulsion or for propulsion and levitation purposes.

7.7. Note on Design Aspects

The design strategy depends upon the fact that the LIM is used as a propulsion means only or as an integrated propulsion-levitation system. The interested reader should consult Reference 6 for details.

7.8. The State Equations

A complete study of LIM transient behavior should include the rectifier-inverter system and the airgap variation. The airgap variation is inevitable due to the track irregularities and to airgap control for levitation. For the rectifier current-inverter scheme,

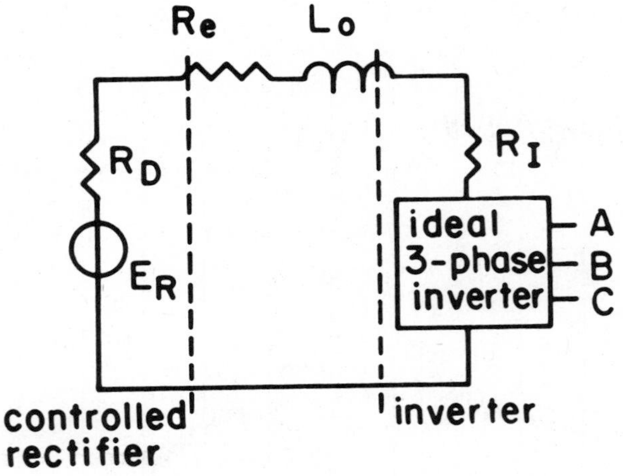

Fig. 7.22. An equivalent circuit.

shown in Fig. 7.10, an equivalent circuit is defined[14] in Fig. 7.22. The phase-controlled rectifier is modeled as a controllable dc voltage, e_R, in series with a resistance, R_R. The parameters of the filter inductor are R_0 and L_0. Also the inverter is modeled as an ideal inverter in series with a resistance, R_I. The commutation delay angle is considered either zero or constant in time. The output of the ideal inverter is a 120° square wave current. The effects of the harmonics of this current may be neglected during transients. To simplify the equations the effect of the currents induced, during the transients, in the secondary segments is neglected. Also, zero saliency is assumed, and $l_e = \tau$.

Considering that the angle between the phase, a, of the armature and the field winding is θ, in a synchronous reference frame, the three-phase armature winding may be replaced by a single winding (along d axes for example) displaced by an angle γ with respect to the field winding (Fig. 7.23) such that

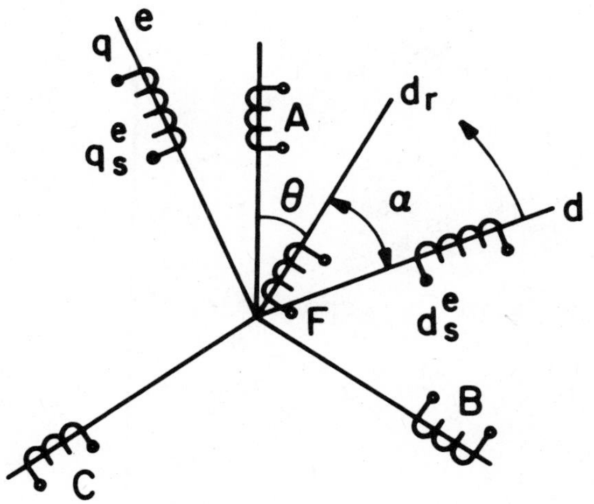

Fig. 7.23. dq-equivalent of 3-phase winding.

$$\theta + \gamma = \omega_1 t \ .$$

The equations in dq-coordinates can be obtained from those in αβ-coordinates (33) by a rotation ($\theta + \gamma$ using the transformation matrix [C]) where

$$C = \begin{bmatrix} \cos(\theta+\gamma) & \sin(\theta+\gamma) & 0 \\ -\sin(\theta+\gamma) & \cos(\theta+\gamma) & 0 \\ 0 & 0 & 1 \end{bmatrix} \tag{57}$$

Equation (33) transformed by the matrix [C] becomes:

$$\begin{bmatrix} v_d^e \\ v_q^e \\ v_F^e \end{bmatrix} = \begin{bmatrix} R_1+pL_1 & \omega_1 L_1 + L_1\dot{\gamma} & \cos\gamma pM_F + \omega_m M_F \cos\gamma \\ -\omega_1 L_1 - L_1 & R_1 + pL_1 & \sin\ pM_F - \omega_m M_F \cos\gamma \\ \cos\gamma pM_F - M_F(\sin\gamma)\dot{\gamma} & -\sin\gamma pM_F + M_F(\cos\gamma)\dot{\gamma} & R_F + pL_F \end{bmatrix} \begin{bmatrix} i_{ds}^e \\ i_{qs}^e \\ i_F^e \end{bmatrix} \tag{58}$$

It should be noted that (58) could have been obtained by inspection using Fig. 7.23. The thrust may be obtained directly from (58), observing that $i_{qs}^e = 0$, by power balance:

$$V_d^e i_{ds}^e = R_1 i_{ds}^{e\,2} + i_{ds}^e pL_1 i_{ds}^e$$

$$+ i_{ds}^e \cos\gamma pM_F i_F^e + \omega_m M_F \sin\gamma i_F^e i_{ds}^e \tag{59}$$

Only the last term is responsible for the thrust F_x:

$$F_x = \frac{\pi}{\tau} M_F i_F^e i_{ds}^e \sin\gamma \tag{60}$$

Now considering that airgap g_0 varies in (59) we obtain:

$$V_d^e i_{ds}^e = R_1 (i_{ds}^e)^2 + (i_{ds}^e)^2 \frac{\partial L_1}{\partial g_0} \frac{dg_0}{dt}$$

$$+ i_{ds}^e L_1 \frac{di_{ds}^e}{dt} + i_{ds}^e \cos\gamma i_F^e \frac{\partial M_F}{\partial g_0} \frac{dg_0}{dt}$$

$$+ i_{ds}^e \cos\gamma M_F \frac{di_F^e}{dt} + \omega_m M_F i_F^e i_{ds}^e \sin\gamma \qquad (61)$$

The second and fourth terms contain the airgap time derivative and are responsible for one component of the normal (attraction force) F_{na}; that is,

$$F_{na} = i_{ds}^{e^2} \frac{\partial L_1}{\partial g_0} + i_{ds}^e i_F^e \frac{\partial M_F}{\partial g_0} \cos\gamma \qquad (62)$$

The other component, F_{np}, of normal force is similarly obtained from the excitation circuit equation

$$F_{nf} = i_{ds}^e i_F^e \frac{\partial M_F}{\partial g_0} \cos \gamma + i_F^{e^2} \frac{\partial L_F}{\partial g_0} \qquad (63)$$

Finally, the third and the fifth terms in (63) represent the magnetic stored-energy time variation. Thus the equations of motion along the two directions are

$$m_1 \left(\frac{\tau}{\pi}\right) \frac{d\omega_m}{dt} = F_x - F_d \qquad (64)$$

$$m_1 \ddot{g}_0 = -(F_{nf} + F_{na}) + m_1 g_g \qquad (65)$$

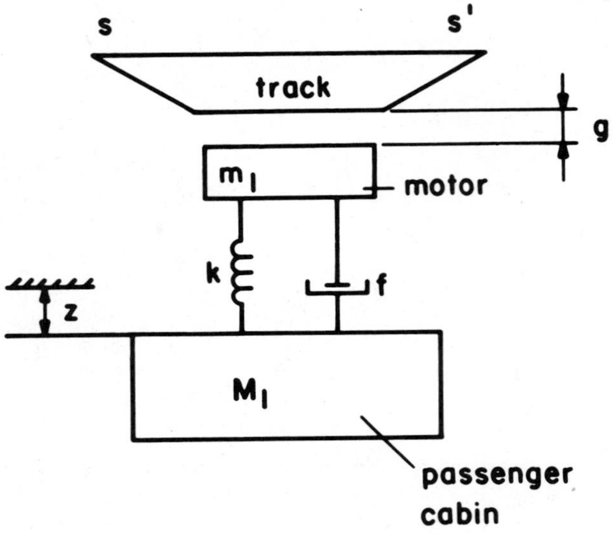

Fig. 7.24. Mechanical analog of a vehicle.

g_g = the acceleration due to gravity. Equation (65) is valid with no additional suspension system when m_1 is a part of the vehicle weight lifted by the motor. In the presence of an additional mechanical suspension system (Fig. 7.24) two more equations are added and (65) becomes

$$m_1 = m_1 + M_1$$

$$m_1\ddot{g}_0 = (m_1 + M_1)g_0 - K(g_0 - g_{00} - z)$$

$$- f(g_0 - z) - (F_{nF} + F_{na}) - \Delta F_{ne} \quad (66)$$

$$M_1\ddot{z} = K(g_0 - g_{00} - z) + f(g_0 - z) \quad (67)$$

where Δ_{ne} = external vertical force (perturbation).

Now considering only the fundamental of the inverter-produced phase currents in the three-phase machine, we find (Fig. 7.21) that the relation between i_{ds} and the rectifier current is

$$i_{ds} = \frac{3\sqrt{2}}{\pi} I_R \tag{68}$$

The power balance in the inverter provides, besides (68), the relation between the inverter voltage, V_I, and V_{ds}:

$$V_{ds} = \frac{\pi}{3\sqrt{2}} V_I \tag{69}$$

with V_I from Fig. 7.22, and

$$V_I = V_R - (R_R + R_I + R_O)I_R - L_O \frac{di_R}{dt} \tag{70}$$

But now if we consider the rectifier voltage, V_R, as a function of the firing angle, α_R, of the phase delay rectifier we have,

$$V_R = \frac{3\sqrt{2}}{\pi} E_s \cos \alpha_R \tag{71}$$

Replacing (71), (68), and (69) in (70) we obtain

$$V_{ds}^e = E_s \cos \alpha_R - R_O' i_{ds}^e - L_O' \frac{di_{ds}^e}{dt} \tag{72}$$

$$R_O' = \frac{\pi^2}{18}(R_R + R_I + R_O) \tag{73}$$

$$L_O' = \frac{\pi^2}{18} L_O \tag{74}$$

Equations (58), (64), (65), and (72) represent the complete set of motor equations valid both for steady state and transients. Some remarks on (58) are in order. As stated above $i^e_{qs} = 0$ but the voltage V^e_{qs} could still be calculated, though in reality its winding does not exist. For steady state the time derivatives are zero in (58), (65), and (72). Also,

$$\frac{dg_e}{dt} = \frac{dz}{dt} = 0$$

Thus the steady-state values of i^e_{ds0}, i^e_{F0}, V^e_{ds0}, and γ_0 could be calculated, if $\alpha_R = \alpha_{R0}$, $\omega_m = \omega_{m0}$, $V_F = V_{F0}$, and $F_d = F_{d0}$ are known, from the equations

$$V^e_{ds0} = R_1 i^e_{ds0} + \omega_{m0} M_{F0} \sin \gamma_0 I_{F0} \tag{75}$$

$$V^e_{F0} = R_F I^e_{F0} \tag{76}$$

$$F_{d0} = \pi \text{Over} \tau \; M_{F0} I^e_{F0} i^e_{ds0} \sin \gamma_0 \tag{77}$$

$$V^e_{ds0} = E_s \cos \alpha_{R0} - R'_0 I^e_{ds0} \tag{78}$$

It is now possible to apply the theory of small deviations to derive the state equations of the motor when used as an integrated propulsion-levitation system:

$$V_{ds} = \Delta V^e_{qs} + V^e_{ds0}; \quad \alpha_R = \alpha_{R0} + \Delta\alpha_R$$

$$i_{ds} = \Delta i^e_{ds} + I_{ds0}; \quad g_0 = \Delta g + g_{00}$$

$$
\begin{bmatrix} v_d^e \\ v_q^e \\ v_f^e \end{bmatrix}
=
\begin{array}{c} \\ d \\ \\ q \\ \\ f \end{array}
\begin{array}{c|c|c}
d & q & f \\
\hline
\begin{array}{l} R_1 + pL_d + \sin^2 \gamma p(L_q - L_d) \\ + \frac{1}{2}\,\omega_m(L_d - L_q)\sin 2 \\ - \frac{1}{2}\,(L_d - L_q)\sin 2\gamma \frac{d\gamma}{dt} \end{array} &
\begin{array}{l} \frac{1}{2} \sin 2\gamma p(L_d - L_q) \\ + L_d\omega_m + \omega_m(L_q - L_d\cos^2\gamma \\ (L_d + (L_q - L_d)\sin^2 \gamma) \frac{d\gamma}{dt} \end{array} &
\begin{array}{l} (\cos \gamma)pM_f' \\ + M_f(\sin \gamma)\omega_m \end{array} \\
\hline
\begin{array}{l} \frac{1}{2} \sin 2\gamma p(L_d - L_q) \\ - L_d\omega_m - \omega_m(L_q - L_d)\sin^2 \\ - [L_d + (L_q - L_d)\cos^2 \gamma] \frac{d}{dt} \end{array} &
\begin{array}{l} R_1 + pL_d + \cos^2 \gamma p(L_q - L_d) \\ - \frac{1}{2}\,\omega_m(L_d - L_q)\sin 2 \\ + \frac{1}{2}\,(L_d - L_q)\sin 2\gamma \frac{d}{dt} \end{array} &
\begin{array}{l} (\sin \gamma)pM_f \\ - M_f'\,\omega_m \cos \gamma \end{array} \\
\hline
\begin{array}{l} (\cos \gamma)pM_f \\ - M_f \sin \gamma \frac{d}{dt} \end{array} &
\begin{array}{l} (\sin \gamma)pM_f \\ + M_f \cos \gamma \frac{d}{dt} \end{array} &
R_f + pL_f
\end{array}
\cdot
\begin{bmatrix} i_d^e \\ i_q^e \\ i_f^e \end{bmatrix}
\qquad (80)
$$

$$p\Delta g = g'; \quad p\Delta z = z'$$

$$i_F = \Delta i_F + I_{F0}; \quad z = \Delta z + z_0$$

$$\omega_m = \Delta\omega_m + \omega_{m0}; \quad F_d = F_{d0} + \Delta f_d \tag{79}$$

Both L_1 and M_F depend upon the airgap and thus when developing (50) we should take care of this aspect. Replacing (79) in (50), (64), (66), (67), and (72) and making use of (75) through (78) we finally obtain, after rearranging the terms, the form shown in (80) below. In the matrix of (80) it should be noticed that $\Delta\gamma$, the variation of internal angle, is either zero or an imposed function of the other variables. In practice γ could be kept constant ($\Delta\gamma = 0$) by special position transducers or varied electronically according to needs.

The state equations evolving from (80) may be written as

$$[V] = [A]p[X] + [B] \tag{81}$$

with [V], [A], [B], and [X] easily identifiable from (80). These equations may be used to study any propulsion and/or levitation control strategies. It is only necessary to add the controller (compensator) equations. It is expected that a simultaneous control of propulsion and levitation stability is necessary for an optimal use of LIM as a propulsion-levitation means. However, when used only as a propulsion means the airgap, g_0, could be considered either constant or variable according to known track irregularities. Finally, we should observe that the thrust may be modified through the rectifier firing angle, α_R, load angle, γ, and levitation current (or voltage), i_F, while the same parameters also determine the attraction (levitation) force, though i_F is preponderant in this latter case.

References

1. E. Rummich, "Linear synchronous machines--Theory and construction," Bul. ASE, Vol. 23, 1972, pp. 1338-1344.

2. E. Levi, "Preliminary design studies of iron-cored synchronously operating linear motor," Polytechnical Institute of Brooklyn, Report No. 76/005, 1976 DOT OR&D, Washington, DC.

3. I. Boldea and S. A. Nasar, "Field winding drag and normal forces of linear synchronous homopolar motors," EME, Vol. 2, No. 2, 1978, pp. 253-268.

4. H. Lorenzen and W. Wild, "The synchronous linear motor" (in German), Internal report, Technical University of Munich, 1976.

5. T. R. Haller and W. R. Mischler, "A comparison of linear induction and linear synchronous motors for high speed transportation," IEEE Trans. Vol. MAG-14, No. 5, 1978, pp. 924-926.

6. I. Boldea and S. A. Nasar, "Linear synchronous homopolar motor--Design procedure for propulsion and levitation," EME, Vol. 4, No. 2-3, 1979, pp. 125-136.

7. I. Boldea and S. A. Nasar, "Thrust and normal force pulsations of current-inverter fed linear inductor motors," EME, Vol. 7, No. 2, 1982.

8. B. I. Ooi, "Homopolar linear synchronous motor dynamics equivalents," _IEEE Trans._, Vol. MAG-13, No. 5, 1977, pp. 1244-1246.

9. I. Boldea, "An experimental study of LSHM" (in English), _Bull. Sci. Techn. Polytechn. Inst. Timisoara_, No. 2, 1978.

10. M. Rentmeister, "Comparison between asynchronous and synchronous linear motors of short stator construction," International Conference on Electric Machines, L3/5, Brussels, Belgium, 1979.

11. G. E. Dawson and E. Unteregelsbacher, "A transverse laminated linear synchronous homopolar machine," IEEE paper CH 1575-0/80/0000-0270.

12. G. R. Slemon and R. P. Bhatia, "Field analysis of the linear homopolar synchronous motor," submitted for publication in _EME_, 1982.

13. G. R. Slemon, "A homopolar linear synchronous motor," International Conference on Electric Machines, Brussels, Belgium, 1979.

14. R. A. Turton an G. R. Slemon, "Stability of synchronous supplied from curent source inverter," _IEEE Trans._, Vol. PAS-98, No. 1, 1979, pp. 181-186.

15. A. El Zawawi, Y. Baudon, and M. Ivanes, "Dynamic analysis of an electromagnetically levitated vehicle using linear synchronous motors," _EME_, Vol. 6, No. 2, 1981, pp. 129-141.

16. I. Boldea and G. Papusoiu, "Brushless linear inductor motor--An experimental study" (in Romanian), Jubiliary Scientific Session, Nov. 1981, Craiova, Romania.

CHAPTER 8

THE ACTIVE GUIDEWAY SYNCHRONOUS MOTOR

8.1. Introduction

The active guideway linear synchronous motor (LASM) represents the linear counterpart of the conventional rotary synchronous motor. The armature winding of the motor is located in a laminated magnetic core installed along the guideway and is fed, section by section, from rectifier-inverter power stations. The frequency and phase of inverter voltage are controlled according to the field winding (located on the moving part) position with respect to the armature axis. From this point of view the LASM power conditioning and control unit resembles that used for rotary synchronous motors. The magnetic field produced by the field winding of an LASM is rather high and consequently the quantity of aluminum used in the armature winding is moderate, generally not higher than that used for the passive-guideway sheet-secondary of the linear induction motor (LIM), 10 to 12 kg/m. A substantial number of laminations are used in the armature magnetic core, 120 to 150 kg/m. However, this core is "used" also for levitation. The absence of high levels of power transfer to the moving part should lead to significant reductions of vehicle weight, lowering the level of the required propulsion force, for a given number of passengers for a vehicle driven by LASMs. The electric power necessary for levitation control and for auxiliary services is rather small in comparison with the propulsion power and is provided by advanced batteries, in parallel with a linear generator (installed in the LASM inductor poles), and a rectifier.

LASM may be applied for remotely controlled high-capacity urban transportation systems as well as for high speed interurban vehicles. We now consider the LASM construction, theory and performance in some detail. The guidance of an LASM vehicle will be provided by separate guidance controlled electromagnets.

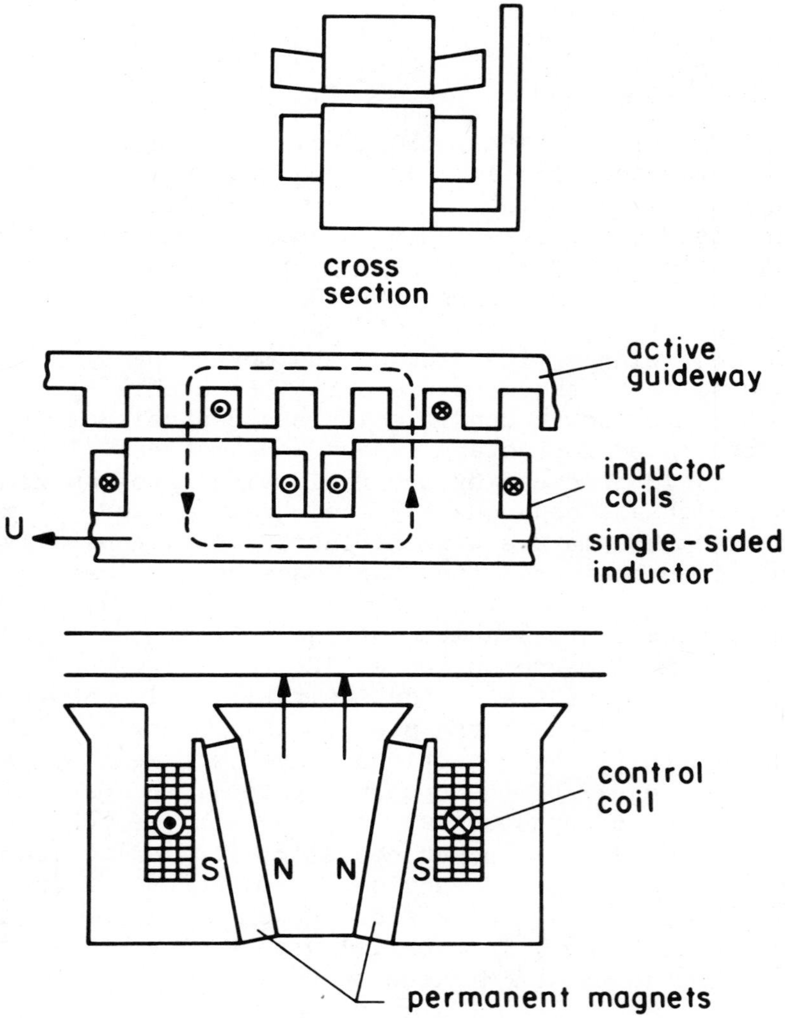

Fig. 8.1. Active guideway single-sided linear synchronous motor. (a) With electromagnets, (b) with permanent magnets.

8.2. LASM Construction

8.2.1. Inductor Structure[1-3]

The LASM heteropolar inductor may be manufactured in two ways: single-sided (Fig. 8.1) and double-sided (Fig. 8.2). To reduce the peak voltage required by the inductor coils in the process of levitation control, the inductor may contain static coils whose magnetomotive force (mmf), $W_{e1}I_e$, is kept constant, and dynamic coils whose mmfs, $N_1 i_c$ and $N_2 i_c$, are fed from an automatically controlled chopper. The static coils may be replaced by permanent magnets[4] (Fig. 8.1b) with the main advantage of reducing the weight of storage batteries on board.

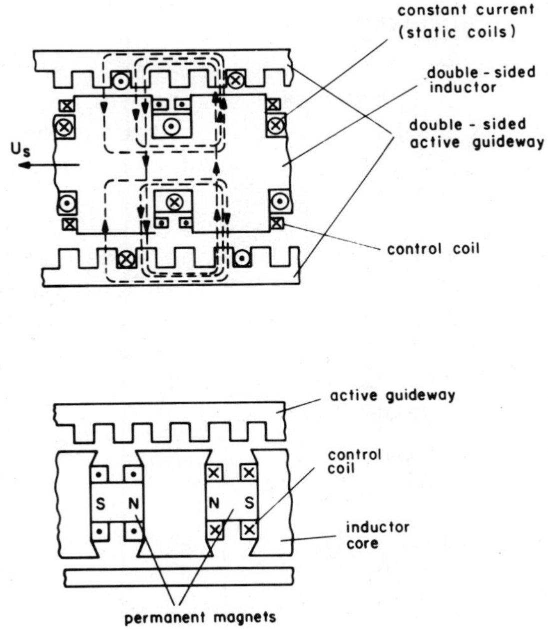

Fig. 8.2. Double-sided LASM. (a) With electromagnets, (b) with permanent magnets.

Low-weight strong permanent magnets are needed, and only samarium-cobalt or ceramic materials seem adequate. These magnets, however, tend to be expensive.

The use of static coils is not necessary for the single-sided construction alternative. The double sided LASM is superior to single-sided one from the point of view of levitation control but requires more inductor core material.[5] The LASM in the single-sided form, however, has been adopted for the world's most advanced program of magnetically levitated trains[6] since the active track is less expensive than that for double-sided inductor case. For this reason we will treat this case in more detail here. The permanent magnet case is not considered because the overall costs and weight are so much higher than those of forced cooled static coils.

8.2.2. Active Guideway Three Phase Armature Winding[7-11]

The three-phase stator is located along the guideway, laterally, on both sides of the vehicle (Fig. 8.3). Single-sided and double-sided configurations are both possible. Several types of windings are feasible, but the single-layer winding made of continuous conductor is adequate (Fig. 8.3). The number of slots per pole and phase is one. For economic reasons the armature continuous conductor is made of aluminum stranded cable, thus reducing the skin effect also. The magnetic core is made from laminations. To reduce the core depth, the pole pitch, τ, of the armature winding is kept under 0.3 m. On the other hand the value of τ is limited by the necessity of reducing the end turns length. However,

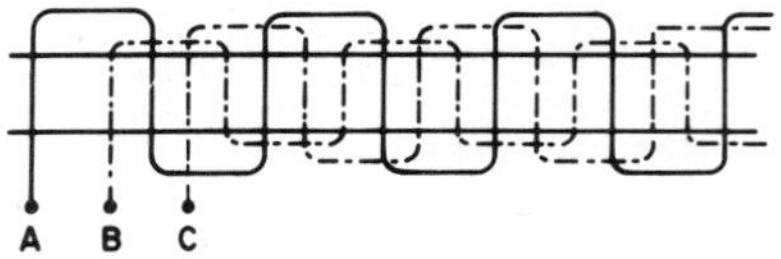

Fig. 8.3. Armature winding.

too small a pole pitch leads to difficulties in winding manufacture. Also, for small pole pitches the frequency required for a given speed is higher, resulting in higher skin effect losses. Higher frequencies lead also to the requirement of using faster thyristors for the inverters located in the power substations placed along the guideway. An involved problem concerning the optimum pole pitch occurs but it is beyond our scope.

8.2.3. The Ground-Based Power Substations

The armature winding is energized section by section. A practical solution[12] is shown in Fig. 8.4. The rectifier-inverter systems are controlled such that the LASM stays continuously locked into synchronism. In a

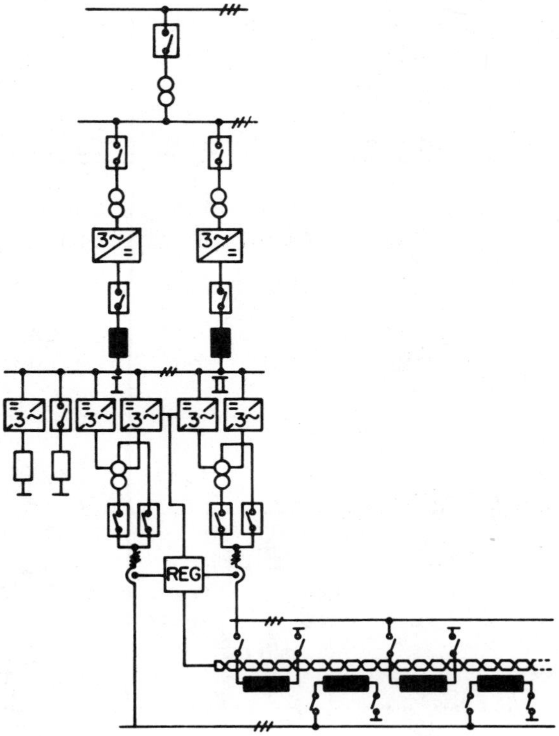

Fig. 8.4. Energized section.

closed-loop control system the frequency is adjusted to speed, thus providing a conventional synchronous motor operation. Position control is also possible. Then the LASM exhibits dc motor-type characteristics.

An inverter may energize more than one section.[12] The inverter commutation from one station to another is managed by fast response static power or vacuum switches[3] (Fig. 8.5). A difficult problem is to determine the optimum length of a section. However, section lengths of 1 to 1.5 km are considered reasonable, especially if an inverter energizes four to six sections. We should observe here that additional cables are necessary to feed more than one section from the same inverter. These rather general data on the section length are justified later. There are also some synchronizing problems when a vehicle passes from one section to another, but the problem is not severe on account of the LASM lagging power factor.

8.2.4. Inductor Introductory Design Guidelines

In general, the field produced by the field winding is significantly higher than the reaction field. Also the mmf's (load) angle may be kept close to 90° by using forced commutation techniques[13-15] in the inverter. Both the field winding and the armature winding

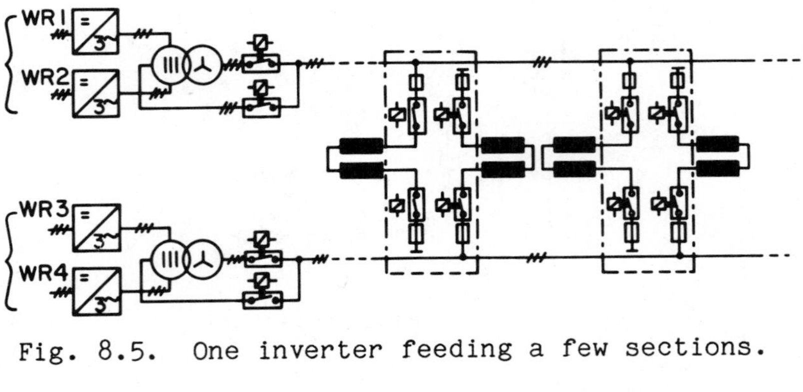

Fig. 8.5. One inverter feeding a few sections.

contribute to the normal levitation force. Because of the armature and inductor slotting the exact calculation of levitation and propulsion force could hardly be done through analytical methods (the inductor slots are used for the linear auxiliary generator). Consequently this aspect has been studied by numerical techniques, obtaining field configurations such as that given in Fig. 8.6.[15] As shown, the field configuration changes with the relative position between the inductor poles and the armature slots, indicating that some pulsations should occur in both the levitation and the propulsion forces (Fig. 8.7).[15]

By carefully adjusting the ratio between the inductor pole shoe length and the pole pitch to about 2/3 and by inclining the pole shoes with respect to the transverse direction by about a quarter of the stator

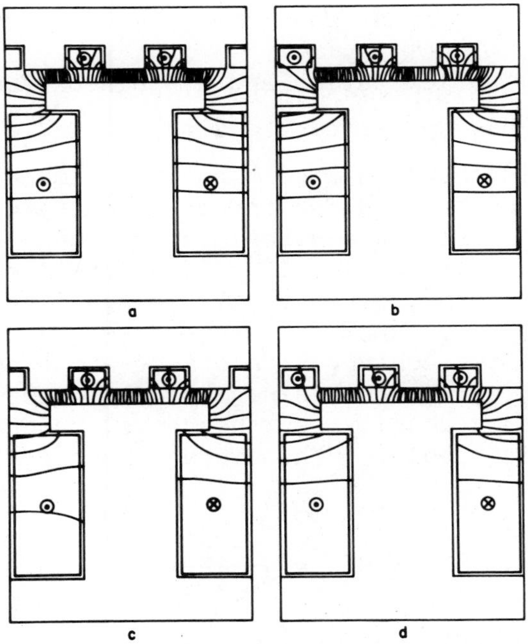

Fig. 8.6. Inductor airgap-field distribution.

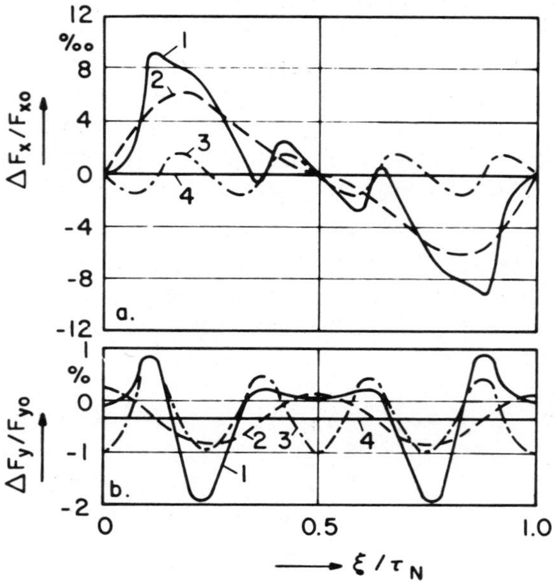

Fig. 8.7. Thrust and normal force pulsations.

slot pitch, the levitation and propulsion forces pulsations may be decreased to almost zero.[15] A practical inductor[12] is shown in Fig. 8.8. This important information may now be used for the preliminary design of the motor by simplified analytical expressions rather than by numerical methods. (For final refinements in the design or performance appraisal, we again rely on numerical approaches.) Thus, for a rough estimate of inductor requirements the contribution of the armature winding to the levitation force is neglected for the time being. The airgap flux density is, therefore, produced by the field winding only. The average airgap flux density, B_f, is

$$B_f = \frac{\mu_o W_f I_f}{g_1 K_c K_s} \tag{1}$$

The Carter coefficient, K_c, accounts for both inductor and armature slotting, and K_s depends on the magnetic

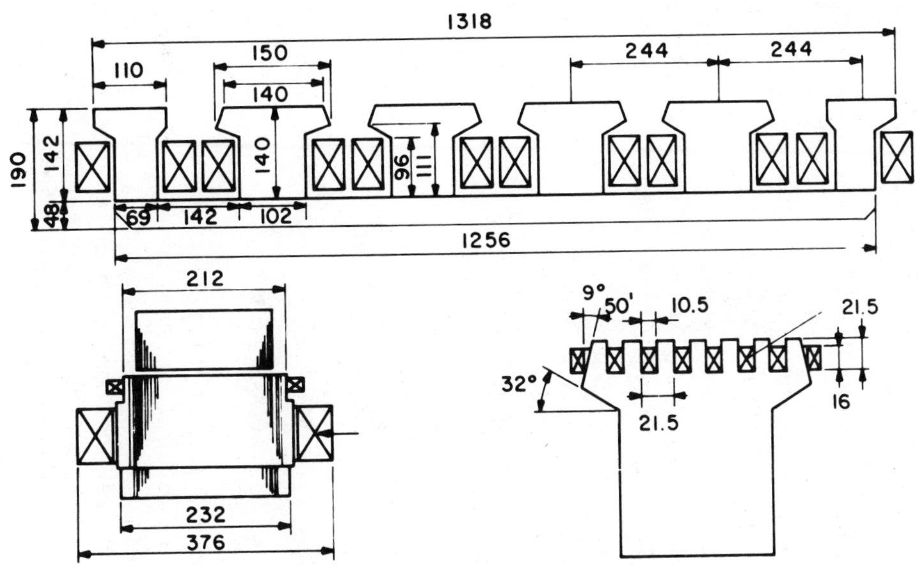

Fig. 8.8. A practical inductor.

core saturation level. In (1) g_1 is the airgap and $W_f I_f$ is the field mmf.

The motor levitation force, F, is approximately given by

$$F = \frac{B_f^2 2p(2/3)\tau l_e}{2\mu_0} \qquad (2)$$

where l_e is the motor width (in the transverse direction). For mechanical reasons g_1 = 10 to 15 mm, and g_1 = 10 mm is considered here as adequate. The airgap flux density B_f ranges from 0.5 to 0.7 T. This limitation is mainly due to the necessity of having wide, small-depth armature slots, resulting in a reasonably low slot leakage. The width l_e can be determined if the weight per vehicle length, G_e, is known.

It is assumed that the LASM is placed on both sides of the vehicle, occupying half of vehicle length (on each side) so that

$$G_e = \frac{2B_f^2}{2\mu_0} \left(\frac{2}{3}\right) l_e \frac{1}{2} \qquad \text{N/m} \qquad (3)$$

In general G_e = 1.5 to 2 x 10^4 N/m, and thus with B_f = 0.65 T, l_e = 0.133 to 0.177 m.

If lower values of flux density are chosen then, of course, the inductor (and armature) width become larger. Also, when such a motor is used only for propulsion purposes the above criteria should not be met when the armature cost limitations become predominant. Although it is true that a higher ratio between l_e and τ would lead to a better usage of armature winding (shorter end connections), the total quantity of copper (aluminum) per meter of track increases and so do the track costs. For transportation purposes values of l_e smaller than

0.20 to 0.22 m are considered adequate. The next step is to determine the pole pitch.

In essence, for a high-speed LASM $V_s \simeq$ 90 to 110 m/sec, the inverter frequency should not go above f_s = 250 Hz (in order to use standard fast thyristors), and thus the pole pitch, τ, is constrained by

$$\tau \geq \frac{V_s}{2f_s} = 0.22 \text{ m} \tag{4}$$

Again, high pole pitch values would mean an unwise use of copper (aluminum) and thus it would be useful to try to fulfill the standard condition known in rotary machines:

$$0.7 < \frac{l_e}{\tau} < 1.6 \tag{5}$$

Now, for transportation purposes ($l_e \simeq 0.2$ m), it results that

$$0.125 < \tau < 0.285 \text{ m} \tag{6}$$

From conditions (4) and (6) it follows that an optimum pole pitch should be about 0.22 m.

Once the values of l_e and τ are obtained, the inductor design follows, roughly, the pattern used in rotary synchronous machines and consequently it is not developed here.

8.3. Parameter Expressions

As for rotary synchronous machines the most important lumped parameters are

1. Phase resistance of the armature, R_1.

2. Phase total leakage reactance, X_t.

3. Longitudinal, X_{dm}, and transverse, X_{qm}, magnetizing reactances.

If 2p' = number of stator poles per track energized section, I_{1r} = rated current, and j_{Al} = the design current density in the armature, R_1 becomes

$$R_1 \simeq \rho_{Al} \frac{(\tau + l_e + 0.06)2p'}{I_{1r}} J_{Al} \qquad (7)$$

where ρ_{Al} is the electrical resistivity of the armature cable. Since a section is energized for a short period of time (while the inductor is within it) a high design current density may be chosen without serious overheating problems. But this value must not be too large, since the armature Joule losses become prohibitively large--though the armature wire costs are reduced this way. For aluminum armature wire, considering the above constraints J_{Al} should be somewhere in the range J_{Al} = 3.5 to 5 A/mm^2.

Now the total leakage reactance (per phase), X_t, consists of three components: X_a, X_{as}, X_e. The expression for X_a is similar to that of rotary synchronous machines adapted to the peculiar type of winding used here (one slot per pole and phase, one turn per coil, one end connection per coil). Hence

$$X_a \simeq \mu_0\omega_1 [(\lambda_c + \lambda_d)l_e + \lambda_e(\tau + 0.06)]2p \qquad (8)$$

where

$$\lambda_c \simeq \frac{h_{s1}}{3b_{s1}} + \frac{0.004}{b_{s1}}; \quad \lambda_d \simeq \frac{5g_1/b_{s1}}{5 + 4g_1/b_{s1}}; \quad \lambda_e \simeq 0.6 \qquad (9)$$

For the rest of the section no inductor is above the energized track. Thus there is flux leakage due to the stator slots, and armature end connections, X_{as}, and also leakage corresponding to the flux paths, above the track, X_e (similar to the case of a rotary machine in the absence of the rotor):

$$X_{as} = \mu_0\omega_1 [\lambda_c l_e + \lambda_e(\tau + 0.06)]2(p' - p) \quad (10)$$

For X_e, the field computations technique in a semi-infinite space above an iron core provides useful information: The equivalent airgap of an open stator winding is roughly τ/π. Thus,

$$X_e = \frac{6\mu_0\omega_1}{\pi^2} \frac{\tau l_e}{\tau/\pi} (p' - p)$$

$$= 24 \times 10^{-7}\omega_1 l_e(p' - p) \quad (11)$$

The total leakage reactance, X_t, is

$$X_t = X_s + X_{as} + X_e \quad (12)$$

The magnetizing reactances may be calculated as in rotary machines to obtain

$$X_{dm} \simeq \frac{6\mu_0\omega_1}{\pi^2} \frac{l_e\tau}{g_1 k_c} pK_{dm} \quad (13)$$

$$X_{qm} = X_{dm} \frac{K_{qm}}{K_{dm}} \quad (14)$$

Now if the interpolar flux paths are neglected, K_{dm} and K_{qm} are given by

$$K_{dm} = 4 \int_0^{\tau_p/\tau} \cos^2 (\frac{\pi}{\tau} x)d(\frac{x}{\tau}) \quad (15)$$

$$K_{qm} = 4 \int_0^{\tau_p/\tau} \sin^2 \left(\frac{\pi}{\tau} x\right) d\left(\frac{x}{\tau}\right) \quad (16)$$

In our case the pole shoe length is $(\tau_p/\tau) \simeq 2/3$ and thus $K_{dm} = 0.948$ and $K_{qm} = 0.3845$. The notable difference between K_{dm} and K_{qm} is offset in the synchronous reactances by the high values of leakage reactance considered for a whole energized section. The motor efficiency (per one energized section) is

$$\eta_t = 1 - \frac{3I_1^2R_1 + p_{i1} + p_{i2}}{F_x U} \quad (17)$$

The iron losses in the active part, p_{i1}, and in the rest of the section, p_{i2}, may be computed as in rotary synchronous machines by replacing the airgap g_1 by τ/π. The thrust F_x expression is still to be determined.

There is one other parameter very useful in the study of transients: the mutual inductance M_f between the field winding and a phase of the armature winding. The expression for M_f may be obtained by dividing the phase flux fundamental to the field current:

$$M_F = \frac{\psi_{ph1}}{I_f \sqrt{2}} \quad (18)$$

But

$$\psi_{ph1} = \frac{2}{\pi} B_{1f} \tau l_e p \quad (19)$$

In (18) and (19), B_{1f} = the fundamental of excitation airgap flux density. And,

$$B_{1f} \simeq \frac{2\sqrt{3}}{\pi} B_f \quad \text{for } \frac{\tau_p}{\tau} = \frac{2}{3} \tag{20}$$

with B_f given by (1). Finally

$$M_f = \frac{4\sqrt{3}\mu_0 \tau l_e p W_f}{\pi^2 \sqrt{2} g_1 K_c K_s} \tag{21}$$

The voltage V_f induced by the field current in an armature phase is

$$V_f = \omega_1 \psi_{ph1} = \omega_1 M_f I_f \tag{22}$$

The phasor diagram of the machine when fed with ideal (sinusoidal) voltages and currents is similar to that of rotary salient pole synchronous machines as given in Fig. 8.9. Thus the thrust, F_x, is

$$F_x = 3\frac{\pi}{\tau}\left[M_f I_f I_q - (L_d - L_q) I_d I_q\right] \tag{23}$$

with

$$L_d = \frac{X_{dm} + X_t}{\omega_1}; \quad L_q = \frac{X_{qm} + X_t}{\omega_1} \tag{24}$$

and X_{dm}, X_{qm}, and X_t are found from (12) through (14). Now, in general, the armature winding is fed from a current inverter with forced (or induced voltage) commutation. The mmf's load angle, θ_0, should be greater than $\pi/2$ for induced voltage commutation (leading power factor angle). But for $\theta_0 \leqq \pi/2$ forced commutation is required (see Fig. 8.9). In a preliminary design, the standard procedures used for rotary synchronous machines may be followed.[16] But it should be taken into account that the LASM is fed from a rectifier-inverter system.

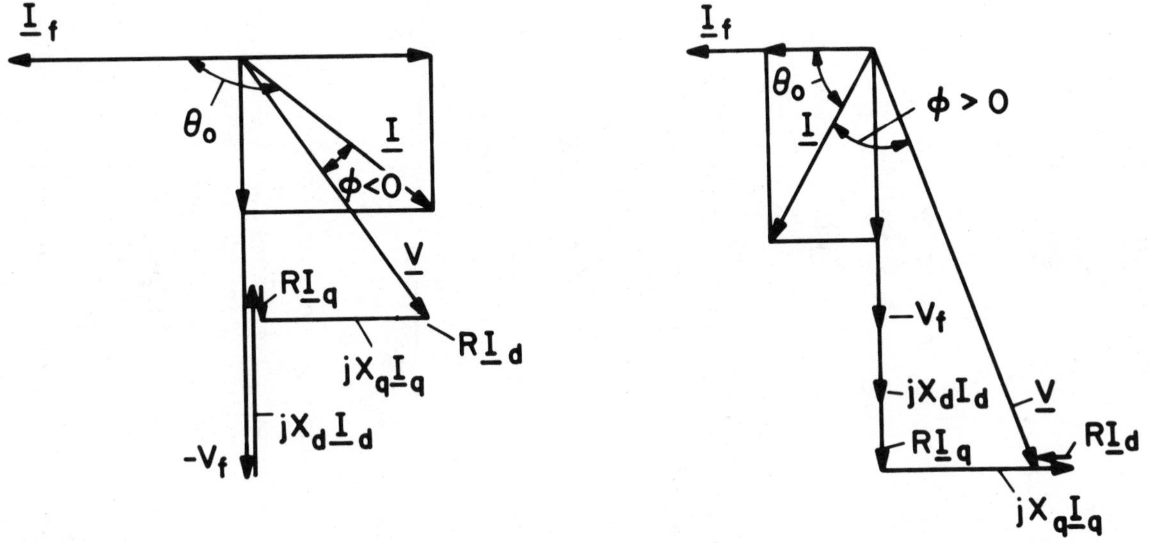

Fig. 8.9. Phasor diagram.

The rating of the machine is somewhat altered by this fact. Before studying this problem let us give a numerical example for performance.

Considering the design limits and expressions developed so far with $\theta_0 = \pi/2$, we may obtain the efficiency and power factor as a function of section length of motors per vehicle length (p'/p). These results are shown in Fig. 8.10.

Reasonable performance levels can be expected as long as the section length does not go beyond 30 times the vehicle length. Thus, for a vehicle of roughly 100 tonnes with 42 m of vehicle length (on each side) occupied by inductors, section lengths of 1000 to 1200 m are adequate. The energy on board of the vehicle (necessary for supplying the field windings of LASMs and other auxiliary services) is obtained by electromagnetic induction in a linear generator placed on the pole shoes of the inductors (see Fig. 8.8).

The pole pitch of the linear generator is half the armature slot-pitch exploiting the slot harmonics of the armature airgap field. The design procedure of the linear generator may again be accomplished by techniques known in rotary synchronous machines.

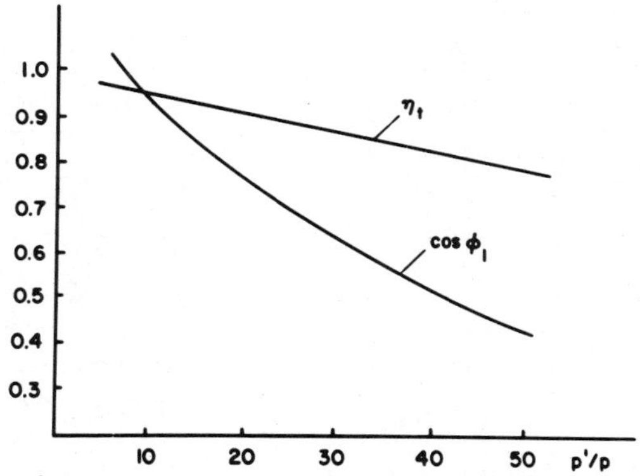

Fig. 8.10. Efficiency η_t and power factor, $\cos\phi_1$ against the ratio between section and LASM lengths.

A practical solution is presented in Reference 11. As should have been noted, there is no place for a damping winding in the inductor structure. The electronic control of the load angle avoids instabilities that occur in conventional synchronous motors. However, the presence of a damping winding on the inductor would have improved to some extent the inverter commutation conditions.[17] On the whole, the linear generator winding placed on the inductor pole shoes provides a greater service.

8.4. The State Equations of LASM

In order to vary the LASM speed within a wide range of speeds, a variable-voltage variable-frequency source is necessary. A bridge rectifier combined with a current-type inverter is typical for this purpose (Fig. 8.11). An equivalent circuit of this power inverter is shown in Fig. 8.12.

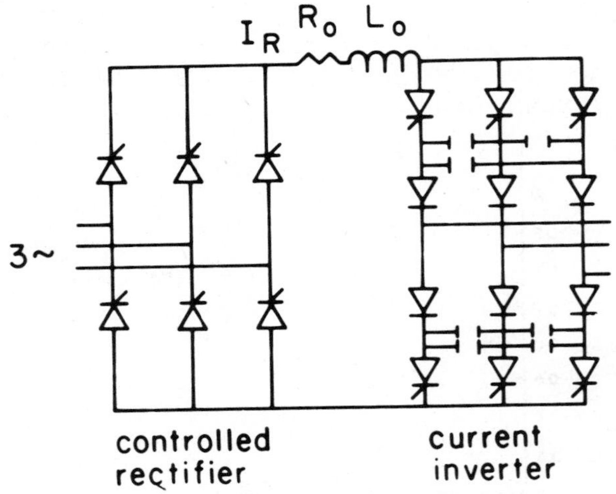

Fig. 8.11. Bridge rectifier-inverter system.

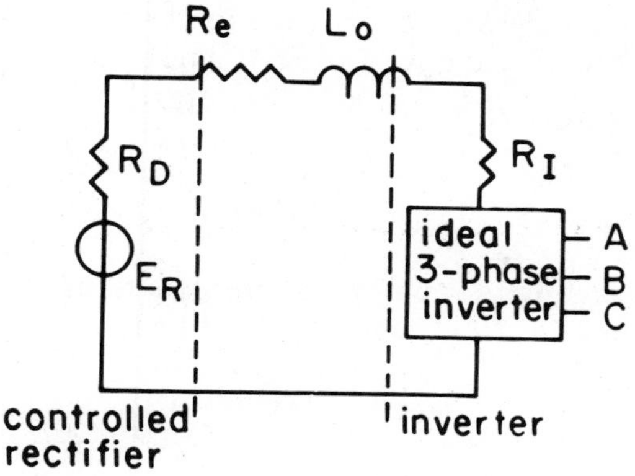

Fig. 8.12. Equivalent circuit of power conditioner.

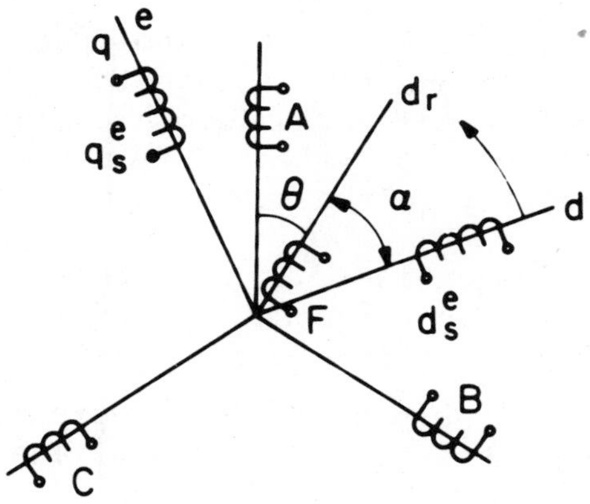

Fig. 8.13. LASM equivalent d_z scheme.

The development of the state equations for the power conditioning and the motor system follows the procedure used for LIMs. There is, however, one notable difference: LASM is a salient pole synchronous machine. Where there is no damper winding, the airgap should be allowed to vary during levitation dynamic control, due to the inevitable track irregularities. The equivalent circuit of the motor in a synchronously rotating reference frame is shown in Fig. 8.13.

The orientation of the synchronously rotating reference system is chosen such that the current in the quadrature axis, q, is equal to zero.

$$\theta + \gamma = \omega_1 t \tag{25}$$

where θ is the angle between the field winding and the armature phase, a, and varies in time. Also,

$$\frac{d\theta}{dt} = \omega_m; \quad \frac{d\gamma}{dt} = \omega_1 - \omega_m \tag{26}$$

where ω_m is the mechanical angular frequency in electric terms.

Equation (26) is valid if a position control for firing the thyristors in the inverter is used. In the absence of damper windings, unless a position control is used the drive will be affected by serious problems of stability especially in the motoring mode.

The machine equations, in a synchronously rotating reference frame, in the absence of damper windings and with $i_q = 0$, may be obtained directly from (10) and (11) of Reference 18 (the procedure for developing them with $L_d = L_q$ has been presented for LIMs). The result is:

The expression for thrust may be obtained by observing the power balance in (27) through isolating the terms not containing the term p, and dividing the result by ω_m the synchronous speed. Hence,

$$F_x = \frac{\pi}{\tau} \Big[\frac{1}{2}(L_d - L_q)\sin 2\gamma\Big(1 - \frac{1}{\omega_m}\frac{d\gamma}{dt}\Big)(i_d^e)^2$$

$$+ M_f' i_d^e i_f^e \sin\gamma - M_f' i_d^e i_f^e \sin\gamma \frac{1}{\omega_m}\frac{d\gamma}{dt} \Big] \tag{28}$$

$$
\begin{bmatrix}
\Delta\alpha E_s \sin\alpha_{R0} \\
\Delta V_F \\
\Delta f_d \\
0 \\
\Delta F_{ne} \\
0 \\
0 \\
0
\end{bmatrix}
=
\begin{bmatrix}
(R_1 + R_0') + (L_{10} + L_0')p & \omega_{m0}M_{F0}\sin\gamma_0 + \cos\gamma_0 M_{F0}p & M_{F0}i_{F0}\sin\gamma_0 & (I_{ds0}\frac{\partial L_1}{\partial g_0} + \cos\gamma_0 I_{F0}\frac{\partial M_F}{\partial g_0})p + \omega_{m0}\frac{\partial M_F}{\partial g_0} I_{F0}\cos\gamma_0 & 0 & 0 & 0 & \omega_{me}M_{F0} I_{F0}\cos\gamma_0 \\
\cos\gamma_0 M_{F0}p & R_F + L_F p & 0 & (\cos\gamma_0 \cdot I_{ds0} \cdot \frac{\partial M_F}{\partial g_0} + I_{F0}\frac{\partial L_F}{\partial g_0})p & 0 & 0 & 0 & -M_F \cdot \sin\gamma_0 \cdot I_{ds0} \cdot p \\
\frac{\pi}{\tau}M_{F0}I_{F0}\sin\gamma_0 & \frac{\pi}{\tau}M_{F0}I_{ds0}\sin\gamma_0 & -(m_1 + M_1)\frac{\tau}{\pi}p & \frac{\pi}{\tau}\frac{\partial M_F}{\partial g_0} I_{F0} I_{ds0} \cdot \sin\gamma_0 & 0 & 0 & 0 & \frac{\pi}{\tau}M_{Fe} \cdot I_{F0}I_{ds0} \cdot \cos\gamma_0 \\
0 & 0 & 0 & -p & 1 & 0 & 0 & 0 \\
I_{ds0}\frac{\partial L_1}{\partial g_0} + I_{F0}\cos\gamma_0 \cdot \frac{\partial M_{F0}}{\partial g_0} & I_{ds0}\frac{\partial M_F}{\partial g_0}\cos\gamma_0 + I_{F0}\frac{\partial L_F}{\partial g_0} & 0 & -K & \begin{smallmatrix}-m_1p \\ -f\end{smallmatrix} & K & f & -I_{ds0}I_{F0} \cdot \sin\frac{\partial M_F}{\partial g_0} \\
0 & 0 & 0 & 0 & 0 & -p & 1 & 0 \\
0 & 0 & 0 & K & f & -K & \begin{smallmatrix}-M_1p \\ f\end{smallmatrix} & 0 \\
0 & 0 & 0 & 0 & 0 & 0 & 0 & 0
\end{bmatrix}
\begin{bmatrix}
\Delta i_{ds} \\
\Delta i_F \\
\Delta\omega_m \\
\Delta g \\
g' \\
z \\
z' \\
\Delta\gamma
\end{bmatrix}
\tag{27}
$$

$$M_f' = M_f \sqrt{\frac{3}{2}}$$

Equation (28) degenerates to the conventional expression for synchronous machines torque if γ = constant, that is for ideal position control of inverter thyristors gating.

The attraction force can also be obtained after multiplying the first equation of (27) by i_d^e and the last by i_f^e, by the addition of the two equations, and by separation of the terms multiplying the airgap time derivative de/dt. It should be noted that M_f', L_d, and L_q are dependent on the airgap, and that by allowing small variations of the airgap, we obtain the normal force from the power balance. Thus, finally,

$$F_n = \sin^2 \gamma \, [\frac{\partial}{\partial e}(L_d - L_q)\,](i_d^e)^2$$

$$+ 2 \frac{\partial M_f'}{\partial e} i_d^e i_f^e \cos \gamma + (i_f^e)^2 \frac{\partial L_f}{\partial e} \qquad (29)$$

In a linear motor the airgap varies due to the track irregularities and (if any) during the magnetic levitation automatic control. Thus, two equations of motion should be accounted for: one along F_x and one along F_n, resulting in

$$\frac{\tau}{\pi} m_e \frac{d\omega_m}{dt} = F_x - F_{dp} \qquad (30)$$

$$m_1 \frac{d^2e}{dt^2} = m_1 g_g - F_n - F_{pn} \qquad (31)$$

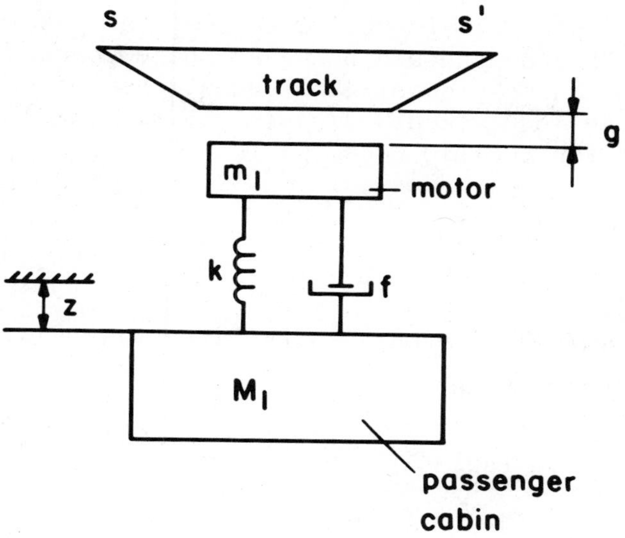

Fig. 8.14. Secondary suspension system of the motor.

where m_e is the mass of the vehicle propelled by the motor and $m_e = m_1$ if no additional mechanical suspension system is used on the vehicle. In general, m_1 is the unsprung weight of the motor. Also, F_p is the longitudinal drag force and F_{pn} the external random vertical perturbation force.

Equation (31) should be replaced by two equations if an additional mechanical suspension system is placed between the motor and the vehicle body, as shown in Fig. 8.14. Hence,

$$M_1\frac{d^2z}{dt^2} = (m_1 + M_1)g_2 - K\,(e - e_0 - z) + f\,[\frac{d(e - z)}{dt}]$$

$$M_1\frac{d^2z}{dt^2} = K(e - e_0 - z)$$

$$+ f\,[\frac{d(e - z)}{dt}]; \quad m_1 + M_1 = m_e \qquad (32)$$

To complete the state equations, the power conditioner equations should be added those corresponding to different control loops. The equations of the power conditioner equivalent circuit (Fig. 8.12) have been developed for the LIM case:

$$V_d^e = E_s \cos\alpha - R_0' i_d^e - L_0' \cdot \frac{di_d^e}{dt} \qquad (33)$$

where $(3\sqrt{2}/pi)\ E_s$ = maximum rectifier voltage; α = rectifier firing angle.

$$R_0' = \frac{\pi^2}{18}(R_0 + R_R + R_I); \quad \text{and } L'_0 = \frac{\pi^2}{18} L_0$$

By considering only the fundamental of the motor phase current, i_d^e (120° rectangular wave), the relation between i_d^e and the rectifier current, i_R, is straightforward, that is,

$$i_d^e = \frac{3\sqrt{2}}{\pi} I_R; \quad V_R = \frac{3\sqrt{2}}{\pi} E_s \cos\alpha \qquad (34)$$

with i_R and V_R are the rectifier current and voltage, respectively.

8.4.1. The Steady State Operating Point

Once the speed, $\omega_m = \omega_{m0}$, rectifier firing angle, $\alpha = \alpha_0$, longitudinal drag, F_{dp}, and field winding supply voltage, $V_f = V_{f0}$, are given, we can calculate the steady-state values of $i_d^e = I_{d0}^e$, $\alpha = \alpha_0$, $i_f^e = I_{f0}^e$, and $V_d^e = V_{d0}^e$ from (27) with p = 0 and $d\gamma/dt = 0$, $i_q^e = 0$:

$$V_{s0}^{e} = E_s \cos \alpha_0 - R_0^{'} I_{d0}^{e}$$

$$V_{f0} = R_F I_{f0}$$

$$F_{dp} = \frac{\pi}{\tau} [\frac{1}{2}(L_d - L_q)(I_{d0}^{e})^2 \sin 2\gamma_0$$

$$+ M_F^{'} I_{d0}^{e} I_{f0}^{e} \sin\gamma_0] \qquad (35)$$

$$Vd0^{e} = R_1 Id0^{e} + \frac{1}{2}(L_d - L_q)\omega_{m0} I_{d0}^{e} \sin 2\gamma_0$$

$$+ M_f \omega_{m0} I_{d0}^{e} \sin\gamma_0$$

Beginning with these values the transients may be studied either directly by a numerical procedure or by small deviation procedure as done for LIMs.

8.5. Conclusion

The LASM study carried out here demonstrates its feasibility for the propulsion and levitation of future transportation systems. Construction and preliminary design guidelines are presented. The state equations of LASM plus the power conditioning unit constitute a good basis for future in-depth studies of LASM transients.

References

1. H. Weh, "Long stator motors and railway technique" (in German), Symposium Motori Lineari, Capri, 1973.

2. H. Weh, "The integration of functions of magnetic levitation and propulsion" (in German), ETZ.A., Vol. 96, No. 9, 1975, pp. 131-135.

3. H. Weh, "Synchronous long stator motor with controlled normal forces" (in German), ETZ.A., Vol. 96, No. 9, 1975, pp. 409-413.

4. H. Weh and M. Shalaby, "Magnetic levitation with controlled permanentic excitation," IEEE Trans., Vol. MAG-13, No. 5, 1977, pp. 1409-1411.

5. H. Weh and H. Mosebach, "Research and development of integrated attraction MAGLEV with active guideway synchronous propulsion at Braunschweig University," Boston, Sept. 1977.

6. H. Alscher and H. G. Raschbichler, "Demonstration facility for magnetic transportation technique at international traffic exhibition (IVA) 1979" (in German). E.T.R., No. 4, 1979, pp. 281-292.

7. D. Frenzel and L. Baur "Goals and main fields of government promotion of research and technological developments for railway systems" (in German), ZEV--Glas. Ann. 105, No. 7-8, 1981, pp. 194-197.

8. V. D. Rogg, "Status of Maglev train development in the German Federal Republic" (in German), ZEV--Glas. Ann., Vol. 105, No. 7-8, 1981, pp. 198-201.

9. E. Eilhaber, "Emsland Transrapid test facility (TVE)" (in German), ZEV--Glas. Ann., Vol. 105, No. 7-8, 1981, pp. 202-204.

10. D. Hilliges, P. Molzer, H. G. Raschbichler and R. Zurek, "The guideway of the Emsland Transrapid test facility (TVE)" (in German), ZEV--Glas. Ann., Vol. 105, No. 7-8, 1981, pp. 205-215.

11. P. Schwarzler, I. Borcherts, G. Steinmetz, P. Knorr, and K. Dreimann, "Transrapid-06 - Concept of the vehicle" (in German), ZEV--Glas. Ann., Vol. 105, No. 7-8, 1981, pp. 216-224.

12. G. P. Parsch and H. G. Raschbichler, "The iron-coned long-stator synchronous motor for the Emsland Transrapid test facility" (in German), ZEV--Glas. Ann., Vol. 105, No. 7-8, 1981, pp. 225-232.

13. W. Luers, "Data processing and safety system" (in German), ZEV--Glas. Ann., Vol. 105, No. 7-8, 1981, pp. 235-239.

14. R. Kretzschmar, "Initial operation of the Emsland Transrapid test facility" (In German), ZEV--Glas. Ann., Vol. 105, No. 7-8, 1981, pp. 240-245.

15. H. May, H. Moseback, and H. Weh, "Pole force oscillations caused by armature slots in the active guideway synchronous motor" (in German), Arch. fur Electrotechn., Vol. 59, 1977, pp. 291-296.

16. J. H. Walker, Large Synchronous Machines, Clarendon Press, Oxford, England, 1981.

17. F. Harashima, H. Naitoh, and T. Hanejoshi, "Dynamic performance of self-controlled synchronous motors fed by current-source inverters," IEEE Trans., Vol. IA-15, No. 1, 1979, pp. 36-46.

18. R. A. Turton and G. R. Slemon, "Stability of synchronous motors supplied from current source inverters," IEEE Trans., Vol. PAS-98, No. 1, 1979, pp. 180-186.

Chapter 9

ACTIVE GUIDEWAY LINEAR SYNCHRONOUS MOTORS WITH SUPERCONDUCTING FIELD WINDING

9.1. Introduction

Propulsion of vehicles with electrodynamic repulsion levitation systems may be provided by double-sided linear induction motors (DSLIMs) or by linear synchronous motors with superconducting field winding (LSCMs). The LSCM has a heteropolar flat field winding placed on the lower part of a Maglev vehicle. A three-phase concrete-embedded armature winding is spread along the guideway and power is supplied section by section from special power substations (Fig. 9.1). The length of a section varies, in general, between 1 and 5 km. The LSCM is thus an active guideway linear motor, eliminating the necessity of high level power transfer to a vehicle.

Apparently, the absence of the on-board power conditioning unit leads to a notable vehicle weight reduction. However, this advantage is counteracted, to some extent, by the cryogenic equipment weight. What follows is devoted primarily to LSCM construction, theory, and performance.

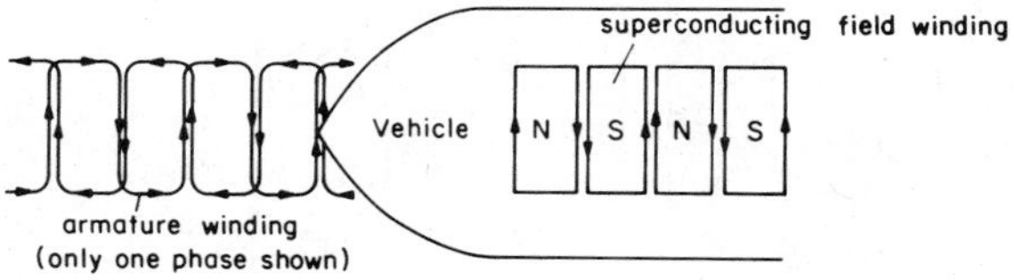

Fig. 9.1. Simple horizontal LSCM.

9.2. Description

Quite a few construction alternatives have been proposed for LSCM to be used for Maglev. Among them three are of greatest importance:

1. Simple-horizontal LSCM[1] (Fig. 9.1).

2. Double-vertical LSCM[2] (Fig. 9.2).

3. Double-horizontal LSCM[3] (Fig. 9.3).

Solutions 1 and 2 provide only propulsion, and solution 3 is able to assure propulsion and guidance by using a ladder (or a guideway made of short-circuited coils) placed between the two twin-armature windings. Furthermore, special LSCM construction variations have been proposed to provide simultaneously propulsion, levitation, and guidance functions[4-14] (Fig. 9.4a and b). These solutions are less effective from the energy-conversion point of view, mainly because levitation and guidance require long superconducting magnets, and propulsion requires shorter magnets. Also, a poorer usage of guideway short-circuited coil material for levitation is made since only one side of the guideway coils is

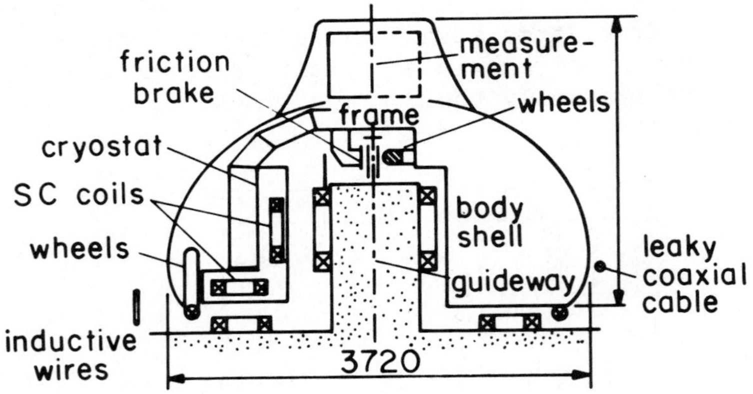

Fig. 9.2. Double-vertical LSCM.

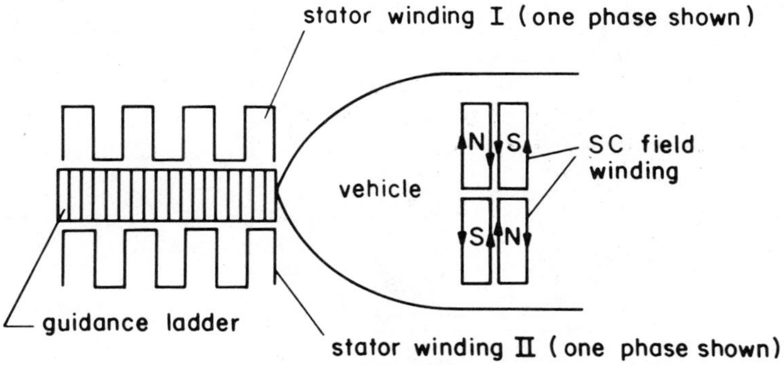

Fig. 9.3. Double-horizontal LSCM.

really active (Fig. 9.4b). It still retains a remarkable simplicity, which may render the solution as practical.

Also, dc current-fed rods in the guideway have been proposed[15] to provide levitation from zero speed, avoiding the short-circuited levitation guideway coils. However, for track energized sections longer than 1.3 km, this solution is energetically inferior to using short-circuited coils. For this reason, it is not treated here. To provide more generality to the mathematical treatment, we explore here the theory of double-horizontal LSCM. (The simple horizontal or the simple vertical[14] LSCM could be treated as a special case of double-horizontal LSCM.)

9.2.1. On-Board Cryogenic System

The cryogenic system[5] has a rather low specific weight and thus seems adequate for Maglev. The tubular dewars of superconducting (SC) coils (of LSCM and levitation coils) are partly filled with atmospheric-pressure liquid helium, and then hermetically sealed. The heat leakage of SC coils during a duty day (for example, 20 hr out of 24) leads to temperature and pressure rise in the dewars. The rate of partial filing

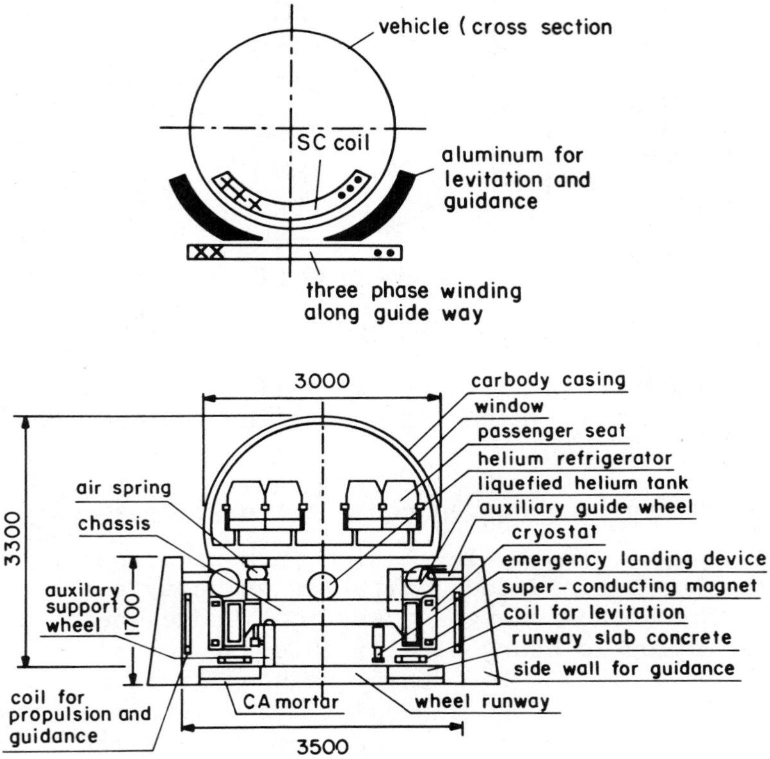

Fig. 9.4. (a) the magneplane, (b) the U guideway LM-500-01.[14]

with liquid helium is chosen such that during a duty day the superconducting state is not endangered. Daily, after the duty hours, the Maglev is serviced and the gaseous helium is replaced by liquid helium.

Liquid nitrogen dewars connected to liquid nitrogen tanks eliminate most of the heat leakage from the SC coils such that, at the end of a work day without accidents, the temperature in the liquid helium dewar is less than 13 K. The capacity of the nitrogen tanks is calculated. To reduce their weight on a vehicle, they may be replaced a few times a day.

Some malfunctions can also occur in the cryogenic system on board, for example (1) an SC coil turns normal, or (2) liquid nitrogen leaks. Under these conditions, the pressure and temperature in the helium dewar rise. The system should be tailored such that the maximum attainable pressure is 20 atm and the maximum temperature, 60 K. Now, by the partial filling of dewars with helium, 50 to 80%, and by an adequate choice of the nitrogen tanks' capacity, these constraints could be achieved in practice. An overpressure valve calibrated for 20 atm is used for protection. Thus, with the exception of an SC coil explosion, even for malfunctions as serious as those pointed out above, the temperature in the helium dewar will not go beyond 60 K. Consequently, the superconducting state will be comfortably restored by the refilling with liquid helium.

9.2.2. Flux Pumps and Cryotrons

It is now accepted that a continuous supply of SC coils with electric current would be rather improper because this would require relatively large power sources on board.[5] Hence, the persistent current mode of SC coils is viewed as a practical solution. For compensating the "current" losses caused by the induced current in coil connectors, armature current harmonic field and diverse Maglev oscillations, the SC coils are fed, when needed, from flux pumps located in the helium dewar.

A flux pump consists of a step-up superconducting transformer and a pair of cryotrons connected as shown in Fig. 9.5. A cryotron[6] is made from a thin superconducting filament (for example, lead, with 4% antimony) that turns normal when an additional magnetic field, produced by a control coil, crosses it (Fig. 9.6). A Hall-type transducer located in the helium dewar measures the SC coil field and controls, by means of an

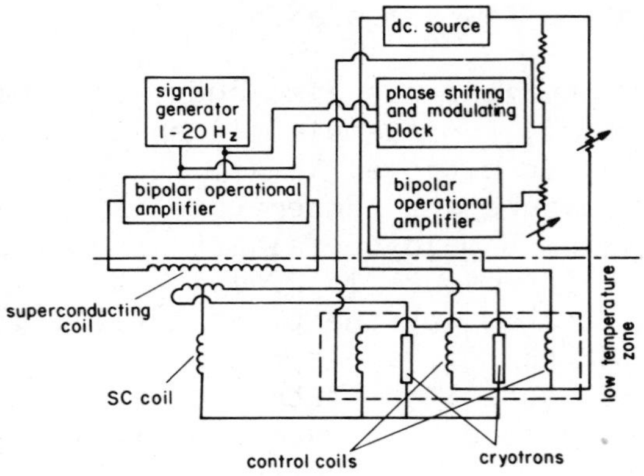

Fig. 9.5. Flux pump.

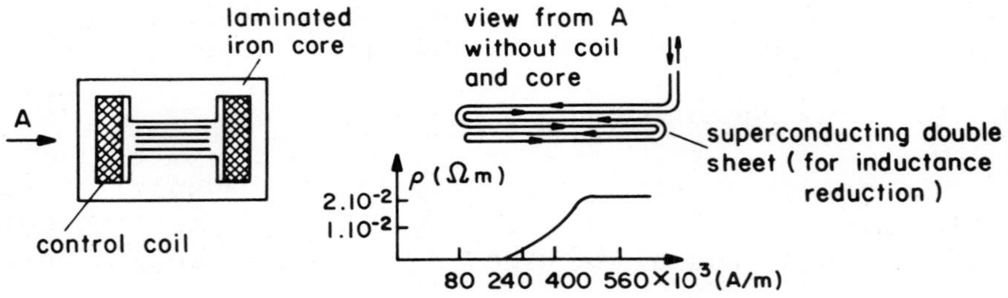

Fig. 9.6. Cryotron.

automatic control system, the voltage or frequency of the flux pump transformer. According to the transformer primary current polarity, the control coils of cryotrons open one or the other of the cryotron gates, thus supplying the SC coil with only one polarity current until the rated value of the SC coil field is reached. The preponderant inductive character of SC coils provides the necessary filtering.

9.2.3. Active Guideway Armature Winding

A few winding lay-outs have been proposed for the LSCM armature.[7] However, if we keep in mind the necessity of reducing armature field harmonics, as well as

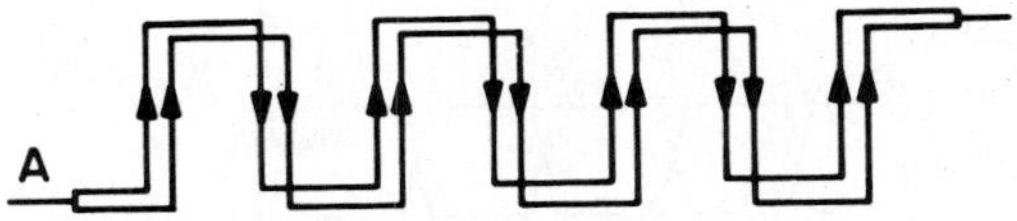

Fig. 9.7. Three-phase armature winding with continuous conductor (only one phase shown).

that of a winding that is easy to mount and less costly, it seems reasonable to choose a one-layer winding with two continuous conductors per pole and phase (Fig. 9.7). To reduce the phase voltage per section, two current paths in parallel are used. The length of a section must be limited[7] if we are to keep the efficiency at a reasonable level. But, the other hand, the length of a section should be as high as possible, to reduce the investment, maintenance, and supervision costs of the complex electronic thyristorized-power substations.[5] Finally, the maximum phase voltage per section is limited to about 4 to 5 kV by the cost of armature conductor isolation and thyristor-equipment protection. These conflicting conditions have led to the rather generally accepted section length[5] of 5 km.

We should recognize here a very complex optimization problem involving costs and energy conversion. Because high-power thyristor equipment is still in the developmental stage, an appraisal of costs would be of little practical use. The star connection of phases should be used because there are two current paths (Fig. 9.8). However, if the two paths are series connected

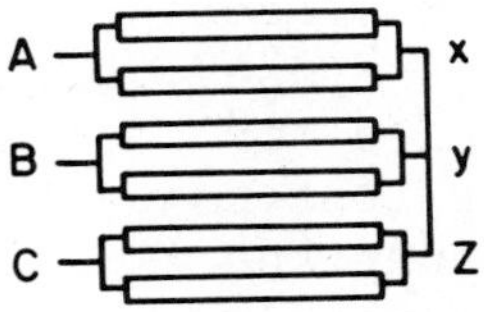

Fig. 9.8. Star connection of phases.

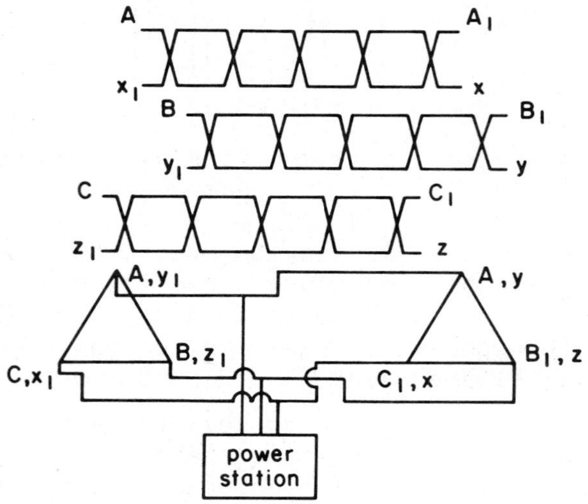

Fig. 9.9. Delta connection of phases.

(Fig. 9.9), the phase voltage is doubled. In this case, the delta connection may be used, together with a synchronized power supply of a section from both ends.[9]

If the feed of a section is made at one end only, a tripolar static switch is necessary at the other end. In the delta connection, the third harmonic armature field reacts with the third harmonic of the field winding field, producing, in general, an additional propulsion force, and reducing, to some extent, the propulsion force pulsations. Thus, at least for high pole pitches, the delta connection may prove better than the star connection.

To reduce the skin effect, the armature conductors are made of thin, transposed, aluminum filaments. The quantity of aluminum used for the armature winding is not much higher than that used for the secondary-sheet of DSLIMs, mainly because the SC coils field is rather strong. After these general construction guidelines, we now consider LSCM theory.

9.3. LSCM Theory

The following aspects of LSCM theory are considered essential, and consequently, are treated here in some detail:

1. SC coils field distribution.

2. Armature winding parameters.

3. Skin effect in armature windings.

4. LSCM propulsion performance.

5. Normal and lateral forces.

6. Guidance force.

7. Thyristorized controlled power for LSCM.

9.4. Field Winding Field Distribution

To calculate the field winding field distribution at points situated in the armature winding plane, we use the expressions for field components of a rectangular coil. The SC coils are considered as filaments. For a double-horizontal LSCM (Fig. 9.10), the Oz component of the flux density produced by the field winding at a point (x, y, z_0), situated in the armature coil plane, is

$$B_z'(x, y, z_0) = B_z(x, y, z_0) - B_z(x, y - 2b, z_0)$$

$$- B_z(x + \tau, y, z_0) + B_z(x + \tau, y - 2b, z_0)$$

$$- B_z(x - \tau, y, z_0) + B_z(x - \tau, y - 2b, z_0) \qquad (1)$$

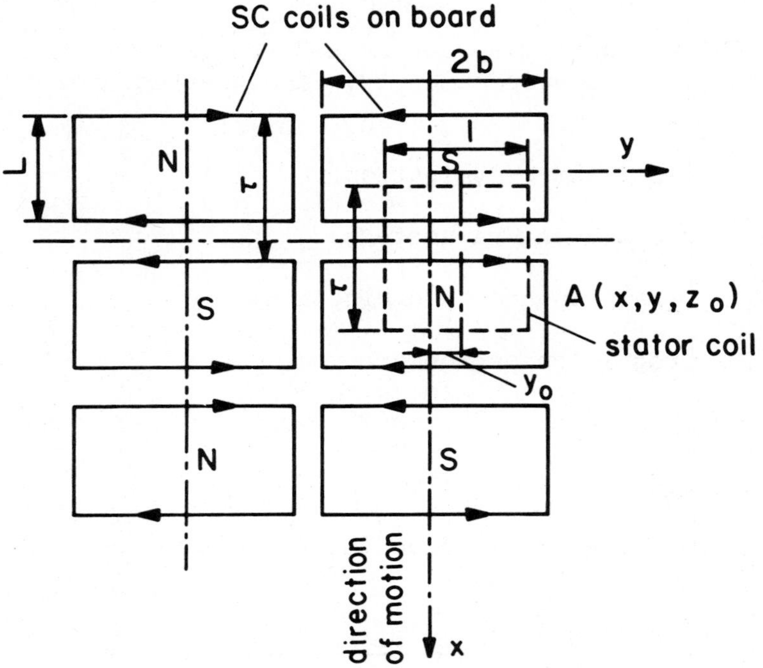

Fig. 9.10. Overview of SC coil.

In (1) the contribution of three neighboring coils has been accounted for. The expression $B_z(x, y, z_0)$ resembles (5) of Chapter 11. The other two components of resultant flux density, $B_x(x, y, z_0)$ and $B_y(x, y, z_0)$ have expressions similar to (1) above. The number of poles is generally high (2p > 10), and thus it is sufficient to study only one pole.

What deserves special attention for propulsion is the magnetic flux variation with respect to the direction of motion:

$$\frac{d\phi_\theta}{dx} = \int_{y_0-l/2}^{y_0+l/2} B_z'(x, y, z_0)\, dy \tag{2}$$

A study of harmonic content of $d\phi_e/dx$, for constant SC coil magnetomotive force (mmf), or for constant weight of SC coils, by varying the coil dimensions, would reveal the solution that provides the maximum fundamental. However, to have the correct conclusion, the weight of aluminum used in the armature winding should also be accounted for as a weighted parameter dependent on traffic density. Such an optimum problem goes beyond our scope here. Instead, a numerical study based on geometrical dimensions of practical interest is carried out.

Thus, for $I_0 = 5 \times 10^5$ ampturns, $2b = 0.9$ m, $L/\tau = 0.5, 0.6, 0.7$, $l = 0.6$ m, $y_0 = 0.2$ m, $z_0 = 2.2$ m, the $d\phi_e/dx = f(x)$ distribution and its harmonic contents have been determined as functions of pole pitch, τ (Fig. 9.11).

It should be noted that the fundamental has reasonable values for $\tau_0 \geqq 1.5$ m and $L/\tau \geqq 0.7$. These results are very close to those of Reference 10, which were cited as having been obtained by accounting also for the minimum armature aluminum weight criterion. The optimum pole pitch strongly depends on the airgap, z_0, and the above value of τ_0 corresponds to $z_0 = 0.22$ m, a value of practical interest.

9.4.1. Armature Winding Parameters

The armature mmf is negligible in comparison with the mmf of the SC coils. Thus, the phase armature inductance can be calculated from the expressions used for long electric lines:

$$L_f = L_t\left(1 - \frac{1}{3} - \frac{1}{6}\right) \qquad (3)$$

where

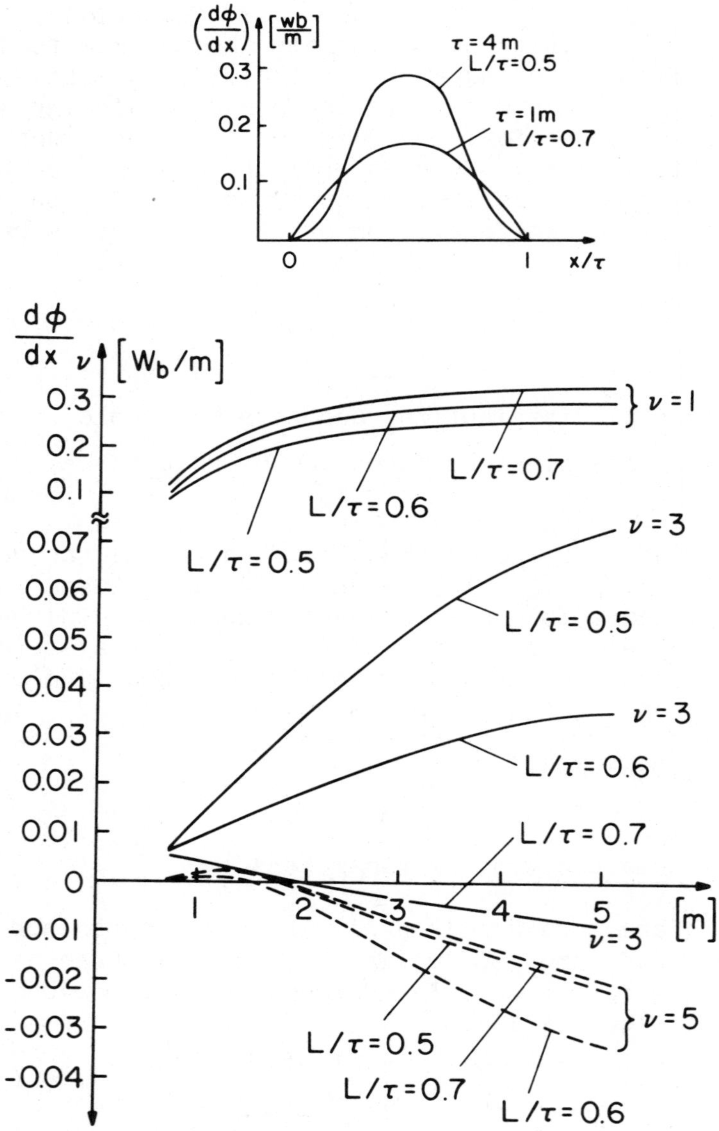

Fig. 9.11. Rate of change of flux with position (a) 0 < x/τ < 10; (b) 1 < x/τ < 5.

$$L_t = \frac{\mu_0}{\pi}(4p'l\ln\{\frac{\tau}{d} + [(\frac{\tau}{d})^2 - 1]^{1/2}\}$$

$$+ 2p'\tau \ln\{\frac{l}{d} + [(\frac{l}{d})^2 - 1)^{1/2}\}) \quad (4)$$

$2p'$ = the number of poles per section; d = the diameter of armature coil conductors; and l = armature coil width (along Oy). In (3), the influence of the other two phases is included. The armature phase resistance per section, R_1, is

$$R_1 \simeq \frac{\rho_{Al}}{2q_{Al}}(l + \tau)\ 2p' \quad (5)$$

where q_{Al} is the armature coil total cross section. The armature skin effects have been neglected so far. Now we return to this problem.

9.4.2. Armature Winding Skin Effects

The field winding field penetrates partially in the armature conductors, producing eddy-current armature

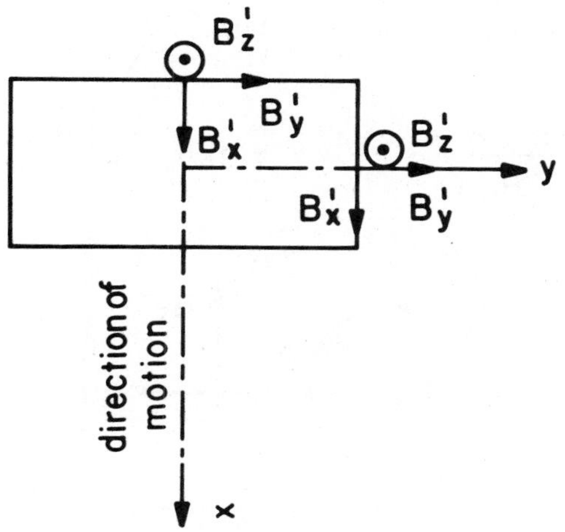

Fig. 9.12. Field winding field components.

losses. On the other hand, the armature ac current produces its own skin effect. The field winding field exhibits two types of components with respect to armature coils sides[11] (Fig. 9.12). These are:

1. A transverse component, B_t, perpendicular to coil side.

2. A longitudinal component, B_l, parallel to coil side.

For the sides AB and A'B' (Fig. 9.12), B_x' and B_z' are transverse, and B_y' is longitudinal. But, for BB' and AA', B_z' and B_y' are transverse and B_x' is longitudinal.

<u>Transverse field</u>, B_t

The transverse field (Fig. 9.13) is assumed sinusoidal in time with respect to armature coils, so that

$$\underline{B}_t = B_{0t}\, e^{j\omega t}; \quad \omega = \frac{\pi}{\tau} U \tag{6}$$

The preponderant induced electric field component, $\underline{E}_x$, satisfies the equation

$$\frac{\partial \underline{E}_x}{\partial y} = - \frac{\partial \underline{B}_t}{\partial t} = - j\omega B_{0t}\, e^{j\omega t} \tag{7}$$

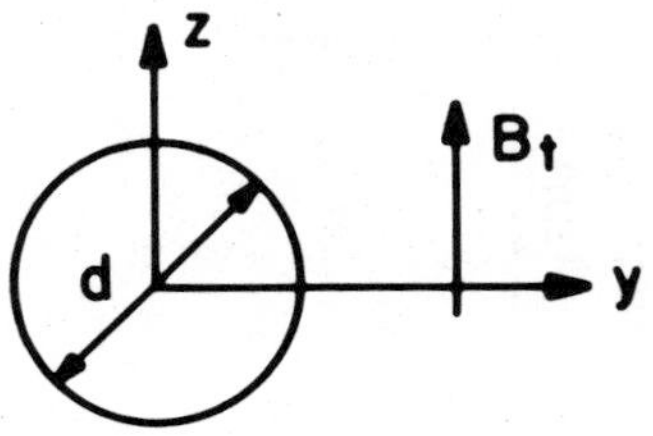

Fig. 9.13. Transverse field, B_t.

and

$$\underline{E}_x = j\omega B_{0t} e^{j\omega t} y \qquad (8)$$

The corresponding current density, $\underline{J}_x$, is

$$\underline{J}_x = \sigma_{Al} \underline{E}_x \qquad (9)$$

The $\underline{J}_y$ component of current density has been neglected since the armature conductor diameter is much smaller than the armature coil length. The Joule losses per conductor unit length, P_t, are

$$P_t = \frac{1}{2} \sigma_{Al} \omega^2 B_{0t}^2 \int_{-[(d/2)^2-z^2]^{1/2}}^{[(d/2)^2-z^2]^{1/2}} \int_{-d/2}^{d/2} y^2 \, dz \, dy \qquad (10)$$

and finally,

$$P_t = \pi \sigma_{Al} \omega^2 B_{0t}^2 \frac{d^4}{128} \quad (W/m) \qquad (11)$$

<u>Longitudinal</u> <u>field</u>, B_1

The axial magnetic field, B_1 (Fig. 9.14), induces circular eddy currents. Their current density final expression, $\underline{J}_\phi$, is written

$$\underline{J}_\phi = - j \frac{\sigma_{Al} \omega B_{01}}{2} e^{j\omega t} r; \quad 0 \leq r \leq \frac{d}{2} \qquad (12)$$

The Joule loss per unit length, P_1, is

$$P_1 = \pi \omega^2 \sigma_{Al} B_{01}^2 \frac{d^4}{256} \quad (W/m) \qquad (13)$$

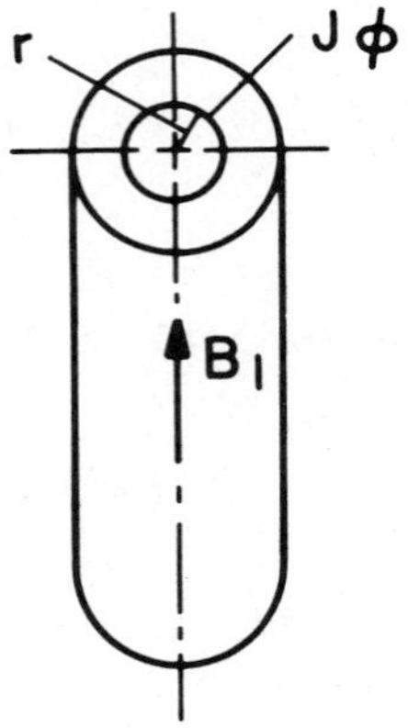

Fig. 9.14. Longitudinal field, B_1.

In the above expression it is implied that the skin effect is relatively small; that is, the depth of field penetration, δ, is greater than the conductor radius, d/2, and,

$$\delta = \sqrt{\frac{2}{\mu_0(\pi/\tau)\sigma_{Al}U}} > \frac{d}{2} \tag{14}$$

To keep the Joule losses within reasonable limits, condition (14) must be fulfilled up to the vehicle maximum speed.

The smaller the pole pitch, for a known speed, the smaller the elementary conductor diameter, d. Thus, the number of thin elementary transposed conductors used in armature coils increases when τ decreases.

To evaluate the skin effect losses (P_t and P_1), B_{0t} and B_{01} must be determined. As a first-order approximation[11]

$$B_{0t} = B_{01} = \frac{\mu_0 K_1 K_{d1}}{\sqrt{2}} e^{-\beta_0 z_0}; \quad \beta_0 = \frac{\pi}{\tau} \tag{15}$$

where

$$K_1 = \frac{4I_0}{\tau} \tag{16}$$

and K_{d1} is the distribution factor of armature winding.

Finally, the number of elementary conductors, n_c', is

$$n_c' = \frac{4I_f}{2j_{Al}\pi d^2} \tag{17}$$

with

$$d \simeq 0.3\delta \text{ to } 0.5\delta \tag{18}$$

In (17) I_f = phase current and j_{Al} = current density. The factor 2 in the denominator of (17) indicates that we have two current paths in parallel. If a more precise calculation of skin effect is necessary, B_x', B_y', and B_z' should be calculated from (1) and used conveniently as B_{01} or B_{0t}.

<u>Distribution of armature current over conductor cross section</u>

The armature current tends to concentrate at the outer surface of the conductor because of skin effect (Fig. 9.15). Proceeding as above, the axial field, $\underline{H}_t$, satisfies the equation

$$\frac{\partial^2 \underline{H}_t}{\partial r^2} + j\omega\sigma_{Al}\mu_0\underline{H}_t = 0 \tag{19}$$

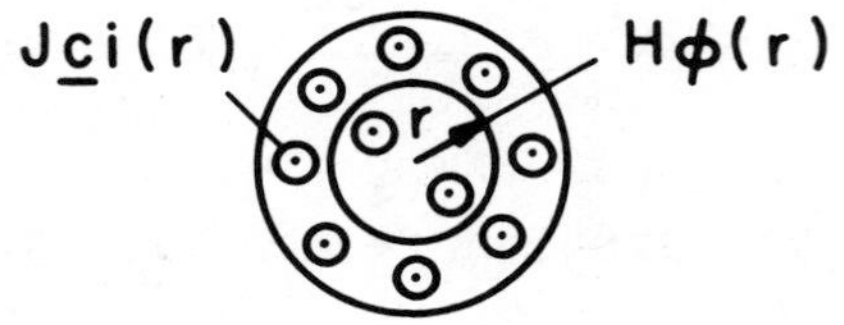

Fig. 9.15. Armature current skin effect.

with the boundary conditions

$$\pi d \underline{H}_t\left(\frac{d}{2}\right) = \frac{I_i}{2n_c'}; \quad \underline{H}_t(0) = 0 \tag{20}$$

where I_i is the ith harmonic of phase current, I_f.

The current density, $\underline{J}_{ci}$, corresponding to $\underline{H}_t$ yields the expression

$$\underline{J}_{ci} = \frac{I_i (j\omega\mu_0 i \sigma_{Al})^{1/2} \cosh[r(j\omega\mu_0\sigma_{Al} i)^{1/2}]}{2\pi n_c' d \sinh[(d/2)(j\omega\mu_0\sigma_{Al} i)^{1/2}]}$$

$$0 \leqq r \leqq \frac{d}{2} \tag{21}$$

Accordingly, the Joule loss per unit length, P_{ci}, is

$$P_{ci} = \frac{1}{\sigma_{Al}} \int_o^{d/2} |\underline{J}_{ci}|^2 2\pi r \, dr \tag{22}$$

As expected, the field penetration depth, δ_{c1}, for the fundamental is the same as δ (14).

Now the total loss due to skin effect, P_{1t}, for the three phases and one section is

$$P_{1t} \simeq 2[(P_t + P_1)2p + 2p' \sum_{i=1,3,5} P_{ci}]\, 6n_c' (\tau + 1) \quad (23)$$

Expression (23) is valid for the type of armature winding presented in Fig. 9.7. The skin effect must be considered in any evaluation of LSCM performance, as shown in the next section.

9.4.3. LSCM Performance

The performance is determined when the motor is fed by a six-pulse current-source inverter. The power angle, θ, pulsates at the instants of commutation. However, to account for the commutation process, the inverter is replaced by an equivalent circuit[12] (presented in Fig. 9.16), where E_0 represents the dc voltage in the dc link of the inverter, L_0 and R_0 are the filtering coil parameters, and

$$R_c = \frac{3\omega L_c}{\pi} \quad (24)$$

represents the commutation phenomenon.

Consequently, the real inverter is replaced by a variable dc voltage source, $V_e \cos \gamma_0$, and a commutation

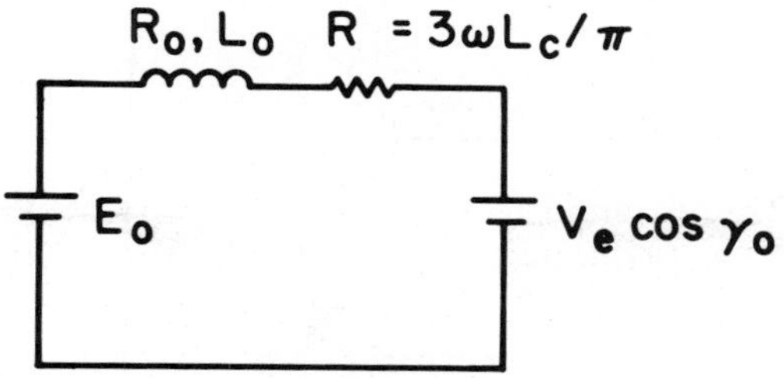

Fig. 9.16. Inverter equivalent circuit.

resistance, R_c; V_e is the phase induced voltage at $\gamma_0 = 0$, and γ_0 is the lead angle of the thyristor's gate pulse. The commutation inductance, L_c, is identical to the phase inductance, L_f, (3).

Under these conditions the propulsion force is given by the expression[12]

$$F_x = \frac{\omega\sqrt{3}}{U} M_{12} I_0 I_f \cos(\theta + \frac{\pi}{3}) \tag{25}$$

with

$$\theta = \omega t - \frac{\pi}{2} - \gamma_0; \quad 0 \leq t \leq \frac{\pi}{3\omega} = T_p \tag{26}$$

and M_{12} = the mutual inductance between field winding and armature phase:

$$M_{12} \simeq \frac{2\tau}{\pi} \frac{d\phi_e}{dx} \frac{2p}{I_0} \tag{27}$$

The propulsion force pulsates in time with the period T_p, which is six times smaller than the phase current period. The average propulsion force, F_{xav}, is

$$F_{xav} = \frac{3}{\pi} \int_0^{\pi/3} F_x \, d\theta = \frac{3\sqrt{3}}{\pi} M_{12} I_0 I_f \frac{\omega}{U} \cos \gamma_0 \tag{28}$$

Efficiency and power factor

Taking into consideration the average propulsion force, we can define a lumped efficiency, η_{av}:

$$\eta_{av} = \frac{1}{1 + P'_{1t}/(F_{xav} U)} \tag{29}$$

where

$$P'_{1t} = P_{1t} + P_{1s}; \quad P_{1s} = \text{cryogenic equipment power.} \qquad (30)$$

To define a power factor we refer to the fundamental. The phase induced voltage $\underline{V}_{e1}$ is

$$-\underline{V}_{e1} = 2U\left(\frac{d\phi_e}{dx}\right)_{1\ max} e^{j\omega t} \qquad (31)$$

while the phase current fundamental, $\underline{I}_1$, is written

$$\underline{I}_1 = I_1 \cdot e^{j(\omega t + \gamma_0)} \qquad (32)$$

The phase equation (Fig. 9.17) is

$$\underline{V}_1 = -V_{e1} + I_1(R_1 + j\omega L_f) \qquad (33)$$

The value of power factor, $\cos \phi_{av}$, may be calculated from the phasor diagram (Fig. 9.17).

For positive values of γ_0 the power factor may exhibit a leading character, while for small values of

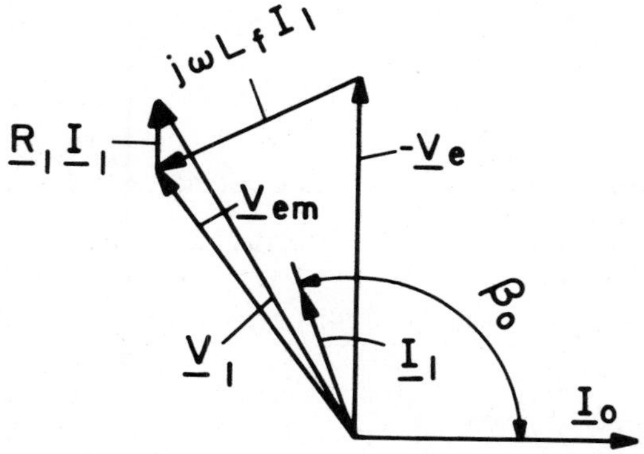

Fig. 9.17. Phasor diagram.

γ_0 the propulsion force becomes high. Thus for a given field winding mmf, armature current and geometric dimensions, a maximization trial of the product $\eta_{av} \cos \phi_{av}$ could be tried with respect to γ_0. However, with γ_0 = 10° to 20°, high performance is obtained.[13] It should also be noted that $\gamma_0 > 0$ corresponds to the unstable side of plot $F_{xa0} = f(\gamma_0)$, and thus rigorous electronic control of stability becomes necessary. To get a feeling for magnitudes, a numerical example of practical interest is worked out.

Numerical example

The following initial data are given: average propulsion force F_{xav} = 40 kN (20 kN per one stator); cruising speed U = 133.3 m/sec; SC coil mmf I_0 = 500 kA; vertical distance between SC coils and guideway z_0 = 0.22 m; pole pitch τ = 1.5 m; SC coil length L_1 = 0.75 m; SC coil width 2b = 0.9 m.

For the armature winding these data are given: paths in parallel = 2; armature coil width l = 0.625 m; lateral asymmetry y_0 = 0.1 m; phase inductance $L_f = 1.28 \cdot 10^{-2}$ H; $\gamma_0 = 20^o$; number of poles 2p = 20.

Solution

With these data, M_{12} is determined from (27), and then from (28) the phase current, I_f, yields

$$I_f = 840 \text{ A (per one stator)}$$

The field depth of penetration, δ, from (14), is δ = 12.5 mm. The diameter of an elementary conductor is selected as d = 6.0 mm.

Thus, for a current density $J_{Al} = 3 \cdot 10^6$ A/m^2, the number of elementary conductors, n_c connected in parallel, is, from (17), $n_c' = 10$. Following the procedure described in Section 9.4.2 the skin effects losses, per unit length of elementary conductor, are determined:

$$P_t = 4.1 \text{ W/m}; \quad P_l = 2.05 \text{ W/m}; \quad P_{cl} = 0.726 \text{ W/m}$$

The efficiency is now $\eta_{av} = 0.888$ from (29).

Finally the average power factor, $\cos \phi_{av}$, and phase voltage fundamental, V_1, follow from (33):

$$\cos \phi_{av} = 0.485; \quad V_1 = 1571.0 \text{ V(RMS)}$$

It is to be noted that the skin effects are rather important even though the elementary conductor diameter is half the field penetration depth. Consequently it is absolutely necessary to use thin elementary transposed conductors in the manufacturing of armature winding. This is true also for the ladder of short-circuited coils secondary of levitation or guidance electrodynamic systems.

The phase voltage V_1 is below 5 kV, and thus the elementary conductors need not be isolated from each other before their transposition. On the other hand, the voltage obtained is below the level of 4 to 5 kV reached by high power static converters. Because of cost considerations we need to find armature winding aluminum weight per unit length of section, M_{Al}:

$$M_{Al} \simeq \frac{2l + 2\tau}{2\tau} \, 6 \, \frac{2I_f}{J_{Al}} \, \gamma_{Al} \quad \text{(kg/m)} \tag{34}$$

Based on our data $M_{Al} = 12.96$ kg/m, which is a reasonably low value. We can conclude that the overall performance of LSCM is satisfactory. Further improvements in the power factor and efficiency could be obtained for the same length of section--5 km--by increasing the mmf in the SC coils to (7 to 9) x 10^5 ampturns.

9.5. Normal and Lateral Forces

The average value of normal force, F_{nav}, can be calculated by a procedure similar to that used for the propulsion force calculation. The only differences are that $(\partial\phi_e/\partial x)$ is replaced by $(\partial\phi_e/\partial z)_{1max}$ and $\cos\gamma_0$ by $\sin\gamma_0$:

$$F_{nav} = \frac{3\sqrt{3}}{\pi} M_{12} I_0 I_f \frac{\omega}{U}\left[\frac{(\partial\phi_e/\partial z_{1max})}{(\partial\phi_e/\partial x)_{1max}}\right] \sin\gamma_0 \qquad (35)$$

In many practical cases we may assume that the magnetic flux of an SC coil varies exponentially with z:

$$\phi_e(z) \simeq \phi_{e0}\, e^{-\pi/\tau\, z} \qquad (36)$$

Thus in a first-order approximation,

$$F_{nav} = F_{xav} \frac{\sin\gamma_0}{\cos\gamma_0} \qquad (37)$$

The lateral average force F_{lav} is

$$F_{lav} = F_{xa} \frac{\sin\gamma_0}{\cos\gamma_0}\left[\frac{(\partial\phi_e/\partial y)_{1max}}{(\partial\phi_e/\partial x)_{1max}}\right] \qquad (38)$$

The computation of the fundamentals of the partial derivatives occurring in (35) and (38) is done on a computer. Both the normal and lateral forces pulsate in time as the propulsion force does:

$$F_n(t) \simeq \frac{\omega\sqrt{3}}{U} M_{12} I_f I_0 \sin(\theta + \frac{\pi}{3}) \qquad (39)$$

$$F_l(t) = \frac{\omega\sqrt{3}}{U} M_{12} I_f I_0 \left[\frac{(\partial\phi_e/\partial y)_{lmax}}{(\partial\phi_e/\partial x)_{lmax}}\right] \sin(\theta + \frac{\pi}{e}) \qquad (40)$$

where

$$\theta = \frac{\pi}{2} - \omega t - \gamma_0 \quad \text{and} \quad 0 \leq t \leq \frac{\pi}{3\omega} \qquad (41)$$

It should be kept in mind that the average normal and lateral forces change their sign at $\gamma_0 = 0$. For $\gamma_0 < 0$ the normal force is attractive while the lateral one has a centralizing (spring) effect. On the contrary when $\gamma_0 > 0$ the normal force is repulsive while the lateral one has a decentralizing character. Both these forces are rather small for low values of γ_0. This is so because both depend upon $\sin\gamma_0$. Moreover, the lateral force is smaller than normal force since the derivative $\partial\phi_e/\partial y$ is small because the SC coil is much wider than the armature coil (2b > 1). However, both these forces should be accounted for in the levitation and guidance systems design.

The double-horizontal LSCM[3] may be supplied with two inverters (one per each stator). Thus it becomes possible for the two "stators" to work at two different modified power angles, $\gamma_{01} \neq \gamma_{02}$, and different phase currents, $I_{f1} \neq I_{f2}$, while maintaining constant propulsion force values. Now if $\gamma_{01} < 0$ and $\gamma_{02} > 0$, so the normal forces of the two "stators" are opposite, developing a torque capable of counteracting some swinging oscillations of the vehicle.

In a similar way the two propulsion forces may be made different from each other, while keeping their sum constant. An adequate torque for winding oscillations damping could thus be created. All these variations of γ_{01}, γ_{02}, I_{f1}, and I_{f2} are electronically monitored using the oscillations parameters (transmitted by radio to substations) as inputs. Thus, it seems possible to avoid some of the on-board control systems used for diverse oscillations damping. A vehicle weight reduction may be expected. Although these theoretical solutions[3] are very tempting, they must be tested on full-scale vehicles before their reliability can be fully established.

9.6. Propulsion Control

As previously mentioned the double-horizontal LSCM could be fed by two six-pulse current-source inverters (Fig. 9.18). The key issue is to adequately control the value of γ_0: Two main solutions have been proposed so far:

1. With on-board position transducer.[5]

2. Indirect control from substation.[10]

Solution 1 implies a telemetric system to transmit the value of $\gamma_0(\beta_0)$ to substations, but is capable of acting from standstill. Solution 2 is more comfortable but, for starting (below 5 to 10 m/sec), a special control solution must be used. In what follows only solution 2 is dealt with in some detail.

Measurement of angle β_0 is carried through by direct measurement of internal power factor angle α (Fig. 9.19) between $\underline{V}_{em}$ and $\underline{I}_1$. For a given ratio of I_1/I_0, the phasor diagram (Fig. 9.19) can be used to determine the dependence between α and $\beta_0(\beta_0 = \pi/2 + \gamma_0)$, the latter being rather independent

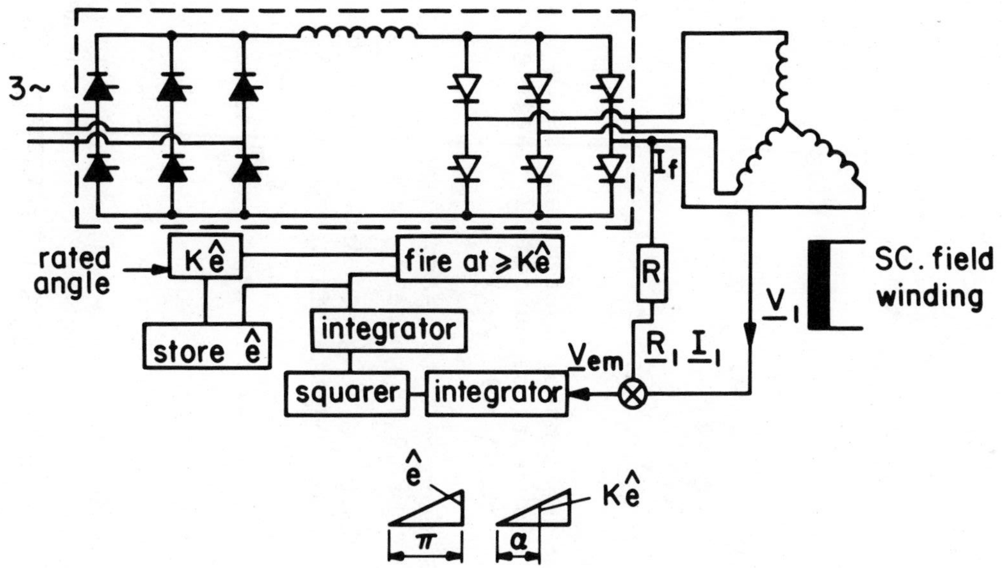

Fig. 9.18. Power conditioning and control unit for LSCM.

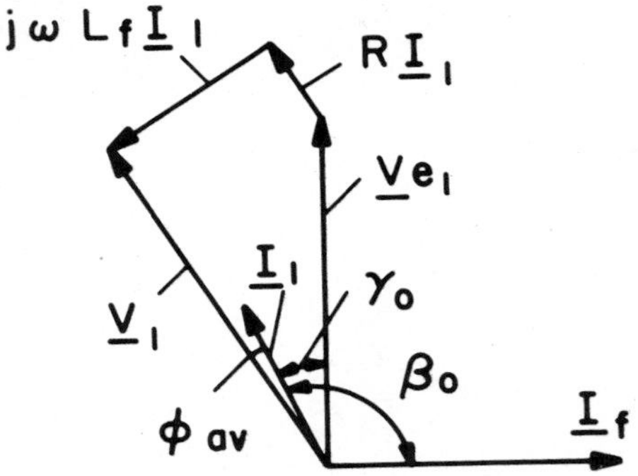

Fig. 9.19. Phasor diagram.

of frequency. Generally α is lagging and may be obtained from the equation

$$\underline{V}_{em} = \underline{V}_1 - \underline{I}_1 R_1 \tag{42}$$

Consequently the control system task is to determine the inverter to inject the current $\underline{I}_1$ lagging with angle α in respect to phase voltage $\underline{V}_1$.

One procedure[10] to accomplish this goal is presented in Fig. 9.18 where $\underline{V}_{em}$ is first integrated, in order to obtain a signal independent of frequency (speed), and then slightly filtered, ψ_m:

$$\psi_m = \int V_{em} \cos(\omega t + \alpha)dt = \frac{V_{em}}{\omega} \sin(\omega t + \alpha) \tag{43}$$

ψ_m is then converted into a rectangular signal and integrated again to yield a ramp signal whose peak value is proportional to $\underline{V}_{em}$ semi-period. In the next period a logic system produces gating pulses when the corresponding ramp signal reaches a given fraction of the peak value previously determined. This fraction corresponds to α, which is known. The gating pulse thus obtained fires the appropriate thyristor of the inverter. The system, however, works only for speeds higher than 8 to 10 m/sec. For starting, the inverter frequency is lowered to zero while the phase currents are kept rather high by increasing the rectifier voltage. Then the frequency is slowly increased as dictated by the speed achieved at that moment, according to a special program, such that β is kept at low values ($\beta = 25^o$ to 30°) that are within the naturally stable zone. Therefore high phase currents are necessary. Automatically, when a speed of 8 to 10 m/sec is reached, the control switches to the cruise control circuits described above. This system of propulsion control also allows for generator operation when regenerative braking is obtained.

For regenerative braking, below 10 m/sec, we return to our starting control logic, appropriately decreasing the frequency while keeping high values of phase current, down to standstill. Consequently we can even obtain a rather precise positioning effect of Maglev, which is necessary in any fully automatized remote-controlled transportation system.

9.7. Guidance Force of LSCM

As mentioned before, the double-horizontal LSCM is able to provide Maglev guidance if an aluminum ladder (or a row of short-circuited coils) is located between the two twin stators embedded in a concrete guideway (Fig. 9.20). The necessity of reducing, to some extent, the drag force associated with guidance force led us to a rather large pole pitch for the LSCM ($\tau = 1.5$ m). Again, an optimum pole pitch problem occurs when the propulsion force of the motor and the drag force of the guidance ladder are both considered. However, this problem is beyond our scope here. The guidance force computation procedure is followed in some detail since the null flux system was not studied earlier.

First, the magnetic flux ψ_0 (x_0, y_0, z_0) in a ladder loop, (Fig. 9.20), is

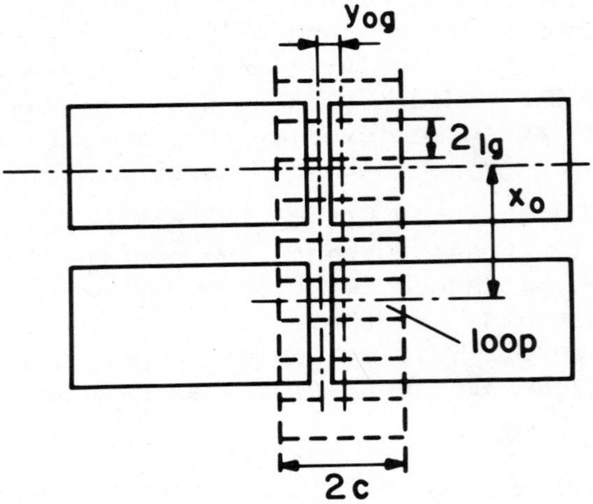

Fig. 9.20. Secondary guideway.

$$\psi_0(x_0,\ y_0,\ z_0)$$

$$= \int_{x_0-l_g}^{x_0+l_g} \int_{-2c-y_0-l/2}^{-c-y_0-l/2} B(x,\ y,\ z_0)\ dx\ dy \qquad (44)$$

Thus the mutual inductance, M_{0v}, between vehicle SC coils and the loop, is

$$M_{0v}(x_0,\ y_0,\ z_0) = \frac{\psi_0(x_0,\ y_0,\ z_0)}{I_0} \qquad (45)$$

The voltage, V_{e0}, induced in a ladder loop is

$$V_{eo}(x_0,\ y_0,\ z_0) = 2U\ \frac{\partial \phi_0}{\partial x} \qquad (46)$$

M_{0v}, as well as V_{e0}, could be decomposed in Fourier series:

$$V_{e0} = \sum_{\nu=1,3,5...} V_{e\nu} \sin[\nu(\omega t + \frac{\pi x_0}{\tau})] \qquad (47)$$

The current $i_0(x_0, t)$ in a loop Q is obtained from (39):

$$i_0(x_0, t) = \sum_{\nu=1,3,5...} V_{e\nu} \sin[\nu(\omega t + \pi/\tau\ x_0) + \phi_\nu]$$

$$\{[2R_1 + 2R_t(1 - \cos(\pi\nu/n))]^2$$

$$+ \nu^2\omega^2[L_s - 2M_{0v} \cdot \cos(\pi\nu/n)]^2\}^{1/2} \qquad (48)$$

For a secondary made of short-circuited coils, the term $2R_L + 2R_t[1 - \cos(\pi\nu/n)]$ in the denominator of (48) is replaced by the coil resistance, R. Also,

$$\tan \phi_\nu = \frac{\nu\omega[L_s - 2M_{0v} \cos(\pi\nu/n)]}{2R_I + 2R_t[1 - \cos(\pi\nu/n)]} \qquad (49)$$

Finally the drag force, F_{dg}, normal force, F_{ng}, and guidance force, F_g, corresponding to guidance-secondary are

$$F_{dg} \simeq 2p \sum_{-n}^{n} I_0 i_0(m1, t) \frac{\partial M_{0v}}{\partial x} \qquad (50)$$

$$F_{ng} \simeq 2p \sum_{-n}^{n} I_0 i_0(m1, t) \frac{\partial M_{0v}}{\partial z} \qquad (51)$$

$$F_g \simeq 2p \sum_{-n}^{n} I_0 i_0(m1,\ t)\ \frac{\partial M_{0v}}{\partial y} \tag{52}$$

These forces pulsate in time. The skin effect may be treated as in Section 9.4.2.

A numerical example

Two loops per pole are selected ($n = 2$). The ladder width is $2c = 0.4$ m. The other data are those used in Section 9.4.3. (It is to be noted that the ladder fits "exactly" between the two stators of Section 9.4.3.) The guidance secondary is made of elementary transposed conductors to reduce the skin effects (Fig. 9.21). From the manufacturing point of view a row of short-circuited coils should be preferred to a ladder secondary.

Using the procedure presented above, the average values of F_{dg}, and F_{ng} are computed as functions of speed, for given values of lateral displacement Y_0 (Fig. 9.22). The guidance force, F_g, depends upon lateral displacement for given speed and shows, in Fig. 9.23, a

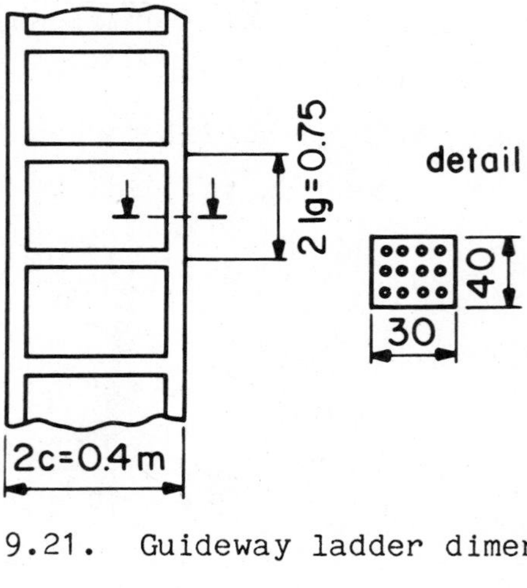

Fig. 9.21. Guideway ladder dimensions.

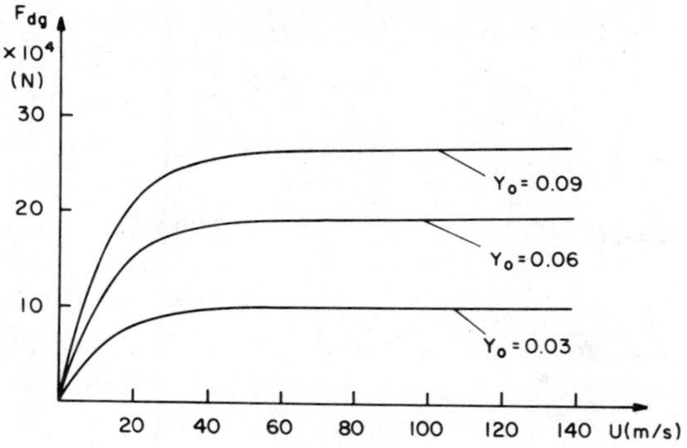

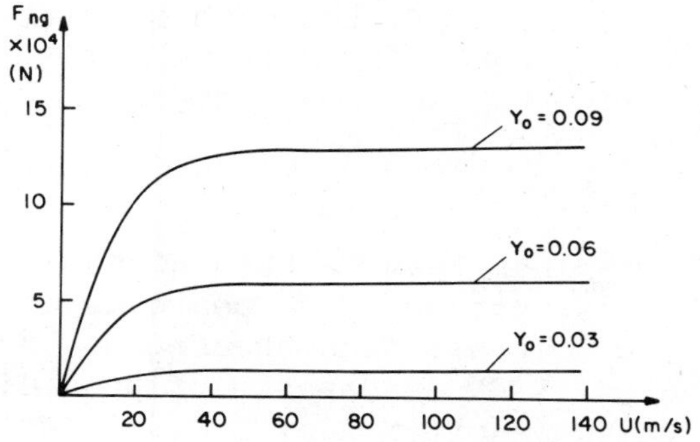

Fig. 9.22. Average forces against speed. (a) F_{dg} = drag, (b) F_{ng} = levitation.

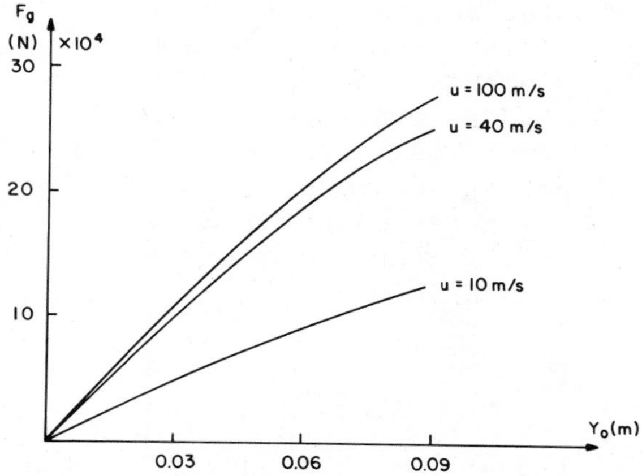

Fig. 9.23. Guidance force, F_g, against lateral displacement, y_o.

linear shape, indicating that the system is very adequate for managing the guidance of Maglev. The rather high value of guidance force obtained justifies the assertion that this system is capable of providing complete guidance of Maglev. The guidance dynamic stability and ride comfort should probably be provided by an active hydraulic secondary system. The occurrence of drag and levitation (normal) forces in the guidance secondary should be accounted for in the design and dynamic behavior study of Maglev propulsion and levitation systems. To simplify the mathematical expressions the interaction between LSCM stators and guidance secondary loops (coils) has been neglected.

9.8. Conclusions

This chapter has been devoted to some of the main aspects of construction and performance of double-horizontal LSCM. It has been theoretically proved that double-horizontal LSCM is capable of providing full propulsion and guidance of Maglev. The performance has been determined for the case when the LSCM is fed from a six-pulse current-source inverter. The overall propulsion performance obtained for energized sections of 5 km is rather satisfactory for reasonably low quantities of aluminum per unit length of guideway (13 kg/m). This simple LSCM represents a particular case of the double-horizontal LSCM, and consequently, the theory presented in this chapter may be directly used in that case. Based on the data presented in this chapter we can conclude that the LSCM represents a promising solution for the electrodynamic guidance and propulsion of high-speed remote-controlled Maglevs.

References

1. Canadian Maglev Group, "Study of magnetic levitation and linear synchronous propulsion," Phase I, Contract Report CIGGT-No. 1, 73, 1, December 1972.

2. S. Yamamura, "Magnetic levitation technology of tracked vehicles: Present status and prospects," IEEE Trans. Vol. MAG-12, No. 6, Nov. 1976, pp. 874-878.

3. S. Kuntz, P. E. Burke, and G. R. Slemon, "Active damping of Maglev using superconducting linear synchronous motors," Int. Quat. Elec. Mach. Electromechan., Vol. 2, No. 4, 1978, pp. 371-385.

4. H. H. Kolm, R. D. Thornton, Y. Iwasa, and W. S. Brown, "The magneplane system," Cyrogenics, July 1975, pp. 377-383.

5. A. R. Eastham (Editor), "Superconducting linear synchronous motor propulsion and magnetic levitation for H.S.G.T," Phase III, Interim report for TDAT Canada, CIGGT Report No. 76, 7, 1976.

6. T. A. Buchhold, "Superconducting machinery," in Applied Superconductivity, (V. L. Newhouse, Editor), Academic Press, New York, 1975, p. 523.

7. R. Yasumochi and H. Mori, "Electromagnetic forces on the linear synchronous motor for magnetically levitated vehicle," International Symposium on Linear Electric Motors, May 1974, Lyon, France.

8. P. C. Sen, "On linear synchronous motor (LSM) for high speed propulsion," IEEE, IAS, Annual meeting, 1975, 10 E.

9. S. A. Nasar and I. Boldea, Linear Motion Electric Machines. Wiley-Interscience, New York, 1976, p. 146.

10. G. R. Slemon, "The Canadian Maglev prospect on high speed interurban transportation," Department of Elec. Eng., University of Toronto, 1975.

11. C. A. Skalski, "The air core linear synchronous motor--An assessment of current development," Report FRA-OR & D 76-260, June 1976, Washington, DC 20590.

12. F. Harashima, H. Naitoh, and T. Haneyoschi, "Dynamic performance of current source inverter-fed synchronous motors," Proc. of International Conference on Power Semiconductor Applications, May 1977, Disney Land Hotel, Tulla, Florida, pp. 462-470.

13. J. Holtz, "Force components and their control for ironless linear synchronous motor" (in German). ETZ-A. Vol. 96, No. 9, 1975, pp. 396-400.

14. T. Ohsuka and Y. Kyotani, "Superconducting Maglev tests," IEEE Trans., Vol. MAG-15, No. 6, 1979, pp. 1416-1421.

15. Y. Amemiya and S. Aiba, "Linear synchronous motors using superconducting coils for both propulsion and suspension," Elec. Eng. Jap., Vol. 99, No. 2, 1979, pp. 59-67.

16. C. P. Parsch and G. Wiegner, "The air-cored linear synchronous motor: The state of the art in Erlangen," L4/3, International Conference on Electric Machines, 1979, Brussels, Belgium.

CHAPTER 10

LINEAR MOTION ELECTROMAGNETIC LEVITATORS

10.1. Introduction

The potential uses of linear motion electromagnetic levitators (LELs) range from standstill applications, such as active magnetic bearings and vibrating tables, to the levitation and guidance of electrical vehicles. LEL construction utilizes electromagnets, with solid or laminated iron secondary (to be lifted or vibrated) and adequate control systems, power amplifiers, and transducers. There are various aspects of LEL, and we deal with the problems that occur when LEL is used for levitation and guidance of vehicles (Maglevs), because the greatest difficulties and the most problems occur in standstill applications.

10.2. Optimal Design of Electromagnets for Maglev Vehicles

The LEL for Maglev vehicles makes use of U-shape electromagnets and the secondary exhibits, in general, one of the three shapes drawn schematically in Fig. 10.1.

In an optimal design of LEL electromagnets two criteria are most important:

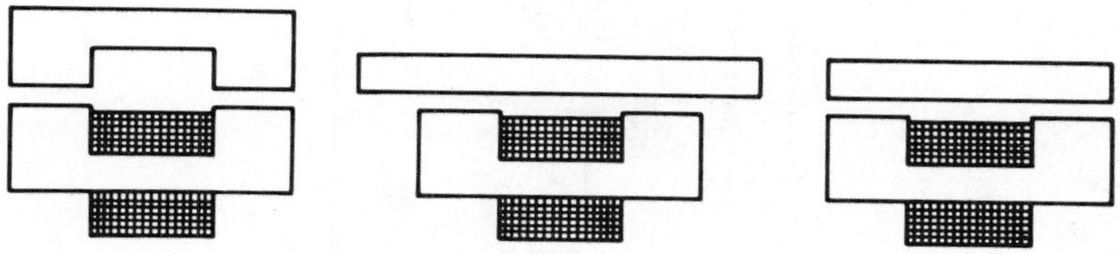

Fig. 10.1. Three different shapes for a secondary in LEL construction.

1. Minimum weight of the electromagnet.[1,2]

2. Minimum average power (including control power) of the electromagnet for a given levitation force.

These two criteria are rather conflicting (as is true for any power electromagnetic system). However, the average power of electromagnets[3] is low compared to the propulsion power (when applied to vehicles); in general, only the minimum weight criterion need be considered. The weight of the electromagnet depends upon its core dimensions and the current density allowed in the coil. At high current densities (above 4 to 6 A/mm^2) a special cooling system should be used. The power and weight of this cooling system should also be accounted for in the total electromagnet weight. On the other hand, for high current densities, the power loss in the electromagnet increases and, thus, for forced-cooled electromagnets the minimum average power cannot be ignored any more. Thus, the optimal design problem gets rather involved.

Here, a simplified method is used. The energy criterion and the presence of a forced-cooling system are neglected and the current density is taken as a parameter. Even under these conditions, for given ampturns, the computation of levitation force is a difficult task, mainly because of large values of the airgap (g_0 = 10 to

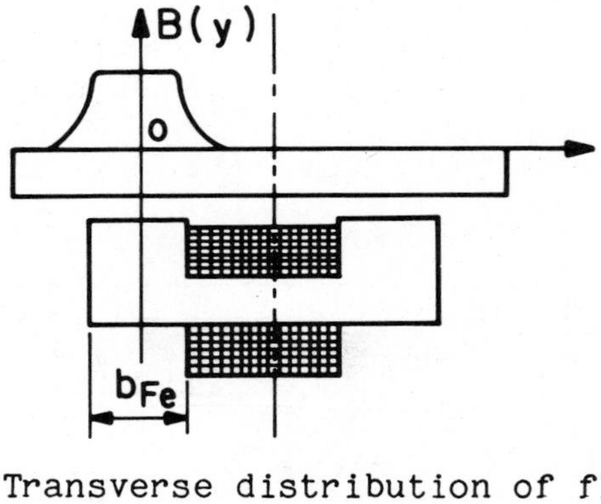

Fig. 10.2. Transverse distribution of flux density.

15 mm). (For standstill applications the airgap may be decreased.) A more precise determination of the field distribution and levitation force can be made by conformal mapping or by two- or three-dimensional dimensional numerical methods. The transverse distribution of airgap flux density obtained from these methods can be approximated by (Fig. 10.2)

$$B_{(y)} = B_0; \quad 0 \leqq y \leqq \frac{b_i}{2}$$

$$B_{(y)} = B_0 \exp\left[\frac{2(y-b_i/2)}{g0}\right]; \quad y \geqq \frac{b_i}{2} \tag{1}$$

The level of magnetic saturation in the magnet core, as well as in the solid iron guideway, should be limited in order to yield fast changes of levitation force by small changes in current, in the process of automatic control of levitation.

However, the magnetic core reluctance should be considered, at least approximately, and thus the core flux density, B_{cm}, away from airgap, is

$$B_{cm} = B_0\left[1 + \frac{2}{b_i} \int_{b_i/2}^{\infty} \exp\left(\frac{-2(y - b_i/2)}{g_0}\right) dx\right] \tag{2}$$

And finally,

$$B_{cm} = B_0 \left(1 + \frac{g_0}{b_i}\right) \tag{3}$$

Moreover, in the transverse plane a leakage flux occurs. The peak flux density B_{maxl}, of this leakage is

$$B_{maxl} = \frac{\mu_0 W_f I_f}{b} \tag{4}$$

The average value of the core flux density, B_i, is

$$B_i = B_{cm} + B_{max1/2} \tag{5}$$

Thus the relationship between the ampturns, $W_f I_f$, and the airgap flux density, B_0, is

$$B_0 = \frac{\mu_0 W_f I_f}{2g_0[1 + \frac{(b + h + 2b_i)}{g_0} \frac{\mu_0}{\mu(B_i)}]} \tag{6}$$

where $\mu(B_i)$ represents the magnetic permeability corresponding to B_i, which is also a function of B_0.

For a given value of ampturns, $W_f I_f$, B_0 can be calculated by means of an iterative procedure if the magnetization curve is stored in the computer memory.

The attraction force per magnet, F_{a0}, is

$$F_{a0} = \frac{4L}{2\mu_0} \int_0^\infty B^2(y)dy \tag{7}$$

And finally,

$$F_{a0} = \frac{2B_0^2 L}{2\mu_0} b_i \left(1 + \frac{2g_0}{b_i}\right) \tag{8}$$

The magnet weight, G_0, is

$$G_0 \simeq L\, [(2(h + b_i)b_i + b_i b)\gamma_i$$

$$+ \frac{2W_f I_f}{J_c} \left(1 + \frac{b/2 + b_i}{L}\right) \gamma_c] \tag{9}$$

where L = the magnet length, J_c = current density, and γ_i,/p γ_c = specific weight of iron and, respectively, copper (aluminum). For a fill factor, K_f, for the core window, its cross section is

$$bh = \frac{W_f I_f}{J_c K_f} \tag{10}$$

10.2.1. Optimal Magnet Selection

As an example we take an LEL used for vehicles of high speed. The initial data are attraction force F_{ao} = 5×10^4 N; rated airgap $g_0 = 1.5 \times 10^2$ m; rated speed U_n = 100 m/sec; core average flux density B_i = 1.55 to 1.65 T; magnet width $L_c = 2b_{Fe} + b \leqq 0.3$ m, in order to keep the guideway costs within reasonable limits.

The procedure used here to select the optimal magnets is divided into two steps:

1. The magnet core width, b_i, is given in the interval $b_i = (4 \text{ to } 6)10^{-2}$ m. Based on the above geometrical restriction we can calculate $b = L_c - 2b_i$. From (3) through (5) we obtain

$$\mu_0 W_f I_f = [B_i - B_0(1 + \frac{g_0}{b_i})]\ 2b \tag{11}$$

For given values of B_i from (10) and (11) we can iteratively calculate B_0 by making use of a real magnetization curve. And $W_f I_f$, h, L, and G_0 are calculated from (11), (10), (8), and (9), respectively. For different values of b_i and J_c the computation process is repeated. The final results are plotted in Fig. 10.3.

Up to this point the end effect has been neglected. The end effect is accounted for as follows:

2. For all the cases studied above, for the rated speed, the resultant attraction force and the drag force due

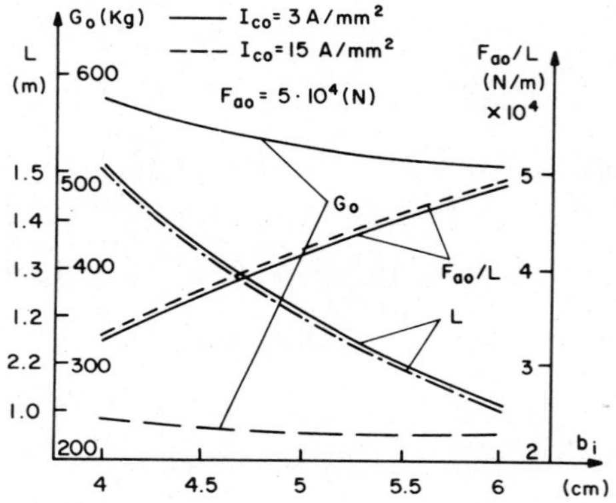

Fig. 10.3. Magnet mass, G, length, L, and levitation force for unit length, F_L/L (without end effect).

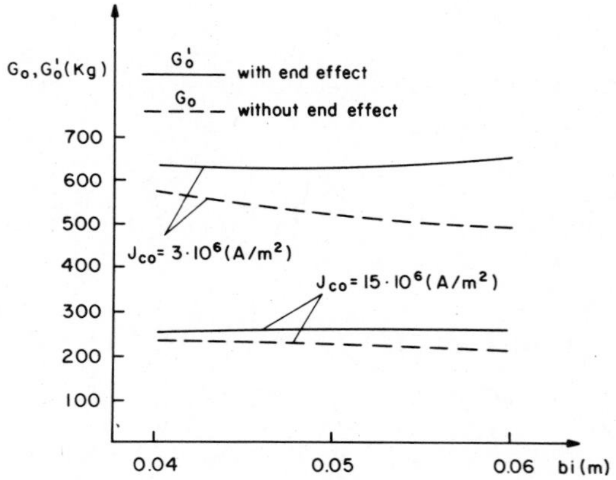

Fig. 10.4. Recalculated magnet mass, G_0', with end effect.

to end effect are calculated. (The end effect is caused by the eddy currents induced in the solid-iron secondary by the magnet coil field. A special procedure to calculate the end effect is developed in Section 10.3.) The ampturns of the magnet are then increased in order to produce the attraction force developed at standstill. Accordingly the core window cross section, bh, is modified, and finally, a new weight, G_0', of the magnet is obtained for a given value of b_i (Fig. 10.4).

When the attraction force and the speed are kept constant a decrease of b_i is accompanied by an increase in the magnet length, L, and thus a reduction of the end effect occurs. However, for mechanical reasons, the magnet length should not exceed 1.5 to 1.8 m. A uniform airgap is difficult to maintain for very long magnets.

To reduce the end effect it has been proposed to form a row of magnets (1.5 to 1.8 m in length). A separate control of these magnets would be necessary in order to obtain a flexible levitation system. On the other hand, the separate control inevitably leads to field discontinuities between magnets, and so they behave more or less as if they were separate. Thus, this proposal is of theoretical interest, and does not result in a significant reduction of end effects. Best results are obtained from magnets 1.5 to 1.8 m long. The end effect calculations (Section 10.3) show that magnets 1.5 to 1.8 m long yield good performance up to 100 to 120 m/sec. When natural cooling is available (J_c = 3 A/mm^2) these magnets could develop an attraction force surpassing by four to six times their own weight. On the other hand, the attraction force of forced-cooled magnets with pure copper coils (J_{c0} = 15 to 20 A/mm^2) is 15 to 18 times the magnet weight. But the average input power is three to four times higher than that of naturally cooled aluminum-coil magnets.

These qualitative remarks demonstrate how difficult it is to choose the current density (or the cooling system). In such an enterprise the specific energy expenditure in kwh per passenger and kilometer and the vehicle total costs will play the key roles.

We now consider the end effect, realizing that this problem does not occur at standstill or in low-speed applications.

10.3. The End Effect in LEL for Maglev Vehicles

When the magnets move at a speed U, with respect to the solid-iron secondary, eddy currents occur in the latter, resulting in a reduction of the attraction force and a production of drag force. A quantitative assessment of these phenomena is absolutely necessary for levitation magnets and propulsion system design.

In the following, the phenomena resulting from the eddy currents induced in the secondary is termed end effect. Recently, some approaches based on Fourier transforms[2,3] and finite element method have been used to calculated the end effect. However, they seem to overestimate the phenomena, as demonstrated by high-speed (90 m/sec) tests.[1] Neglecting the real field penetration depth in the secondary causes discrepancy between theory and tests. We obtain the depth of field penetration in secondary iteratively in a new approximate approach.[15]

10.3.1. Assumptions and Field Equations

We assume a sinusoidal distribution of induced currents along the transverse direction. And, the real transverse distribution of magnet ampturns is approximated by an equivalent rectangular one (Fig. 10.5) where the equivalent magnet shoe width, $2a_e$, is

$$2a_e = b_i\left(1 + \frac{g_0}{b_i}\right) \tag{12}$$

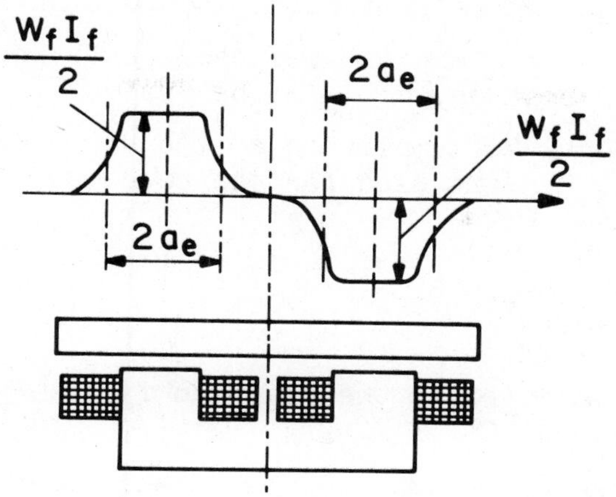

Fig. 10.5. Transverse mathematical model.

The secondary width, 2c, is

$$2c = L_1 + b_i \tag{13}$$

The airgap flux density distribution along Ox (Fig. 10.6) is

$$B_0(x) = 0; \quad x < 0 \tag{14}$$

$$B(x) = B_0; \qquad 0 \leqq x \leqq L \tag{15}$$

$$B(x) = 0; \qquad x > L \tag{16}$$

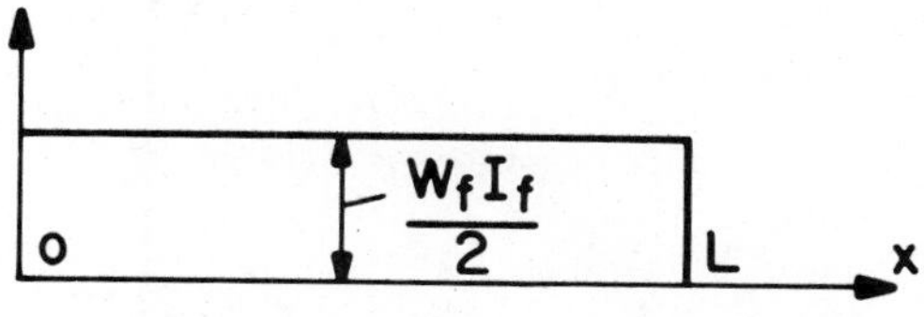

Fig. 10.6. Logitudinal mathematical model.

where B_0 has been calculated as in Section 10.2. The induced currents exhibit two components, and the field depth of penetration, d_i, will be calculated later in this section. The amplitude of the airgap flux density fundamental, B_{1y}, as used for the calculation of induced currents, is

$$B_{1y} = B_0 \frac{4}{\pi} \sin \frac{\pi}{2c} a_e = B_0 K_y \tag{17}$$

Under these conditions the reaction field, H_r, satisfies the equation

$$\frac{\partial^2 H_r}{\partial x^2} + \frac{\partial^2 H_r}{\partial y^2} - \mu_0 \sigma_i U \frac{d_i}{g_e} \frac{\partial H_r}{\partial x} = 0 \tag{18}$$

The solution to this equation takes the forms

$$H_{r0}(x,y) = A_0 e^{\gamma_1 x} \cos \frac{\pi}{2c} y; \quad x \leq 0 \tag{19}$$

$$H_{r1}(x, y) = [A_1 e^{\gamma_1 (x-L)} + B_1 e^{\gamma_2 x}] \cos \frac{\pi}{2c} y;$$

$$0 \leq x \leq L \tag{20}$$

$$H_{r2}(x, y) = B_2 e^{\gamma_2 (x-L)} \cos \frac{\pi}{2c} y; \quad x \geq L \tag{21}$$

where

$$\gamma_{1,2} = \frac{C_u}{2} \pm [(\frac{C_u}{2})^2 + (\frac{\pi}{2c})^2]^{1/2} \tag{22}$$

with

$$C_u = \frac{\mu_0 \sigma_i d_i U}{g_e} \tag{23}$$

For small speeds (0 to 10 m/sec),

$$\frac{C_u}{2} << \frac{\pi}{2c} \quad \text{and thus} \quad \gamma_1 \simeq |\gamma_2| \simeq \frac{\pi}{2c} \tag{24}$$

And at high speeds (60 to 100 m/sec),

$$\frac{C_u}{2} >> \frac{\pi}{2c} \quad \text{and consequently} \quad \gamma_1 >> |\gamma_2| \tag{25}$$

The integration constants, A_0, A_1, B_1, B_2, are obtained from boundary conditions:

$$\begin{aligned} (\mu_0 H_{r0})_{\substack{x=0\\y=0}} &= B_0 K_y + (\mu_0 H_{r1})_{\substack{x=0\\y=0}} \\ (\mu_0 H_{r2})_{\substack{x=L\\y=0}} &= B_0 K_y + (\mu_0 H_{r1})_{\substack{x=L\\y=0}} \end{aligned} \tag{26}$$

$$\begin{aligned} \left(\frac{\partial H_{r0}}{\partial x}\right)_{x=0} &= \left(\frac{\partial H_{r1}}{\partial x}\right)_{x=0} \\ \left(\frac{\partial H_{r1}}{\partial x}\right)_{x=L} &= \left(\frac{\partial H_{r2}}{\partial x}\right)_{x=L} \end{aligned} \tag{27}$$

Finally we get

$$B_1 = \frac{K_y B_0}{\gamma_2/\gamma_1 - 1)}$$

$$A_1 = \frac{K_y B_0}{\gamma_1/\gamma_2 - 1} \qquad (28)$$

$$A_0 = B_0 + A_1 e^{-\gamma_1 L} + B_1$$

$$B_2 = B_0 + A_1 + B_1 e^{\gamma_2 L} \qquad (29)$$

At high speeds B_1 approaches $-K_y B_0$ and A_1 approaches zero.

The theory allows the consideration of harmonics in the transverse direction, with $K_{y\nu} = (4/\pi\nu)\sin(\pi\nu a_e/2c)$ and $\nu\pi/2c$ instead of $\pi/2c$ in (19) through (22). However, for preliminary design purposes only the fundamental component is adequate.

<u>Depth of field penetration</u>

Taking into account that $x = Ut$ we can define two time constants, T_1 and T_2:

$$T_1^{-1} = \frac{U\gamma_1}{2\pi} \quad \text{and} \quad T_2^{-1} = \frac{U|\gamma_2|}{2\pi}; \quad T_1^{-1} >> T_2^{-1} \qquad (30)$$

Their depths of penetration, d_{i1} and d_{i2}, are

$$d_{i1} = \left(\frac{2}{\mu_0 2\pi T_1^{-1}\sigma_i}\right)^{1/2} \quad \text{and} \quad d_{i2} = \left(\frac{2}{\mu_0 2\pi T_2^{-1}\sigma_i}\right)^{1/2} \qquad (31)$$

For high speeds $d_{i2} >> d_{iz}$. Consequently the reaction field component containing γ_1 attenuates quickly at the exit end of the magnet. Thus, along the magnet length the resultant flux density is reduced by eddy-currents reaction. In reality the permeability of secondary varies along Ox and Oy. This fact could be accounted for only by numerical methods. For simplicity an average permeability is defined. It corresponds to an average secondary flux density, B_{xm}:

$$B_{xm} = \frac{\left(\frac{A_0 + B_2}{2}\right) b_i}{d_0}; \quad B_0 = \frac{\mu_0 W_f I_f}{2g_e} \tag{32}$$

that is,

$$B_{xm} = [B_0 + \frac{A_1}{2}(1 + e^{-\gamma_1 L}) + \frac{B_1}{2}(1 + e^{\gamma_2 L})] \frac{b_i}{d_0} \tag{33}$$

where the equivalent airgap g_e accounts also for the secondary saturation:

$$g_e = g_0[1 + \frac{(b + b_i)b_i}{g_0 \ d_0} \frac{\mu_0}{\mu(B_{xm})}] \tag{34}$$

Using an iterative procedure, for known speed U and amp-turns, the value of average permeability $\mu(B_{xm})$ is calculated. Then the reaction field distribution can be determined.

10.3.2. Numerical Example

A magnet with the following data is considered: $b_i = 0.04$ m, $B_0 = 1$T, $b = 0.12$ m, $g_0 = 0.015$ m, secondary thickness $d_0 = 0.025$ m, iron conductivity $\sigma_i = 3.52 \times 10^6 \ (\Omega m)^{-1}$.

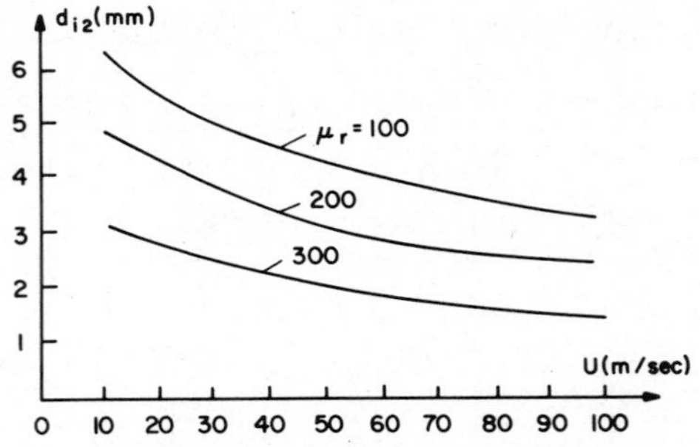

Fig. 10.7. Depth of field penetration, against speed.

Based on these data the depth of penetration, d_{i2} (31), is calculated iteratively, for given values of permeability, μ, as function of speed (Fig. 10.7). Notice that the depth of penetration d_{i2} is much smaller than the secondary thickness, d_0, at least at medium and high speeds. Consequently, the skin effect in the secondary should be accounted for in end effect calculations.

10.4. Levitation and Magnetic Drag Forces

The levitation (attraction) force per magnet, F_n, is

$$F_n = \frac{2}{\mu_0} \int_0^{b_i/2+g_0/2} dy \int_{-\infty}^{\infty} [B(x, y) + \mu_0 H_r(x, y)]^2 \, dx \qquad (35)$$

On the other hand, the drag force, F_d, may be determined from secondary Joule losses:

$$F_d = \frac{1}{\sigma_i U} \int_{-c}^{c} \int_{-\infty}^{\infty} (J_x^2 + J_y^2) d_{i2} \, dx \, dy \qquad (36)$$

If the integral is limited to the interval 0 to L, F_d becomes

$$F_d = \frac{d_{i2}}{\sigma_i U}\left(\frac{g_e}{d_{i2}}\right)^2\left[\gamma_2^2 + \left(\frac{\pi}{2c}\right)^2\right] B_1^2 K_y^2 c\,\frac{e^{\gamma_2 L} - 1}{2\gamma_2 \mu_0^2} \qquad (37)$$

where

$$J_x d_{i2} = g_e \frac{\partial H_r(x, y)}{\partial y}; \quad J_y d_{i2} = g_e \frac{\partial H_r(x, y)}{\partial x} \qquad (38)$$

In (21) the airgap, g_e, was kept constant for $x > L$, but in reality here the field attenuates rapidly, and perhaps a higher value, $g_e' = (2g_e$ to $4g_e$, would yield better results. In fact a precise distribution of magnetic field behind the magnet $x > L$ could be obtained only by numerical methods. For long magnets the fast attenuation of magnetic field, for $x > L$, makes it possible to neglect the end effect in this zone, at least for preliminary design purposes.

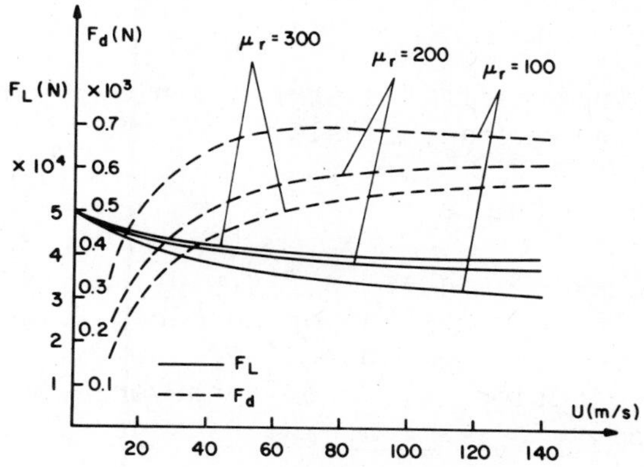

Fig. 10.8. Levitation and drag forces, against speed.

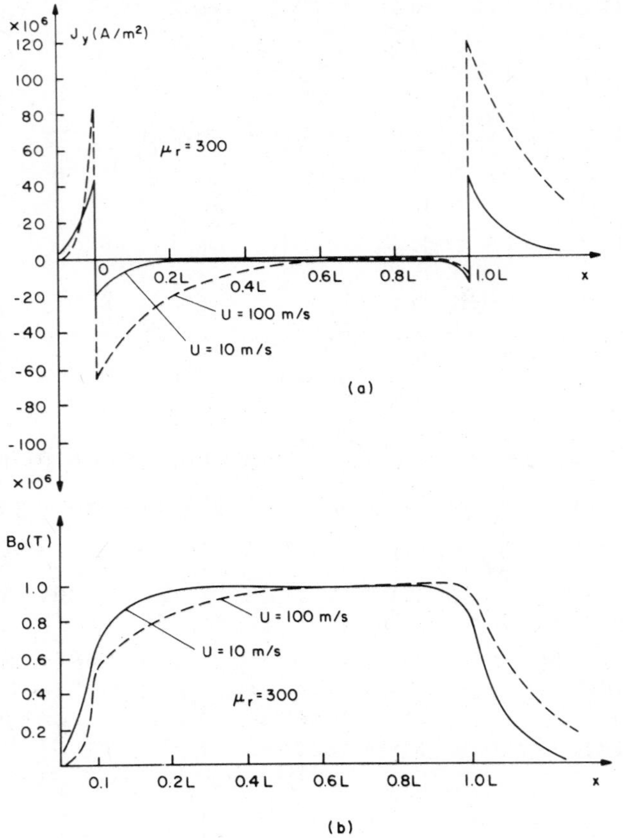

Fig. 10.9. Longitudinal distribution of flux density and current density.

<u>Numerical example</u>

For the same data as in Fig. 10.7 and for a magnet length L = 1 m, the levitation and drag forces were calculated as functions of speed (Fig. 10.8). The longitudinal distribution of the transverse component of current density, J_y, and airgap flux density have been calculated and plotted in Fig. 10.9 for two different speeds. It is remarkable (Fig. 10.8) that end effect reduces the levitation force by 25 to 35% at high speeds

(90 to 100 m/sec). Thus, to keep the levitation force constant, an appreciable increase in the magnet current is necessary.

The drag force is very small as compared to levitation force but represents about 5 to 10% reduction in the propulsion force required during vehicle starting at an acceleration $a = 1\ m/sec^2$. Consequently the drag force must be considered in the propulsion system of the vehicle.

In Sections 10.2 through 10.3, end effect has already been considered in the optimal LEL design. It is understood that for standstill or low-speed applications the end effect does not occur and the optimal design of LEL is greatly simplified.

10.5. Control and Dynamics of LELs

The LEL dynamics could best be investigated by studying the magnet-guideway subsystem with one degree of freedom. The dynamics of electromagnetically levitated vehicles involve five degrees of freedom. However, it has been demonstrated,[16] in theory and practice, that by using adequate secondary and tertiary passive mechanical suspension systems, each LEL could be controlled separately. Standstill applications of LEL

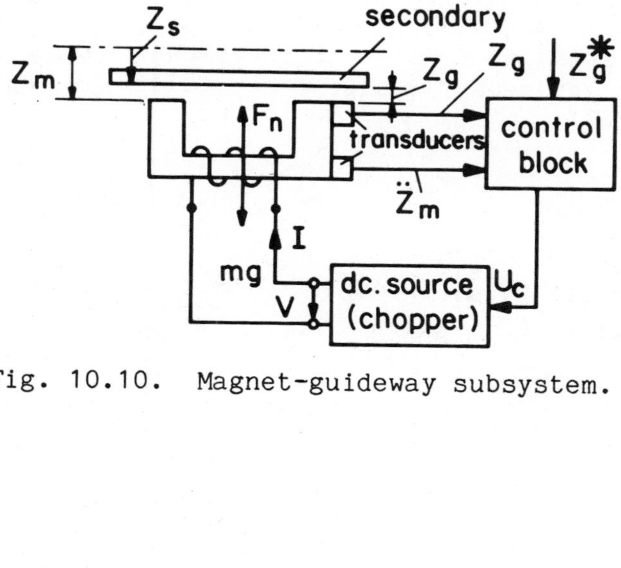

Fig. 10.10. Magnet-guideway subsystem.

also allow a separation of levitation and guidance functions, thus allowing separate LEL control as well.

10.5.1. One Magnet-Guideway Subsystem

The automatic control of the magnet-guideway subsystem (Fig. 10.10) must maintain the airgap, Z_g, as close as possible to its rated (imposed) value, Z_g^*, in the presence of the following perturbations: F_{ext} (external force) and Z_s (guideway irregularities) such that a reasonable degree of ride comfort is provided at a reasonable energy consumption.

After a careful study of various controlled systems proposed so far,[5,7,16,18] we prefer, for both generality and overall quality, a state control system whose state measured variables are: the airgap, Z_g, and the absolute normal acceleration of LEL $\ddot{Z}_m$. A transistorized chopper is used as a power supply. The chopper voltage, V_c, is made a linear function of these variables and LEL speed, through a state observer, such that the airgap, Z_g, is kept, through dynamic equilibrium, close to its rated (imposed) value, Z_g^*.

10.5.2. Basic Equations

The attraction force, F_n, between the LEL (magnet) and the guideway yields the following approximate expression:

$$F_n = \alpha(Z_g, U)\left(\frac{i}{Z_g}\right)^2 \tag{39}$$

The coefficient, α, increases with Z_g, according to (8), and decreases with speed because of the end effect (35).

If these two aspects are neglected:

$$\alpha_0 = \mu_0 W_f^2 b_i L \tag{40}$$

However, if we consider them, the levitation force decreases linearly with speed (Fig. 10.8):

$$\alpha(Z_g, U) \simeq \alpha_0 (1 + \frac{Z_g}{b_i}) (1 - \beta_r U) \tag{41}$$

The coefficient β_r could be obtained from (35), or Fig. 10.8, and is zero for standstill applications. For simplicity, in (41) Z_g will be replaced by Z_{g0}.

The circuit equation of the magnet is

$$V_f = Ri + \frac{d\psi}{dt}; \quad \text{where } \psi = Li \text{ and } L = L_\sigma + \frac{L_p}{Z_g} \tag{42}$$

L_σ being the leakage reactance considered independent of airgap.

The equation of motion becomes

$$m\ddot{Z}_m = mg + F_{ext} - F_n \tag{43}$$

From (39), (42), and (43), it becomes clear that the magnet-guideway system is not linear. Nevertheless, because the changes in the airgap during automatic control are rather small, the above system of equations may be linearized around the rated airgap, Z_{g0}, and the current, I_n.

Let us denote by x_1 and i_1 the deviations in the airgap and current, respectively, from their rated values.

By linearizing (39) we obtain

$$F_n = \alpha_1(i_1 - \beta_1 x_1);$$

$$\text{where } \beta_1 = \frac{I_n}{Z_{g0}} \text{ and } \alpha_1 = 2\alpha \frac{I_n}{Z_{g0}^2} \tag{44}$$

Except for airgap and speed, α_1 and β_1 depend on the eddy currents induced in the guideway during the control process.[8] However, for simplicity this effect is neglected here.

Recombining (42) through (44) we get

$$V_e = Ri_1 + L_\sigma \frac{di_1}{dt} - (L_0 - L_\sigma)\, \beta_1 \frac{dx_1}{dt};$$

$$\text{where } L_0 = L(Z_{g0}) \tag{45}$$

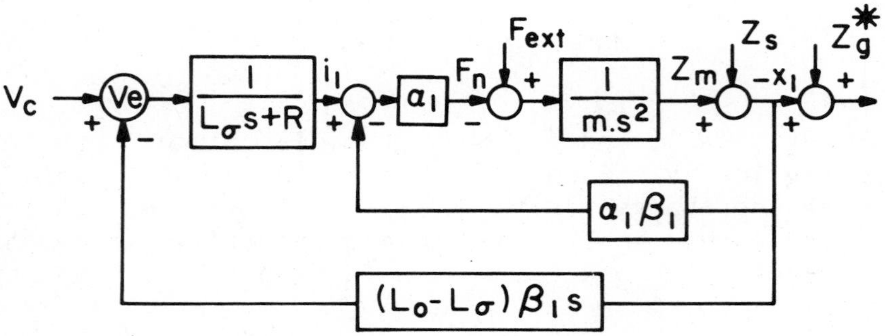

Fig. 10.11. Block diagram for small perturbations.

Thus, the block diagram of the magnet-guideway subsystems, for small perturbations, is as shown in Fig. 10.11. The transfer function between the magnet input voltage, V_e, and the airgap deviation, X_1, is

$$\frac{X_1(s)}{V_e(s)} = \frac{-\alpha_1/mL_0}{s^3 + \frac{Rs^2}{L_0} - \frac{L_\sigma}{L_0}\frac{\alpha_1\beta_1 s}{m} - \frac{R\alpha_1\beta_1}{mL_0}} \quad (46)$$

Based on a stability criterion it may be shown that the magnet-guideway subsystem is unstable. Hence, the use of a closed-loop control system becomes necessary.

10.5.3. Optimal Control System Design Procedure[9]

Equations (44) and (45) may be written in matrix form as

$$\dot{\underline{X}}_1 = \underline{A}\underline{X}_1 + \underline{B}\underline{Y} + \underline{P}_1\underline{Z}; \quad \underline{Y} = \underline{C}\underline{X}_1 + \underline{P}_2\underline{Z} \quad (47)$$

where $\underline{X}_1$ = the state vector,

$$\underline{X}_1 = [Z_g \dot{Z}_g \ddot{Z}_m]^T \quad (48)$$

$\underline{Z}$ = perturbation vector,

$$\underline{Z} = [F_{ext} \dot{F}_{ext} \ddot{Z}_s]^T \quad (48')$$

$\underline{Y}$ = output variables vector,

$$\underline{Y} = [Z_g \ddot{Z}_m]^T \quad (49)$$

$\ddot{Z}_s$ = the vertical acceleration of the guideway position, and $\underline{A}$, $\underline{B}$, $\underline{C}$, $\underline{P}_1$, $\underline{P}_2$ are

$$\underline{A} = \begin{bmatrix} 0 & 1 & 0 \\ 0 & 0 & 1 \\ \frac{C_g}{m} & \frac{(C_g - C_I K'_g / K_i)}{m} & \frac{-1}{T} \end{bmatrix} ; \quad \underline{B} = \begin{bmatrix} 0 \\ 0 \\ \frac{-C_I}{mk_i} \end{bmatrix} ; \quad \underline{C} = \begin{bmatrix} 1 & 0 & 0 \\ 0 & 0 & 1 \end{bmatrix}$$

$$\underline{P}_1 = \begin{bmatrix} 0 & 0 & 0 \\ 0 & 0 & -1 \\ \frac{1}{mT} & 1/m & 0 \end{bmatrix} ; \quad \underline{P}_2 = 0 \tag{50}$$

where

$$C_g = -\frac{\partial F_n}{\partial Z_g} = \alpha_1 \beta_1 ; \quad C_I = \frac{\partial F_n}{\partial I} = \alpha_1 \tag{51}$$

$$K_I = \frac{\partial \psi}{\partial I} = L_0 ; \quad K_g' = \frac{\partial \psi}{\partial Z_g} ; \quad T = \frac{K_I}{R} ; \quad \psi = L_0 I$$

All these coefficients are determined for Z_{g0} and I_n. The usual perturbations consist of external forces (caused generally by wind), F_{ext}, and the guideway irregularities Z_s, $\ddot{Z}_s$.

There are a number of approaches to the algorithmic design of an automatic control system. Here, we adopt a state control system optimized according to an integral squared criterion given by

$$I(V_e) = \frac{1}{2} \int_0^\infty [q_g^2 Z_g^2(t) + q_a^2 \ddot{Z}_m^2(t) + V_e^2(t)] \, dt \tag{52}$$

Minimizing the integral (52) a control law is obtained:

$$V_e = K_g Z_g + K_u \dot{Z}_g + K_a \ddot{Z}_m = \underline{K}^T \underline{X}_1 \quad (53)$$

where $\underline{K}^T$ is the transposed matrix of compensator vector:

$$\underline{K}^T = [K_g K_u K_a]^T \quad (54)$$

The compensator components K_g, K_u, and K_a depend on q_g and q_a determined from the minimization of integral (52), while also observing the control system saturation limits. The squared criterion provides a weighted lumped optimum. The first term of integral (52) aims at a reduction in the airgap deviations in order to avoid magnet-guideway collisions. The second term leads to a reduction in the absolute acceleration of the magnet, resulting in an acceptable ride comfort. Finally, the third term sets a limit on V_e, thus reducing the control power.

The compensator vector is obtained from the equation:

$$\underline{K}^T = - R^{-1} B^T \underline{P} = -b'[P_{13} P_{23} P_{33}]^T \quad (55)$$

where

$$\underline{R} = [1]; \quad b' = - \frac{C_I}{mK_I} \quad \text{and} \quad \underline{P} = \begin{bmatrix} p_{11} & p_{12} & p_{13} \\ p_{12} & p_{22} & p_{23} \\ p_{13} & p_{23} & p_{33} \end{bmatrix}$$

The matrix $\underline{P}$ is the solution of Riccati equation:

$$\underline{PA} + \underline{A}^T\underline{P} - \underline{PBB}^T\underline{P} = \underline{-Q}; \quad \text{where } \underline{Q} = \text{diag}\,[q_g^2,\ 0,\ q_a^2] \tag{56}$$

Consequently, when $\underline{Q}$ is known, the compensator, K^T, and the control law could be determined from (56).

To determine $\underline{Q}$ we proceed as follows. The matrix equation (56) is decomposed into six equations with the unknowns (p_{11}, p_{12}, p_{13}, p_{22}, p_{23}, p_{33}). By using (55) and by eliminating p_{12} an equivalent system of three equations with three unknowns is finally obtained:

$$K_g^2 + \frac{2a_1}{b} K_g = q_g^2$$

$$K_a^2 + \frac{2a_3}{b'} K_a + \frac{2}{b'} K_u = q_a^2 \tag{57}$$

$$-\frac{2a_3}{b'} K_g - \frac{2a_1}{b'} K_a - 2K_gK_a + \frac{2a_2}{b'} K_u + K_u^2 = 0$$

where

$$a_1 = \frac{C_g}{mT}; \quad a_2 = \frac{C_gT_d}{mT}; \quad a_3 = -\frac{1}{T}; \quad T_d = T - \frac{C_IK_g'}{C_gR}$$

By applying any of the algebraic stability criteria two more conditions are obtained. Hence

$$K_g \geqq 2K_{g0}$$

$$(K_a + K_{a0})(K_u - K_{u0}) + \frac{(K_g - K_{g0})}{b'} \geq 0 \qquad (58)$$

where

$$K_{g0} = -\frac{C_g}{mTb'}; \quad K_{u0} = \frac{C_g T_d}{mTb'}; \quad K_{a0} = -\frac{1}{b'T}$$

The limits on the amplifiers lead to new restrictions:

$$K_g \leq K_{gmax}; \quad K_u \leq K_{umax}; \quad K_a \leq K_{amax} \qquad (59)$$

The constants K_{gmax}, K_{umax}, and K_{amax} are determined by accounting for the performances of the primary transducers and measuring bridges performance, the saturation levels of the amplifiers, and the limits introduced by the chopper that supplies the magnet with electric power.

The values of q_g and q_a for which the real solution of system (57) fulfills conditions (58) represent a domain qualitatively shown in Fig. 10.12, for two dif-

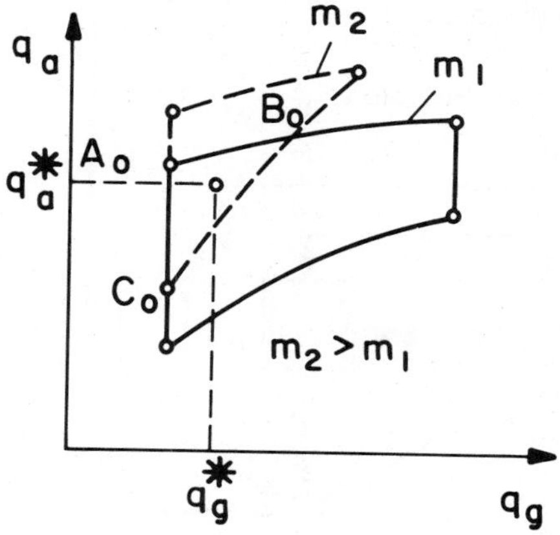

Fig. 10.12. The domain of q_g and q_a.

ferent values of m. For high values of m the domains tend to become triangular in Fig. 10.12.

For a given Maglev the minimum mass, m_1 (without passengers), and the maximum, m_2 (with passengers), are known. Thus the triangle $A_0B_0C_0$ can be determined. For a point inside $A_0B_0C_0$, for any m ($m_1 \leqq m \leqq m_2$), the system is stable and the optimal control problem has a solution. By choosing a pair of values, q_g and q_a, inside $A_0B_0C_0$ we can now calculate K_g, K_u, K_a from (57). Finally, if the dynamic performance of the automatic control system proves to be unsatisfactory, the compensator coefficients may be corrected by imposing new restrictions such as gain response to airgap step change.

It should, however, be kept in mind that the central zone of $A_0B_0C_0$ (Fig. 10.12) acceptably satisfies the optimum criterion as well as the stability reserve, having a sufficient provision for the trial and error tuning of the control system blocks.

10.5.4. Control Schemes

Although the control law is by now determined, its transposition into a block scheme is not yet possible because the state variable, $\dot{Z}_g$, is not directly measur-

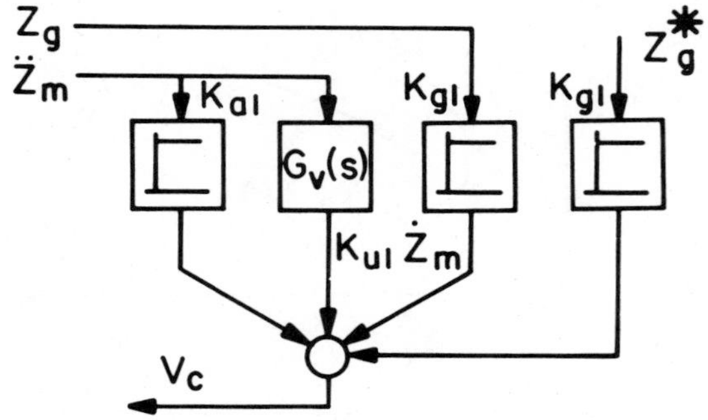

Fig. 10.13. Typical control block.

able. In References 7 and 10 two schemes have been presented to circumvent this difficulty. Basically, they can be reduced to the scheme presented in Fig. 10.13, where K_s represents the transfer coefficient of the magnet chopper. (The chopper is considered as a noninertial element.)

$$K_{g1} = \frac{K_g}{K_s}$$

$$K_{u1} = \frac{K_u}{K_s}$$

$$K_{a1} = \frac{K_a}{K_s} \tag{60}$$

In Fig. 10.13 $G_v(s)$ represents first an ideal integrator:

$$G_v(s) = \frac{K_{ul}}{s} \tag{61}$$

and the second alternative refers to a first-order transfer element with a high time constant, τ_I,

$$G_v(s) = \frac{K_{u1}\tau_I}{\tau_I s + 1} \tag{62}$$

One drawback of the first solution, (61), is the fact that the system lies, theoretically, at the margin of stability, having the following characteristic equation:

$$s|s\underline{I} - \underline{A} - \underline{BK}^T| = 0 \tag{63}$$

In the second case, (62), the system is stable but has a small phase reserve. However, the principal deficiency of these solutions occurs in the system behavior with respect to the guideway irregularities. This fact is predictable because the acceleration $\ddot{Z}_g$ is replaced by $\ddot{Z}_m$ and thus false controls are possible.

As an example, for the first solution, when $\dot{Z}_m = \int_0^t \ddot{Z}_m(t)\, dt$, the following difference occurs between $\dot{Z}_g$ and $\dot{Z}_m$, for zero initial conditions,

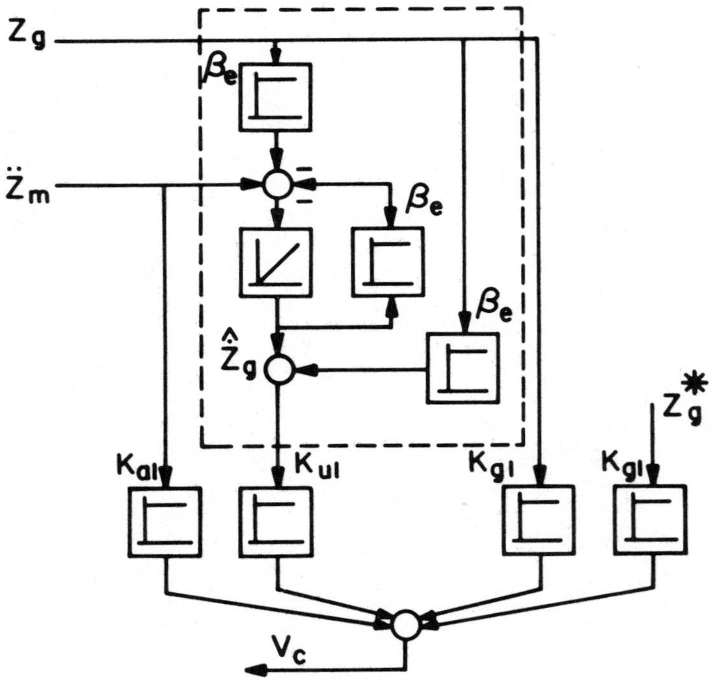

Fig. 10.14. Control scheme with state observer.

$$\dot{Z}_g(t) - \dot{Z}_m(t) = - \int_0^t U_{-1}(t - \tau)\ddot{Z}_s(\tau)\, d\tau \qquad (64)$$

where $U_{-1}(t)$ is the step unity function.

To eliminate this disadvantage, in References 9 and 16 a few control schemes with state observers have been proposed. One of them is presented in Fig. 10.14. In these schemes the state variable $\dot{Z}_g$ is estimated, by means of a low-order Leunberger-type state observer, as $\hat{Z}_g$. The simple state observer implies only the choice of s_{0b} pole:

$$s_{0b} = -\beta_0$$

because

$$\hat{Z}_g(s) = \frac{\ddot{Z}_m(s)}{s + \beta_0} + \frac{s\beta_0}{s + \beta_0}\dot{Z}_g(s) \qquad (65)$$

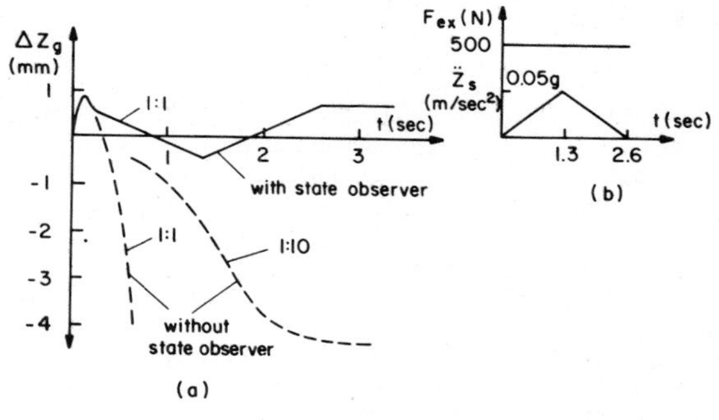

Fig. 10.15. System response at perturbation.

In (64) false controls occur when $\ddot{Z}_s \neq 0$. This situation occurs frequently in practice when ascending a hill or for guideway deformations caused by an attraction force, or by the weight of the vehicle.

The response of a control system,[9] with and without an state observer, to a strong perturbation caused by an external force, F_{ext}, and an acceleration, $\ddot{Z}_s$ (Fig. 10.15), shows the necessity of using a state observer. In fact, in the absence of the state observer the great changes in the airgap, and consequently in the control voltage of the magnet, render the system unpracticable.

The principal data of the example given in Fig. 10.15 are: $m = 200$ kg, $Z_{g0} = 1 \times 10^{-2}$ m, $T = 0.053$ sec, $R = 1.532\ \Omega$, $K_I = 8.21 \times 10^{-2}$ Hz, $C_I = 236.35\ \mathrm{NA}^{-1}$, $C_g = 9.6 \times 10^5 (\mathrm{Nm})^{-1}$. The controlled LEL dynamic behavior is now explored.

10.5.5. Automatic Control System Performance

The control system performance computation is based upon its transfer functions: $G_1(s)$, between guideway position, $Z_s(s)$, and airgap deviation, $X_1(s)$, is

$$G_1(s) = \frac{X_1(s)}{Z_s(s)} = \frac{-[s^3 + (\frac{R}{L_0} + \frac{\alpha_1 K_s K_a}{mL_0})s^2 + \frac{K_s \alpha_1 K_u}{L_0 m} s]}{D(s)} \tag{66}$$

$G_2(s)$, between perturbation force, F_{ext}, and airgap change, $X_1(s)$, is

$$G_2(s) = \frac{X_1(s)}{F_{ext}(s)} = \frac{\frac{1}{m}(s + R/L_0)}{D(s)} \tag{67}$$

$G_3(s)$, between controlled voltage, $V_e(s)$, and the air-gap, $Z_g(s)$, is

$$G_3(s) = \frac{Z_g(s)}{V_e(s)} = \frac{-\alpha_1}{mL_0 D(s)} \tag{68}$$

The polynomial D(s) is given by

$$D(s) = s^3 + \left(\frac{R}{L_0} + \frac{\alpha_1 K_s K_a}{mL_0}\right) s^2 + \left(\frac{\alpha_1 K_s K_u}{mL_0} - \frac{\alpha_1 \beta_1 L_o}{mL_0}\right) s$$

$$+ \frac{\alpha_1 K_s K_g}{mL_0} - \frac{\alpha_1 \beta_1 R}{mL_0} \tag{69}$$

These three transfer functions can be used to illustrate the time variation of different parameters, for known perturbations. However, there are only three essential practical aspects related to the performance:

1. Magnet-guideway collision avoidance condition.

2. Average control power.

3. Ride comfort.

The guideway irregularities can be expressed by the power spectral density (PSD) of guideway deviations from rectiliniaty; PSD, $\phi_z(\omega)$, may be written as

$$\phi_z(\omega) = \frac{AU}{\omega^2} \tag{70}$$

where U = vehicle speed, ω = angular pulsation, and A = guideway constant.

The squared average value, $\overline{X}_1$, of airgap deviations caused by the guideway irregularities, in the presence of the control system, is

$$\overline{X}_1 = \left[\int_{0.2\pi}^{\omega_0'} |G_1(j\omega)|^2 \, \phi_z(\omega) \, d\omega \right]^{1/2} \tag{71}$$

The quantity $\overline{X}_1$ may be considered as a standard value,[11] and thus the condition of collision avoidance becomes

$$\overline{X}_1 < Z_{g0} \tag{72}$$

<u>Average control power</u>

In order to determine the average control power for a guideway of known irregularities, it is first necessary to calculate the squared average voltage, σ_v, and current, σ_{ii}, deviations:

$$\sigma_v = \left[\int_{0.2\pi}^{\omega_0'} \left| \frac{G_1(j\omega)}{G_3(j\omega)} \right|^2 \phi_z(\omega) \, d\omega \right]^{1/2} \tag{73}$$

$$\sigma_{ii} = \left[\int_{0.2\pi}^{\omega_0'} \left| G_1(j\omega) \cdot \frac{[G_3^{-1}(j\omega) + (L_0 - L_\sigma)\, \alpha_1 j\omega]}{R + j\omega L_0} \right|^2 \phi_z(\omega) \, d\omega \right]^{1/2} \tag{74}$$

Thus the average control power, P_c, is

$$P_c = (V_n + \sigma_v)(I_n + \sigma_{ii}) \tag{75}$$

The knowledge of voltage and current variations in the control process is necessary for designing the magnet-chopper subsystem.

Ride comfort

As already mentioned, the ride comfort may be defined using the power spectral density, $\phi_{\ddot{z}_m}(\omega)$, of accelerations:

$$\phi_{\ddot{z}_m}(\omega) = \omega^4 |G_1(j\omega) + 1|^2 \phi_z(\omega) \tag{76}$$

If $\phi_{\ddot{z}_m}(\omega)$ lies below the Janeway curve for all frequencies, then the ride comfort is considered satisfactory. The higher the vehicle speed and frequency, the more difficult it is to fulfill Janeway conditions. Thus the ride comfort should be checked up to maximum speed.

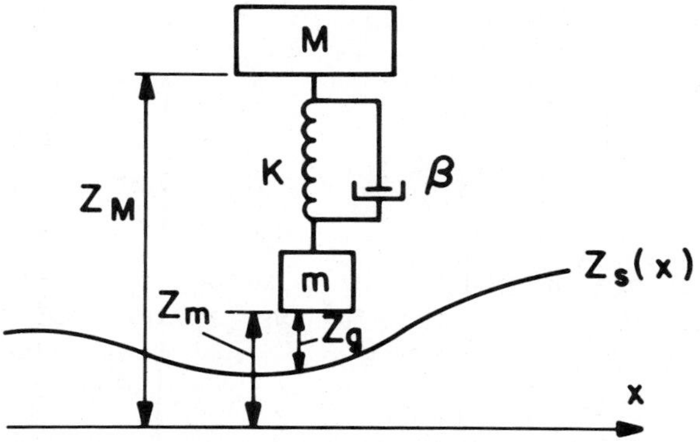

Fig. 10.16. Secondary suspension system.

10.6. Secondary Suspension System

At high speeds, an acceptable ride comfort is not generally achieved. Therefore, a secondary system should be used or the control system should be "forced." In the latter case the average control power increases prohibitively, whereas in the former the weight and cost of secondary suspension system is the necessary burden. Suppose a secondary suspension system is added (Fig. 10.16). In this case there are two equations of motion:

$$MZ_M = Mg - K\ (Z_M - Z_s - Z_g - l_0) - \beta(\dot{Z}_M - \dot{Z}_m) \quad (77)$$

where l_0 is the distance between unspringed mass, m, and springed mass, M, when spring K and damper β are unloaded, and

$$Mg = K(Z_{M0} - Z_{g0} - l_0) \quad (78)$$

where Z_{M0} = rated height of mass M and Z_{g0} = rated air-gap.

For the mass, m, with no spring (77) becomes

$$m\ddot{Z}_m = mg + K(Z_M - Z_s - Z_g - l_0)$$

$$+ \beta(\dot{Z}_M - \dot{Z}_m) - F_n + F_{ext} \quad (79)$$

The problem may be solved in a manner similar to the way as it has been solved in the absence of a secondary suspension system. In this case smaller airgap variations ($\overline{X}_1$ = 0.15 to $0.25Z_{g0}$ may be obtained, and thus, a guideway with higher irregularities (and lower costs) may be followed tightly by an electromagnetic suspension. The ride comfort will be checked now for Z_M, that is, at the secondary suspension system level. For an efficient usage of a secondary suspension system, $m/M < 0.2$ to 0.3, a condition that is generally fulfilled in practice.

10.7. Mechanical Drag Power Caused by Guideway Irregularities

In the process of airgap automatic control, an average drag power, P_m, occurs:

$$P_m = F_n \dot{Z}_s \tag{80}$$

It is now demonstrated that this power must be provided by the propulsion system.

By decomposing the attraction force, F_n, into two components (Fig. 10.17) a drag component, F_{dm}, occurs, which is given by

$$F_{dm} = F_n \tan \alpha' \tag{81}$$

where

$$\tan \alpha' = \frac{dZ_s}{dx} = \frac{1}{U}\frac{dZ_s}{dt} = \frac{1}{U}\dot{Z}_s \tag{82}$$

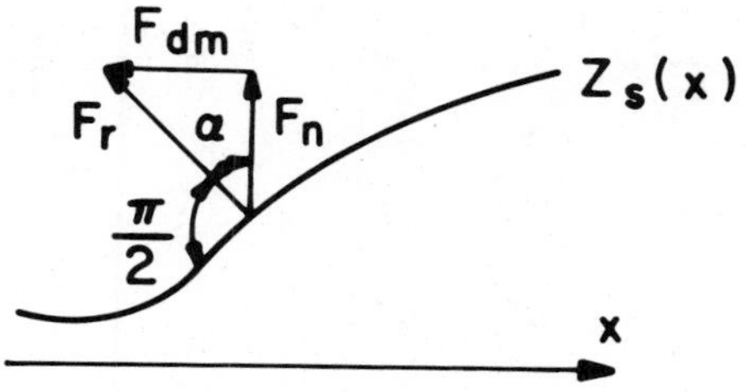

Fig. 10.17. Control drag power.

After the control system design has been terminated this control drag power must be calculated and accounted for in the propulsion system design. However it seems that in most cases P_m = 0.08 to 0.2 kW per tonne of vehicle, and thus the linear motor[10] is not overloaded by very much.

In the following, a numerical example of a 50-tonne Maglev control system is worked out.

10.8. Control System Performance

Suppose a 50-tonne, 100 m/sec Maglev has a rated airgap $Z_{g0} = 1.5 \times 10^{-2}$ m. Following the optimal magnet design procedure (Section 10.2) we finally select magnets of 1.5 m in length and $b_i = 0.04$ m for a current density $J_c = 15 \times 10^6$ A/m^2.

The attraction force per magnet, at 100 m/sec, accounting for end effect, is $F_n = 5 \times 10^4$ N. The required ampturns, magnet rated voltage, and current are

$$W_f I_f = 3.58 \times 10^4; \quad V_n = 200 \text{ V}; \quad I_n = 200 \text{ A}.$$

The drag force per magnet (Section 10.3) $F_d = 7 \times 10^3$ N. There are 10 magnets of 300 kg each (m_e = 300 kg).

The coil resistance R = 1 Ω, and the inductance is

$$L_0 = L + \frac{L_p}{Z_{g0}} = 2.497 \times 10^{-3} + \frac{1.43 \times 10^{-3}}{Z_g};$$

$$L_0 = L(Z_{g0}) = 0.1 \text{ H}$$

From (37), $\alpha(U, Z_{g0})$ is $\alpha(100, 0.015) = 8.39 \times 10^{-4}$ (Ωm). Thus α_1 and β_1, (41), are

$$\alpha_1 = 1492.0 \text{ N/A}; \quad \beta_1 = 13300.0 \text{ A/m}$$

Based on these preliminary calculations the control system is designed. Finally its dynamic performance is explored in some detail.

10.8.1. Levitation Control System Transfer Functions

The mass, m in (41) is the levitated mass since no secondary suspension system is considered. Making use of Section 10.5 the optimal control system is designed. Finally, for

$$q_a = 308.58\text{V} \cdot \text{sec}^2/\text{m}; \quad q_g = 1.65 \times 10^5 \text{V/m}$$

by using (57) we get

$$k_g = 1.8 \times 10^5 \text{ V/m}; \quad K_u = 1.067 \times 10^4 \text{ V} \cdot \text{sec/m};$$

$$K_a = 340.0 \text{ V sec}^2/\text{m}.$$

The three G_s transfer functions are now calculated from (66) through (69):

$$G_1(s) = \frac{-(s^3 + 964.1s^2 + 3.18 \times 10^4 s)}{D(s)}$$

$$G_2(s) = \frac{0.2 \times 10^{-3}s + 2.1 \times 10^{-3}}{D(s)}$$

$$G_3(s) = \frac{-2.98}{D(s)}$$

$$D(s) = s^3 + 964.1s^2 + 3.179 \times 10^4 s + 4.96 \times 10^5$$

10.8.2. Control Performance

For $A = 1.5 \times 10^{-6}$ m (corresponding to a good conventional welded guideway) the mean squared airgap deviation, $\overline{X}_1$ in (71), is

$$\overline{X}_1 = [\int_{0.2\pi}^{\omega_0'} |G_1(j\omega)|^2 \, \phi_z(\omega)d\omega]^{1/2}$$

$$= 4.61 \times 10^3 \text{m} < \frac{Z_{g0}}{3} = 5 \times 10^{-3} \text{ m}$$

$$f_0 = \frac{\omega_0'}{2}\pi; \quad f_0 = 25 \text{ Hz}$$

We may conclude that the Maglev avoids the collision with the guideway for 100 m/sec and, consequently, this is true for smaller speeds also. However, the mean squared deviations of voltage and current are high:

$$\sigma_v = 4692.3 \text{ V}$$

$$\sigma_{ii} = 670.1 \text{ A}$$

Thus the control power is unacceptably high.

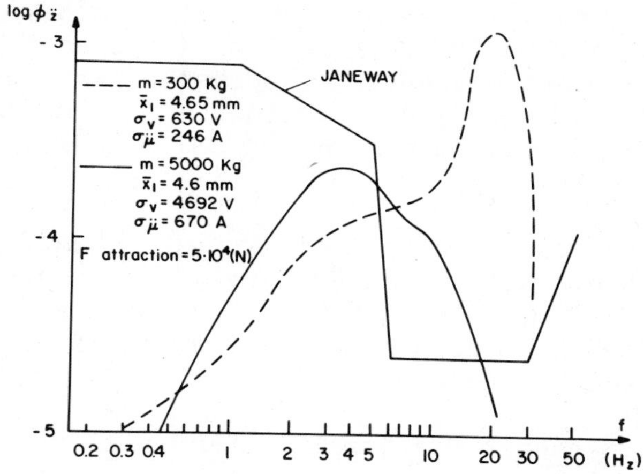

Fig. 10.18. Ride comfort.

If now we consider only the weight of the magnets, m_e, 300 kg (that is, if a secondary suspension system is used), we obtain

$q_a' = 2.73$ V • sec^2/m; $q_g' = 5.83 \times 10^4$ V/m; $K_g' = 7.4 \times 10^4$; $K_u' = 5600.0$ $K_a' = 14.8$; $\overline{X}_1 = 4.65 \times 10^{-3}$ m; $\sigma_V = 630.8$ V; $\sigma_{ii} = 246.8$ A These are acceptable values.

10.8.3. Ride Comfort

The ride comfort is calculated by the function $\phi_{\ddot{z}_m}(\omega)$ as given by (76). It is evident (Fig. 10.18) that, at 100 m/sec, the controlled electromagnetic suspension without secondary suspension system is not capable of providing a satisfactory ride comfort for frequencies in excess of 5 Hz, in spite of the very high control power involved.

On the other hand, if a secondary system is used (m_e = 300 kg instead of 5000 kg), the magnet ride comfort is about the same, but for reasonable control power levels. A passive secondary suspension system (SSS) suffices since, for frequencies smaller than 5 Hz, the controlled electromagnetic suspension (EMS) is able to provide a satisfactory ride comfort. A complete study of the joint action of EMS and SS is, however, beyond our scope.

The transient airgap variation at standstill from "waiting position" to rated airgap in a controlled EMS may be handled by (43) through (45) by means of a numerical procedure since the linearization is not possible for large airgap changes.

10.9. The Magnetic Bearings--A Standstill LEL

For high speeds, conventional bearings for rotating shafts imply higher energy consumption than active magnetic bearings,[28] which are, in fact, a set of separately controlled linear motion levitators. The maximum speed attainable is three to five times higher with active magnetic bearings than with conventional bearings.

10.9.1. Construction Features

An active magnetic bearing is composed of a stator and rotor, sensor coils, a feedback loop and a power amplifier as shown in Fig. 10.19. In order to reduce the eddy-current losses the iron cores of both stator magnets and rotor are made of thin laminations. Also, the four magnets of the stator and the inductive gap sensors may be enclosed in a typical induction machine stator slotted core. Two such stators at each end of the shaft are necessary for suspension and guidance (along the vertical and radial-horizontal directions). Two more LELs are necessary along the axial direction.

In principle, the design aspects of the magnet, transducers power amplifiers and the control loops are similar to those encountered in LEL for Maglev vehicles. However, there is no such problem as ride comfort here.

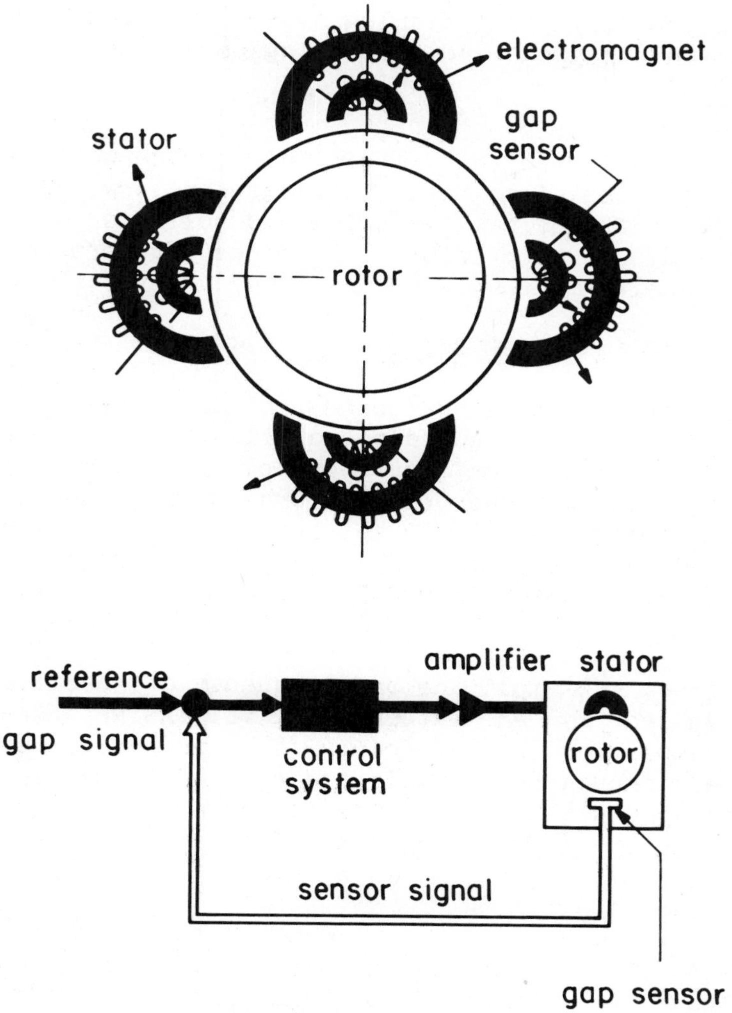

Fig. 10.19. Magnetic bearing.

Consequently, the control loop implies only the displacement and its time derivative (as damping component) as input signals for the compensator that controls the power amplifier voltage. Because of the fact that at very high speeds the time allowed is in the order of milliseconds (50 to 500 Hz band span), only power transistor choppers could be used. The potential assets of active magnetic bearing have prompted research efforts in the area and promising results have been obtained.[28-30]

It is expected that within this decade we will see industrial use of magnetic bearings for very high-speed rotating shafts of many grinding and polishing machines, vacuum pumps, compressors, turbines, generators, and centrifuges.

10.10. A Note on LEL Oscillators

Vibrating tables are used in industry for many applications. For sustained oscillations whose amplitude is less than 5 mm and frequency less than 50 Hz (for transistor power amplifiers) up to 500 Hz (for thyristor amplifiers), LEL may be successfully used. As discussed earlier, the system should be stabilized as for the LEL for Maglev vehicles, but there are two major differences.

First, the amplitude and frequency of oscillations are imposed by an external ac voltage acting on the power amplifier gating circuits. Secondly, special measures should be taken in the control system in order to provide a high degree of fidelity for a minimum distortion in the input voltage oscillations.

10.11. Conclusion

The LEL construction, theory, and design fundamentals have been presented for three main types of applications: magnetically levitated vehicles, magnetic bearings, and vibrating tables. There are, however, many aspects not treated here and the interested reader is urged to consult the references cited at the end of this chapter. The very promising results obtained in applying the LEL to vehicles and magnetic bearings of fast heavy shafts should prompt more in-depth investigations in the field in the future.

References

1. H. Kemper "Suspension systems through electromagnetic forces, A possible approach to basically new transportation technologies" (in German), ETZ, No. 4, 1933, pp. 391-395.

2. H. Kemper "Electrical railroad vehicles magnetically guided" (in German), ETZ-A, No. 1, 1953, pp. 11-14.

3. P. N. Nave, "Maglev test facilities at M.B.B. Munich," High Speed Ground Transp. J., Vol. 8, No. 3, 1974.

4. G. Bohn, P. Romstedt, W. Rothmayer, and P. Schwarzler "A contribution to magnetic levitation technology," Proceedings of the 4th ICEC, Eindhoven, 1972, pp. 202-208.

5. R. H. Bocherts and L. C. Davis, "Lift and drag forces for the attraction electromagnetic suspension systems," Ford Motor Company, Sci. Res. Staff, 1976.

6. S. Yamamura and T. Ito, "Analysis of speed characteristics of the attractive electromagnet for the magnetic levitation vehicles," Elect. Eng. Jap., Vol. 95, No. 162, 1975, pp. 84-89.

7. I. Boldea and S. A. Nasar, "Field winding drag and normal forces of linear synchronous homopolar motors," Int. Quat. Elect. Mach. Electromechan., Vol. 2, No. 3, 1978, pp. 253-268.

8. P. Appun and H. J. Von Thun, "An electromagnetic levitation and guidance system for guided high speed vehicles" (in German), Elektr. Bahnen, Vol. 46, No. 4, 1975, pp. 861-864.

9. F. Matsumura and S. Yamada, "A method to control the suspension system utilizing magnetic attraction force," Elec. Eng. in Jap., Vol. 94, No. 6, 1974, pp. 50-57.

10. S. Yamamura, "Performance analysis of electromagnetically levitated vehicle," World Electrotechnical Congress, Moscow, June 20-25, 1977.

11. G. Bohn, "Calculation of frequency responses of electromagnetic levitation magnets," IEEE Trans., Vol. MAG-13, No. 5, 1977, pp. 1412-1414.

12. T. Dragomir and D. Ciortuz, "Automatic control of electromagnetic levitation test facility" (in Romanian), Presented at IPA Conference, Bucharest, Romania, 1978.

13. J. R. Reitz, "Preliminary design studies of magnetic suspension for high speed ground transportation," Rep. FRA-RT-73-27, 1973, Washington, DC, 20590, pp. 103-127.

14. B. V. Jayawant et al., "Development of a 1-ton magnetically suspended vehicle using controlled d.c. electromagnets," Proc. IEE, Vol. 123, 1976, pp. 941-948.

15. B. V. Jayawant and P. K. Sinha, "Low-speed vehicle dynamics and ride quality using controlled d.c. electromagnets," Automatica, Vol. 13, 1977, pp. 605-610.

16. Y. Jizo, T. Yamada and M. Ivamoto, "Analysis of lift and drag forces in electromagnetic levitation systems," Elec. Eng. Jap., Vol. 97, No. 2, 1977, pp. 94-100.

17. A. Shunichi, "3 dimensional magnetic field calculation of the levitation magnet for HSST by the finite element method," IEEE Trans., Vol. MAG-16, No. 5, 1980, pp. 725-727.

18. I. Boldea, "Optimal design of attraction levitation magnets including the end effect," EME, Vol. 6, 1981, pp. 57-66.

19. F. Matsumura and S. Tachimori, "Magnetic suspension system suitable for wide range operation," El. Eng., Vol. 99, No. 1, 1979, pp. 29-35.

20. E. Gottzein and W. Cramer, "Critical evaluation of multivariable control techniques based on Maglev vehicle design," Proceedings of 4th Symposium on Multivariable Technological Systems, IFAC 77.

21. T. Hoshino, N. Sato, Y. Hayashi, and Y. Ogura, "An application of the observer to the attractive-type magnetic levitation," Elec. Eng. in Jap., Vol. 99, No. 4, 1979, pp. 107-115.

22. Y. Hayashi, N. Sato, Y. Ogura and T. Hoshimo, "Effects of exciting frequency on the characteristics of attractive-type magnetics levitation," Elec. Eng. Jap., Vol. 99, No. 3, 1979, pp. 66-73.

23. E. Gottzein and B. Lange, "Magnetic suspension control systems for the MBB high speed train," 5th IFAC Symposium, Genoa, Italy, 1973.

24. H. Alscher and H. G. Raschbichler, "A demonstration system for magnetic levitation technology at the International Transportation Exhibition (IVA), 1979" (in German), Eisenbahn Technische Rundsch., No. 4, 1979, pp. 281-292.

25. S. Nakamura and A. Hayashi, "Development of HSST system," presented at International Transportation Exhibition (IVA), Hamburg, West Germany, 1979.

26. V. D. Nagoysk (Editor), "Problems of magnetic levitation equipment for high speed ground transportation" (in Russian), brochure, printed by the Moskow Institute of Railroad Transportation Engineers, 1977, 115 pages.

27. B. V. Jayawant, "Passenger carrying vehicles using controlled d.c. electromagnets," L 7/2, Proc. International Conference on Electric Machines, Brussels, 1977.

28. H. Haberman and G. L. Liard, "Practical magnetic bearings," IEEE Spectrum, No. 9, 1979, pp. 26-30.

29. P. L. Geary, Magnetic and Electric Suspensions, British Sci. Instr. Res. Assoc., 1964, London.

30. R. H. Frazier, P. I. Gilinson, and G. A. Oberbeck, Magnetic and Electric Suspensions, MIT Press, Cambridge, MA, 1974.

CHAPTER 11

LINEAR MOTION ELECTRODYNAMIC LEVITATORS

11.1. Introduction

Among the electrodynamic levitation systems[1] (EDS), the normal flux type with aluminum sheet or ladder secondary is considered most suited for high-speed vehicles (Fig. 11.1). (The secondary made of short-circuited coils is a particular case of the ladder secondary.) Two criteria are defined here to evaluate the quality of an EDS. They are: levitation goodness factor, G_e, and levitation rigidity, S_1.

From the multitude of problems related to EDS, the following are considered important and, consequently, are treated in this chapter:

1. Definition of G_e and S_1.

2. Quasi-static performance calculations.

3. Study of dynamic behavior (vertical oscillations attenuation).

11.2. Goodness Factor and Rigidity

The principle of operation of EDS consists of the interaction between the superconducting (SC) coils placed on a vehicle and the currents induced through the vehicle motion in a conducting secondary placed along the guideway. Besides the levitation force, F_L, a drag force, F_d, is also produced (Fig. 11.2). The SC coils produce a flux density of about 0.3 to 0.4 T at the secondary surface. The weight of the SC coils is at least five times smaller than their levitation force. The SC coils are placed in a dewar kept at very low temperatures, 4 to 15 K.

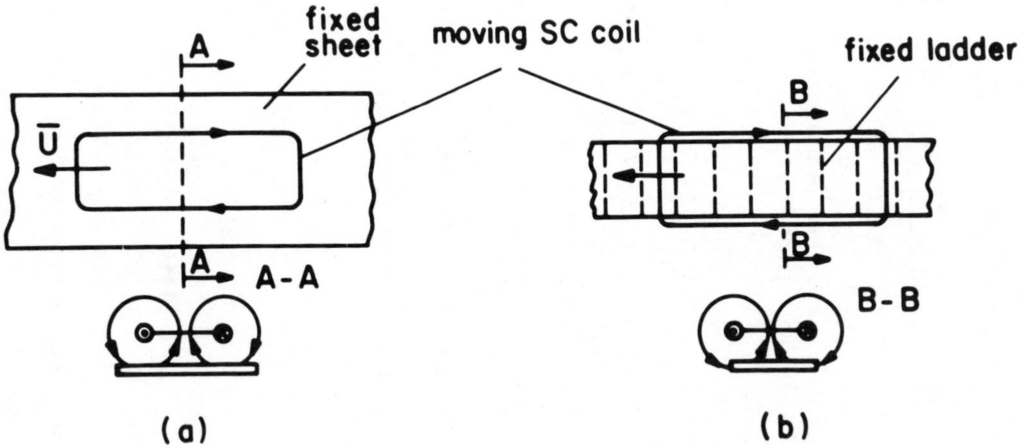

Fig. 11.1. Normal flux EDS. (a) With sheet secondary, (b) with ladder secondary.

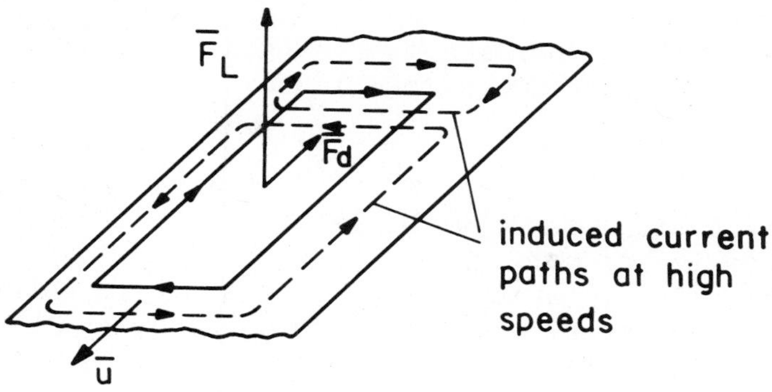

Fig. 11.2. EDS principle.

Under these conditions the Nb_3-Ti or Nb_3-Sn conductors of the SC coils, made of very thin transposed filaments and embedded in a copper matrix, lose drastically their resistivity, thus becoming superconductors. For a standard 50 tonne vehicle the losses in the levitation SC coils are less than 100 W. However, a special refrigeration system is necessary to maintain the superconducting state. As a cooling medium, liquid helium in the supercritical state is used. Advanced liquid helium refrigerators require only about 250 W input power for a watt of SC loss removed from the dewar.

If we denote by P_1 the SC refrigerator input power, and by P_c the levitation control power, then the levitation goodness factor, G_e, is defined by

$$G_e = \frac{F_L}{F_d + (P_1 + P_c)/U} \qquad (1)$$

where U is the vehicle speed.

At the rated speed the levitation force, F_L, "cancels" the vehicle weight. The greater the G_e, the higher is the overall efficiency. However, in the EDS design, the minimum SC coil weight, for a given levitation force, should also have to be taken into consideration. The minimum weight criterion does not necessarily result in maximum levitation goodness factor.

Also of great importance is the rate of increase of the levitation force with speed, since at small speeds the SC coils are not generally able to levitate the vehicle. Thus, a system of aircraft-type retractable wheels is necessary. On the other hand, the dynamic behavior of EDS is greatly affected by its rigidity, S_1:

$$S_1 = \frac{-(\partial F_L/\partial_z)}{F_L} \qquad (2)$$

The higher the S_1, the more rigid the system is; that is, the guideway irregularities are tightly followed and thus the ride comfort gets poorer. Since the levitation force increases when the vertical distance, z, between the vehicle and the guideway decreases, S_1 is positive and, consequently, the EDS system is statically stable. Finally, a high goodness factor, G_e, results in a high rigidity, S_1, and thus, a global compromise (optimum) proves necessary. Even from these introductory remarks, the complexity of the EDS optimal design is evident.

11.3. SC Coils--Construction Guidelines

The structure of a typical levitation SC coil is presented in Fig. 11.3. In essence, an SC coil system consists of a superconducting coil, liquid helium dewar, fiberglass evacuated isolation, supports, and electrical connections. A vacuum zone enclosed by nonmagnetic steel walls thermally isolates the helium dewar from the environment.

The helium dewar and its SC coils are gradually brought to very low temperatures, 4.2 to 15 K, using,

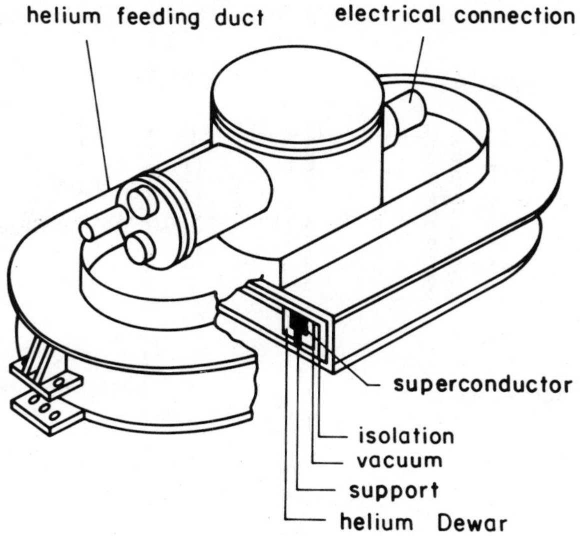

Fig. 11.3. Levitation SC coil.

first, liquid nitrogen. After the liquid nitrogen has evaporated, liquid helium is poured until the desired temperature is obtained. Only after this operation is the electric current gradually "introduced" in the SC coil which is finally short-circuited.[2]

To reduce the weight of SC coils to about 10 to 15% of their levitation force, high equivalent current densities (300 to 400 A/mm^2) are to be obtained by using intrinsically stabilized superconductors impregnated in epoxy resins.[3] In principle, the electric scheme of an SC coil appears as shown in Fig. 11.4. To reduce the losses, a thermal switchgear is mounted right in the vacuum zone. In order to reduce the refrigerator weight on the vehicle it is adequate to introduce liquid helium in excess in the dewar so that no refilling, for one day service, is necessary. Special measures should be taken in case of different malfunctions.

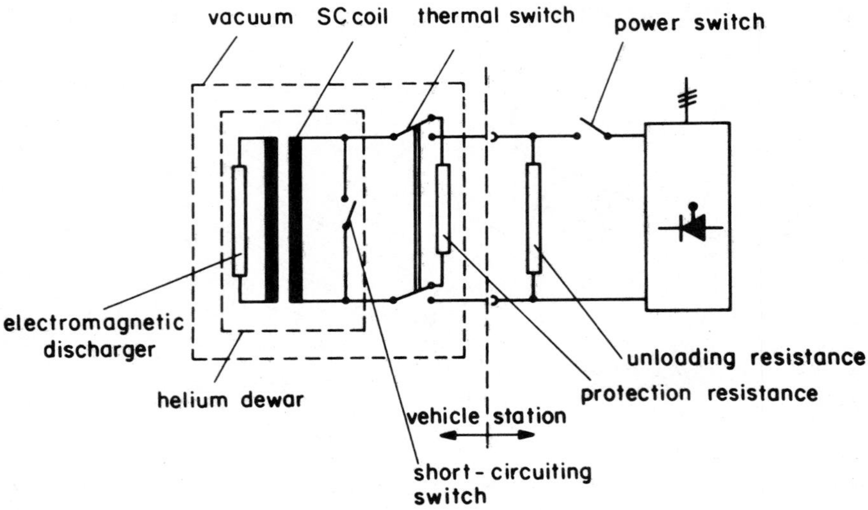

Fig. 11.4. Electric scheme.

11.4. Study of Sheet-Secondary EDS

The EDS with a sheet secondary is easier and less costly to manufacture than the ladder-secondary EDS. A precise study of the sheet-secondary EDS performance is difficult, mainly because of the continuous structure of the secondary. Though the sheet-secondary EDS with double magnets[4] has been proposed for integrated levitation and guidance, it would require very powerful SC coils and manifest high drag forces because of its inherently low sheet width. See Fig. 11.5. Moreover, an increase in the guidance force would be accompanied by a reduction of levitation force. Additional vehicle oscillations would thus be produced. For these strong reasons we believe that levitation and guidance should be managed through separate SC coils. For analysis, in this case the sheet could be considered to be infinitely wide, the width being once or twice the width of the SC coil.

Studies have shown that the ratio between the SC coil length, L, and its width, 2b, should be above 5 ($L/2b \geq 5$) in order to insure a high levitation goodness

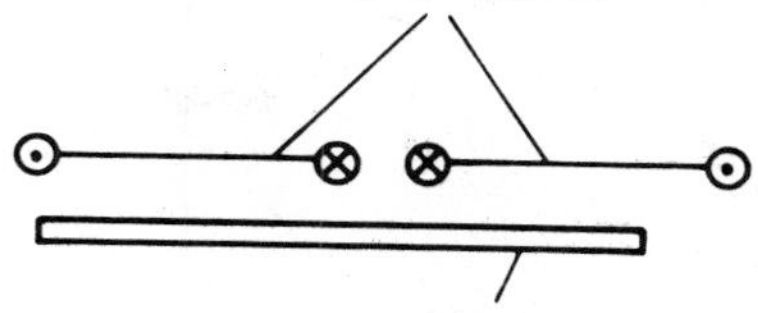

Fig. 11.5. Integrated levitation-guidance EDS.

factor.[6] On the other hand, to keep the secondary coils and vehicle width within reasonable limits, the width of the SC coils should not exceed 0.5 to 0.8 m. Finally, the longer the SC coils, the higher the speed at which the maximum levitation force is attained, though the ratio F_L/F_d is higher.[6] For these reasons, we conclude that SC coils of lengths L = 2.5 to 3 m with a positioning pitch L_v = L + 0.3 to 0.5 m are most suited for high speed vehicles.[6]

11.4.1. SC Coils Field Distribution

It is generally accepted that the cross section of the sides of the SC coils is negligible as compared to its dimensions. The three components of flux density (Fig. 11.6) produced by a rectangular SC coil are[7]

$$B_x = \frac{\mu_0 I_0 z}{4\pi}\left\{ \frac{1}{(x - L/2)^2 + z^2}\right.$$

$$\cdot \left[\frac{y + b}{[(x - \frac{L}{2})^2 + (y + b)^2 + z^2]^{1/2}}\right.$$

$$\left. - \frac{y - b}{[(x - L/2)^2 + (y - b)^2 + z^2]^{1/2}} \right]$$

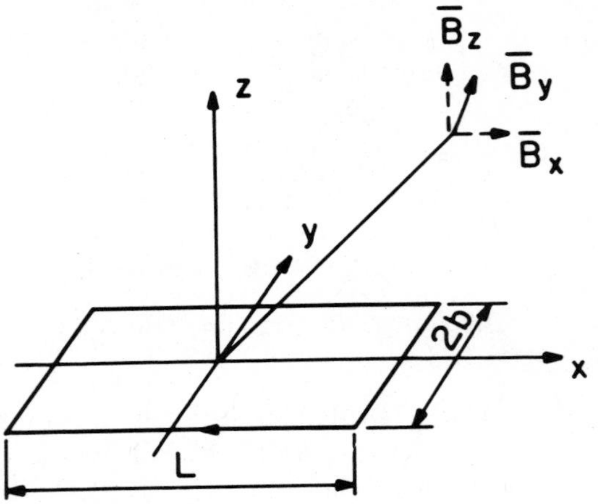

Fig. 11.6. Magnetic field of rectangular coil.

$$- \frac{1}{(x + L/2)^2 + z^2}$$

$$\left[\frac{y + b}{[(x + L/2)^2 + (y + b)^2 + z^2]^{1/2}} \right.$$

$$\left. - \frac{y - b}{[(x + L/2)^2 + (y-b)^2 + z^2]^{1/2}} \right] \Big\} \qquad (3)$$

$$B_y = \frac{\mu_0 I_0 z}{4\pi} \Big\{ \frac{1}{(y - b)^2 + z^2}$$

$$\cdot \left[\frac{(x + L/2}{[(x + L/2)^2 + (y - b)^2 + z^2]^{1/2}} \right.$$

$$\left. - \frac{x - L/2}{[(x - L/2)^2 + (y - b)^2 + z^2]^{1/2}} \right]$$

$$- \frac{1}{(y + b)^2 + z^2}[\frac{x + L/2}{[(x + L/2)^2 + (y + b)^2 + z^2]^{1/2}}$$

$$- \frac{x - L/2}{[(x-L/2)^2+(y+b)^2+z^2]^{1/2}}]\} \qquad (4)$$

$$B_z = \frac{\mu_0 I_0}{4\pi} \{ \frac{y + b}{(y + b)^2 + z^2}$$

$$[\frac{x + L/2}{[((x + L)/2)^2 + (y + b)^2 + z^2]^{1/2}}$$

$$- \frac{x - L/2}{[(x - L/2)^2 + (y + b)^2 + z^2]}] - \frac{y - b}{(y - b)^2 + z^2}$$

$$\cdot [\frac{x + L/2}{[(x + L/2)^2 + (y - b)^2 + z^2]^{1/2}}$$

$$- \frac{x - L/2}{[(x - L/2)^2 + (y - b)^2 + z^2]^{1/2}}] \} \qquad (5)$$

The B_z component distribution along Ox and Oy, respectively, is presented in Figs. 11.7 and 11.8 for L = 2.7 m, L_v = 3 m, 2b = 0.5 m, I_0 = 3 x 10^5 ampturns, and z_0 = 0.2 m and z_0 = 0.3 m. The contribution of the neighboring coils has been also accounted for by a simple displacement of the origin of the reference frame.

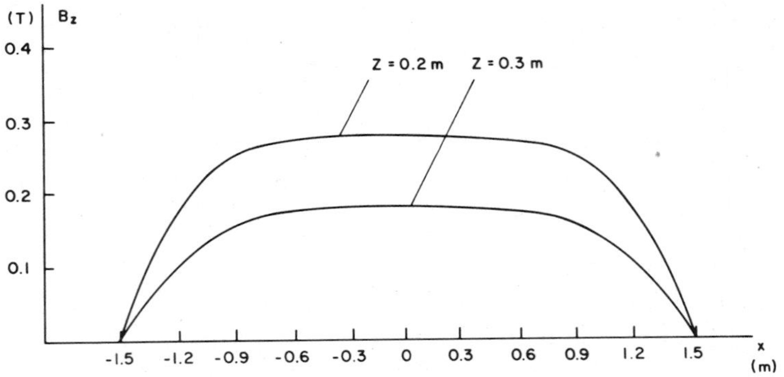

Fig. 11.7. Longitudinal distribution of flux density, B_z.

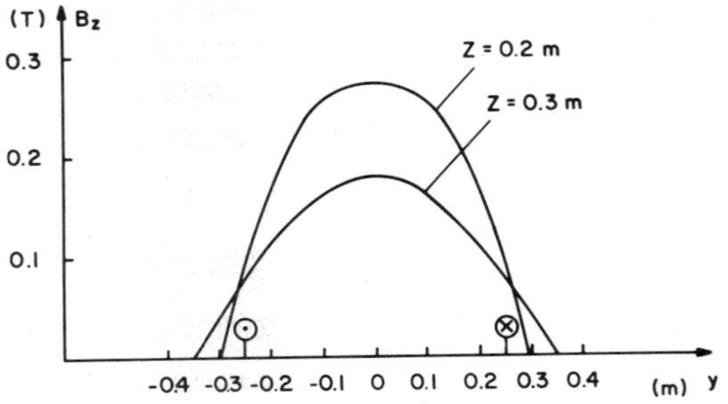

Fig. 11.8. Transverse distribution of flux density, B_z.

It is to be observed that along the transverse direction (Oy) the flux density distribution is almost sinusoidal, and if the sheet width is properly chosen, the fundamental component suffices for design studies. In the longitudinal direction, however, the periodic distribution created by a long row of SC coils should be decomposed in harmonics and, at least, the fundamental, third, and, fifth harmonics should be taken into account.

The condition of a long row of SC coils is fulfilled in practice since, for a 50-tonne typical vehicle, at least six coils of 2.7 x 0.5 m on each side are necessary. In order to account for the end effect, a more precise study, where the SC coils are considered as separate and superposition is then used, is necessary.[5] But for practical vehicles, this is hardly necessary for design purposes.

11.4.2. Field Equations

According to the above-mentioned assumptions, to describe the field distribution, a double Fourier series is used. We retain only the fundamental along the transverse direction, Oy, and the first, third, and fifth harmonics along the direction of motion, Ox. This is termed here the technical field theory of EDS. Under these conditions the three components of the fields of the SC coils become

$$\underline{H}_{x0}(x, y, z)$$

$$= \sum_{\nu=1,3,5} C_{\nu 0} j \frac{\pi\nu}{L_V} e^{-\gamma_{\nu 0}(z-z_0)} \cos\left(\frac{\pi}{2c} y\right) e^{(j\pi\nu/L_V)x} \tag{6}$$

$$\underline{H}_{y0}(x, y, z)$$

$$= \sum_{\nu=1,3,5} C_{\nu 0} \frac{\pi}{2c} e^{-\gamma_{\nu 0}(z-z_0)} \sin\left(\frac{\pi}{2c} y\right) e^{(j\pi\nu/L_V)x} \tag{7}$$

$$H_{z0}(x, y, z)$$

$$= \sum_{\nu=1,3,5} C_{\nu 0} \gamma_{\nu 0} e^{-\gamma_{\nu 0}(z-z_0)} \cos\left(\frac{\pi}{2c} y\right) e^{(j\pi\nu/L_V)x} \tag{8}$$

where

$$\gamma_{\nu 0} = [(\frac{\pi}{2c})^2 + (\frac{\pi\nu}{L_v})^2]^{1/2} \tag{9}$$

Expressions (6) through (8) are solutions to Laplace's equation. The constant $C_{\nu 0}$ is determined by numerical integration of (5).

For values of z = 0.1 to 0.3 m the exponential distribution of the field along z, as used in (6) through (8), causes negligible errors (below 2%) as compared with more precise expressions, (5). Within the secondary sheet, the reaction field of induced currents fulfills Poisson's equation:

$$\frac{\partial^2 \underline{H}_{rz}}{\partial x^2} + \frac{\partial^2 \underline{H}_{rz}}{\partial y^2} + \frac{\partial^2 \underline{H}_{rx}}{\partial z^2} - jU \frac{\pi\nu}{L_v} \sigma_{Al}\mu_0 \underline{H}_{rz}$$

$$= jU \frac{\pi\nu}{L_v} \sigma_{Al} \underline{H}_{z0}\, e^{(j\pi\nu/L_v)x} \tag{10}$$

In (10) it is implied that the field variation of the SC coil, along the sheet thickness (20 to 25 mm), may be neglected.

The solution of (10) is

$$\underline{H}_{rz}(x, y, z) = \sum_{\nu=1,3,5} \underline{C}_v(\nu) H_{z0}^{\nu}(x, y, z_0)$$

$$\cdot\, e^{-\gamma_\nu (z-z_0)} \cos(\frac{\pi}{2c}\, y) e^{j\pi\nu x/L_v} \tag{11}$$

In a similar manner we get

$$\underline{H}_{rx}(x,\ y,\ z) = \sum_{\nu=1,3,5} \underline{C}_{v}(\nu) H^{\nu}_{x0} x,\ y,\ z_0)$$

$$\cdot\ e^{\underline{\gamma}_{\nu}(z-z_0)} \cos(\frac{\pi}{2c}\ y) e^{j\pi\nu x/L_v} \tag{12}$$

$$\underline{H}_{ry}(x,\ y,\ z) = \sum_{\nu=1,3,5} \underline{C}_{v}(\nu) H^{\nu}_{y0}(x,\ y,\ z_0)$$

$$\cdot\ e^{-\underline{\gamma}_{\nu}(z-z_0)} \sin(\frac{\pi}{2c}\ y) e^{j\pi\nu x/L_v} \tag{13}$$

where

$$\underline{\gamma}^2_{\nu} = \gamma^2_{\nu 0} + jU \frac{\pi\nu}{L_v} \mu_0 \sigma_{Al};\quad \underline{C}_{\nu}(\nu) = \frac{-jU\nu\pi\mu_0\sigma_{Al}}{L_v \underline{\gamma}^2_y} \tag{14}$$

where H_{x0}, H_{y0}, and H_{z0} are the amplitudes of the harmonics of the SC coil field components, as available from (6) through (8).

The induced current density components are obtained by making use of the law of magnetic circuits. Hence,

$$\underline{J}_x = -\frac{\pi}{2c} \sum_{\nu=1,3,5} \underline{C}_v(\nu)(\underline{\gamma}_{\nu} - \gamma_{\nu 0})$$

$$C_{\nu 0} e^{-\gamma_{\nu}(z-z_0)} \sin(\frac{\pi y}{2c}) e^{j\pi\nu x/L_v} \tag{15}$$

$$\underline{J}_y = -\frac{j\pi}{L_v} \sum_{\nu=1,3,5} \nu \underline{C}_v(\nu)(\underline{\gamma}_\nu - \gamma_{\nu 0})$$

$$C_{\nu 0} e^{-\gamma_\nu (z-z_0)} \cos(\frac{\pi y}{2c}) e^{j\pi\nu x/L_v} \quad (16)$$

In order to get a feeling for the magnitudes, the distribution of the current density components is calculated and plotted in Fig. 11.9 for the following numerical data $L = 2.7$ m, $L_v = 3$ m, $I_0 = 3 \times 10^5$ ampturns, $2b = 0.5$ m, $z_0 = 0.25$ m, $2c = 0.7$ m; $d_{Al} = 25$ mm.

For the same two speeds (10 and 100 m/sec) the distribution of the amplitude of the current density, $\underline{J}_y$, along the secondary sheet thickness is given in Fig. 11.10. The significant variation of the current density, especially at high speeds, shows that this phenomenon should always be taken into consideration in any engineering performance computation procedure.

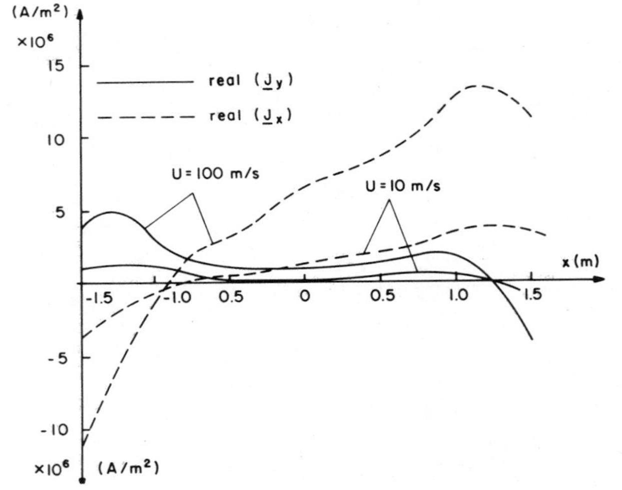

Fig. 11.9. Current density components.

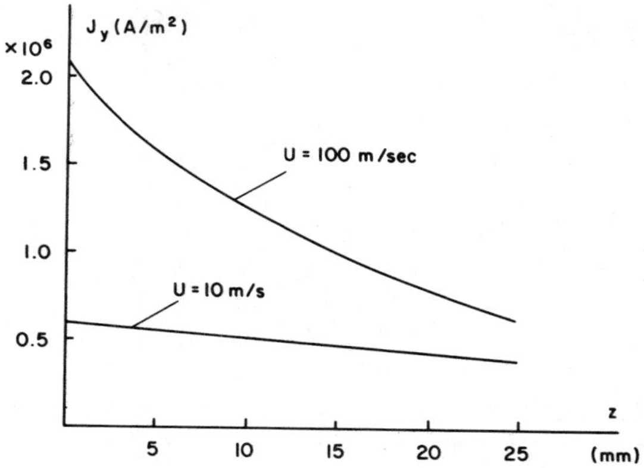

Fig. 11.10. Skin effect in secondary sheet.

Levitation and drag forces

By volume integration of specific forces, the levitation force, F_L, for a vehicle with N SC coils, is given by

$$F_L(U) = \frac{\mu_0}{2} N \,\text{Real}\Big[\int_0^{L_v} \int_{-c}^{c} \int_{z_0}^{z_0+d_{Al}} (\underline{J}_x \underline{H}^*_{y0} - J_y \underline{H}^*_{x0})\,dx\;dy\;dz \Big] \qquad (17)$$

$$F_L(U) = \frac{\mu_0 N L_v c}{2} \sum_{\nu=1,3,5} c^2_{\nu 0}\, \gamma^2_{\nu 0}\, \text{Real}\Big[\underline{C}_v(\nu)\, \frac{\underline{\gamma}_\nu - \gamma_{\nu 0}}{\underline{\gamma}_\nu} \cdot (1 - e^{-\underline{\gamma}_\nu d_{Al}})\Big] \qquad (18)$$

The drag force, F_d, is

$$F_d(U) = \frac{\mu_0 N}{2} \text{Real}[\int_0^{L_v} \int_{-c}^{c} \int_{z_0}^{z_0+d_{Al}} (\underline{H}^*_{z0} \underline{J}_y) dx\, dy\, dz] \quad (19)$$

and, finally,

$$F_d(U) = \frac{\mu_0 N c \pi}{2} \sum_{\nu=1,3,5} \nu C^2_{\nu 0} \gamma_{\nu 0} \text{Real}[j \frac{\underline{\gamma}_\nu - \gamma_{\nu 0}}{\underline{\gamma}_\nu}$$

$$\cdot (1 - e^{-\underline{\gamma}_\nu d_{Al}}) \underline{C}_v(\nu)] \quad (20)$$

For the same data as in Figs. 11.9 and 11.10, and N = 12, levitation force, $F_L(U)$, drag force, $F_d(U)$, and drag power, $P_2(U) = F_u(U) \cdot U)$ are calculated and plotted in Fig. 11.11 as functions of speed. It should be noticed that the peak drag force occurs at about 40 m/sec, and has a value of about 15% of vehicle weight. Hence, in order to accelerate the vehicle with less energy, the secondary sheet will not be mounted along

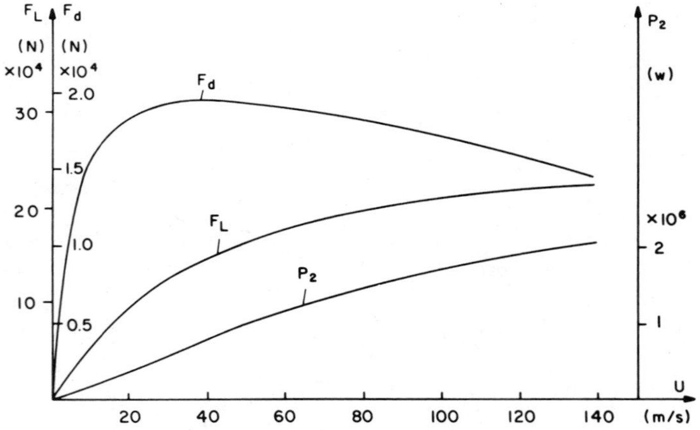

Fig. 11.11. Levitation and guidance forces against speed.

the first 1 to 2 km from stop stations, where the vehicle suspension is provided by light aircraft-type rubber retractable wheels. The image levitation force is not reached in practice because of secondary-sheet skin effect. However, the levitation forces and levitation height may be properly matched so that above 70 m/sec the vehicle is completely levitated by SC coils.

11.4.3. Secondary-Sheet Heating Problems

Secondary-sheet overheating, especially in the neighborhood of stop stations, could limit the interval, T, between two consecutive vehicles. The calculations of heating should be made for the first few meters of secondary sheet mounted at the exit of the stop station, because the most difficult situation occurs in this vicinity.

The computation of head time, T, is made as in linear induction machines (LIMs) for transportation, the only difference being that the time, t_t, within which the vehicle SC coils pass over a point of the secondary sheet is determined from the equation

$$\frac{at_t^2}{2} + U_1 t_t = \frac{NL_v}{2} \qquad (21)$$

For the data given in this section with $a = 1$ m/sec^2, $U_1 = 30$ m/sec and $N = 12$, $T = 3.2$ sec. Thus, in practice, the secondary overheating does not set any limit on transport capacity.

11.4.4. Sheet-Secondary Equivalent Time Constant

An equivalent time constant, T_{d2}, of the secondary sheet may be obtained from the expression

$$T_{d2} = \frac{2W_m}{P_2}; \quad P_2 = F_d U \qquad (22)$$

where W_m is the magnetic energy of reaction field and P_2 is the Joule losses in the secondary sheet. W_m results from the integral

$$W_m = \frac{N}{2}\frac{\mu_0}{4}\int_0^{L_v}\int_{-c}^{c}\int_{-\infty}^{\infty}(|\underline{H}_{rx}|^2 + |\underline{H}_{ry}|^2 + |\underline{H}_{rz}|^2)\,dx\,dy\,dz \tag{23}$$

Finally, we get

$$W_m = \frac{NcL_v\mu_0}{4}\sum_{\nu=1,3,5}\underline{C}_v(\nu)^2[(H_{x0}^{\nu})^2 + (H_{y0}^{\nu})^2 + (H_{z0}^{\nu})^2]_{z=z_0}\left(\frac{1 + e^{-2\gamma_{\nu 0}d_{Al}}}{2\gamma_{\nu 0}} + \frac{1 - e^{-2\gamma_{\nu r}d_{Al}}}{2\gamma_{\nu r}}\right) \tag{24}$$

where $\gamma_{\nu r} = \text{Real}(\underline{\gamma}_\nu)$.

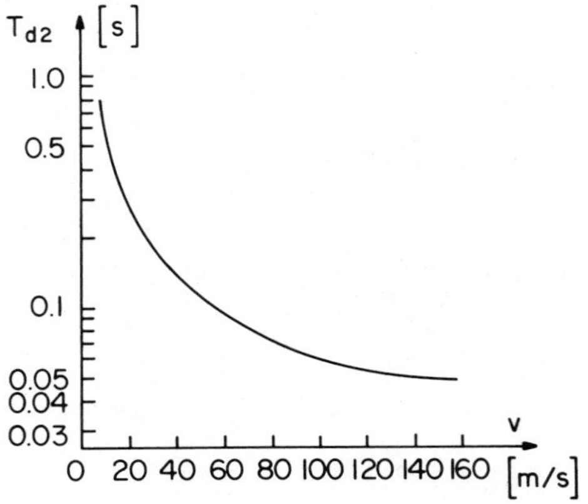

Fig. 11.12. Time T_d versus velocity.

The time, T_{d2}, depends considerably upon SC coil positioning pitch, L_v, and the speed (Fig. 11.12). It is well understood that the conductivity of the sheet, its thickness, and its width have a notable influence on T_2, but for economical reasons, they vary little $2c$ = 0.7 to 1 m, d_{Al} = 20 to 25 mm

11.4.5. Sheet-Secondary Skin Effect

The first three harmonics (accounted for) lead to specific field penetration conventional depths, δ_ν:

$$\delta_\nu = \text{Real}\,\frac{1}{\left[\,(\pi/2c)^2 + (\pi\nu/L_v)^2 + jU\sigma_{Al}\mu_0(\pi\nu/L_v)\right]^{1/2}} \quad (25)$$

Their dependence upon speed, for $2c = 0.7$ m, $L_v = 3$ m, $\sigma_{Al} = 3 \times 10^7 (\Omega\text{m})^{-1}$, is plotted in Fig. 11.13.

It should be noticed that, at least for the third and fifth harmonics, at high speeds, the depth of penetration is appreciably smaller than secondary-sheet thickness (d_{Al} = 25 mm). Thus the skin effect cannot be neglected in the performance calculations. A

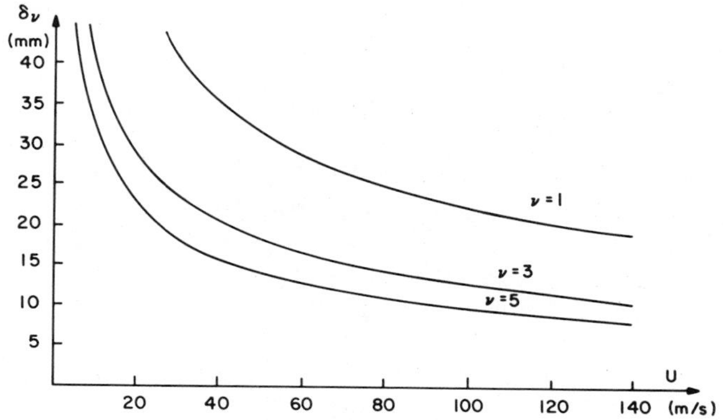

Fig. 11.13. Field penetration depth.

secondary-sheet guidance EDS may be similarly calculated, resulting again in an additional drag force that is to be considered in the propulsion system design.

In general, even for strong lateral winds, the guidance force does not exceed half of the vehicle weight. The two lateral guidance EDSs work differentially, resulting in rigid guidance characteristics. A detailed study of guidance EDS statics and dynamics for an entire vehicle is beyond the scope of this book, but it may be managed with the methods developed here for levitation EDS. In the following the ladder-secondary EDS performance is calculated.

11.5. Ladder-Secondary EDS

The literature[5] mentions the superiority of ladder secondary over sheet secondary. This superiority consists of a smaller drag force at a high speed, while the image (maximum) levitation force is attained at smaller speeds (30 to 40 m/sec). The secondary made of short-circuited coils has about the same qualities and is mathematically a special case of the ladder secondary. However, because of secondary discreteness, the levitation and drag forces pulsate in time. AC currents may, this way, be induced in SC coils, endangering their superconducting state. For this reason the levitation force fluctuations should generally be reduced to below 1 to 2%.

11.5.1. Induced Currents and Forces

Let l_s be the ladder loop length, and n be their number along the SC coils of pitch, L_v, such that

$$n = \frac{L_v}{l_s} \tag{26}$$

Let M_{oi} be the mutual inductance between two neighboring loops Fig. 11.14. In practice, it is adequate to

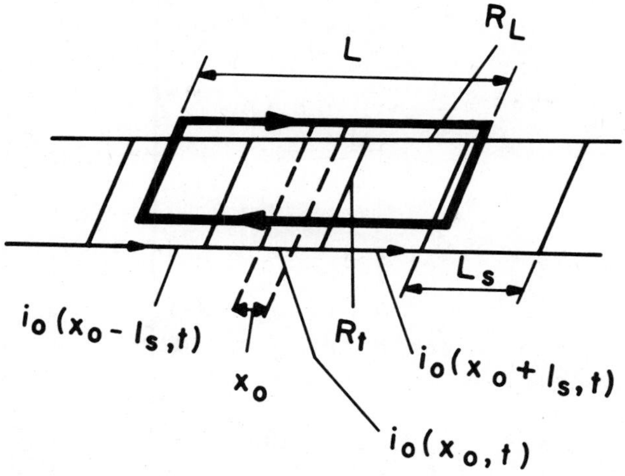

Fig. 11.14. Ladder secondary.

consider the reciprocal influence of neighboring loops. Thus, the complete circuit equation of the loop o is

$$L_s \frac{d}{dt} i(x_0, t) + R_t[2i(x_0, t)$$

$$- i(x_0 - l_s, t) - i(x_0 + l_s, t)] + 2R_L i(x_0, t)$$

$$- M_{oi} \frac{d}{dt} [i(x_0 + l_s, t) + i(x_0 - l_s, t)]$$

$$= - \frac{d}{dt} \phi(x_0, t) = V_e(x_0, t) \qquad (27)$$

where ϕ is the magnetic flux of SC coils in loop o and L_s is the self-inductance of a loop.

The induced voltage $V_e(x_0, t)$ is given by

$$V_e(x_0, t) = - 2U \frac{d\phi}{dx} (x_0, t) \qquad (28)$$

where:

$$\frac{d\phi}{dx} = \int_{-b}^{b} [B_z(x, z_0)$$

$$- B_z(x + L_v, z_0) - B_z(x - L_v, z_0)]\, dy \tag{29}$$

Thus only three SC coils are considered to produce a nonnegligible flux density at a point in the secondary.

For the short-circuited secondary coils, the only difference is that the second term in (27) becomes $2R_t i(x_0, t)$.

The voltage $V_e(x_0, t)$ induced in the loop o, as well as the corresponding current, $i(x_0, t)$, may be decomposed into harmonics:

$$V_e(x_0, t) = \sum_\nu V_{e\nu} \sin[\nu(\omega t + \frac{\pi x_0}{L_v})] \tag{30}$$

$$i(x_0, t) = \sum_\nu i_\nu \sin[\nu(\omega t + \frac{\pi x_0}{L_v}) - \phi_\nu] \tag{31}$$

where

$$\omega = \frac{\pi U}{L_v} \tag{32}$$

After some manipulations, the loop current $i(x_0, t)$ gets the expression

$$i(x_0, t) = \sum_{\nu} V_{e\nu} \sin[\nu(\omega t + \pi x_0/L_v) + \phi_\nu]/\{[2R_l$$

$$+ 2R_t(1 - \cos(\pi\nu/n))]^2 + \nu^2\omega^2[L_s - 2M_{oi} \cos(\pi\nu/n)]^2\}^{1/2} \tag{33}$$

with

$$\tan \phi_\nu = \frac{\nu\omega[L_s - 2M_{oi} \cos\pi\nu/n]}{2\{R_l + R_t(1 - \cos(\pi\nu/n)]\}} \tag{34}$$

The mutual inductance M_{oi} is given by

$$M_{oi} \simeq \frac{1}{I_0} \int_{-b}^{b} dy \int_{l_s/2}^{3l_s/2} [B_z(x, y, z)]_{z=0} \, dx \tag{35}$$

The expression $(B_z(x, y, z))_{z=0}$ is given by (5) where L is replaced by l_s. Finally, the self-inductance, L_s, of a loop is obtained from

$$L_s \simeq \frac{1}{I_0} \int_{-b}^{b} dy \int_{-l_s/2+0.02}^{l_s/2+0.02} [B_z(x, y, z)]_{z=0} \, dx \tag{36}$$

The integrals (35) and (36) are solved generally by numerical integration.

An equivalent time constant, T_s, may be defined such that for the fundamental T_{s1} is

$$T_{s1} = \frac{L_s - 2M_{oi}\cos(\pi\nu/n)}{2\{R_1 + R_t[1 - \cos(\pi\nu/n)]\}} \tag{37}$$

Finally, the levitation force, F_L, and drag force, F_d, are, respectively,

$$F_L = N \sum_{m=-\infty}^{\infty} I_0 i(ml_s + x_{01}, t) \frac{\partial M_m}{\partial z} \tag{38}$$

and

$$F_d = N \sum_{m=-\infty}^{\infty} I_0 i(ml_s + x_{01}, t) \frac{\partial M_m}{\partial x} \tag{39}$$

where M_m is the mutual inductance between an SC coil and the mth loop of the ladder, and is given by

$$M_m = \frac{1}{I_0} \int_{-b}^{b} dy \int_{x_0 - l_s/2}^{x_0 + l_s/2} [\, B_{z0}(x, y, z_0)$$

$$- B_{z0}(x - L_v, y, z_0) - B_{z0}(x + L_v, y, z_0)]\, dx \tag{40}$$

In (40) $x_0 = x_{01} + ml_s$, and x_{01} represents the displacement of the center of the oth loop from the SC coil center, at t = 0. It should be noted that both forces pulsate in time (or with x_{01}).

11.5.2. Numerical Example

A numerical example having the following initial data is now considered: $L_v = 3$ m, $2b = 0.5$ m, $I_0 = 3 \times 10^5$ ampturns, N = 12 SC coils per vehicle, $z_0 = 0.3$ m, n

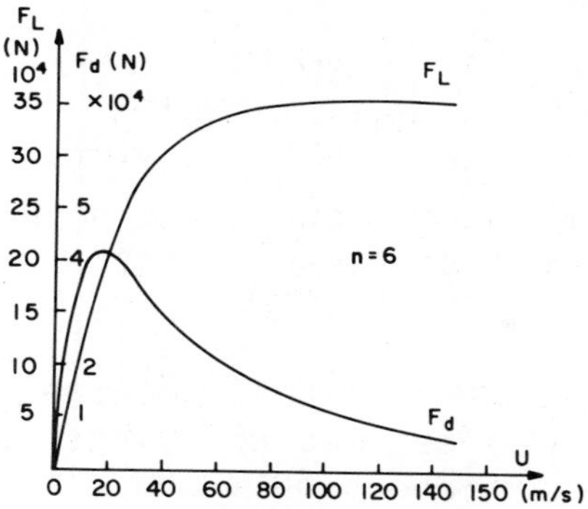

Fig. 11.15. Average forces against speed.

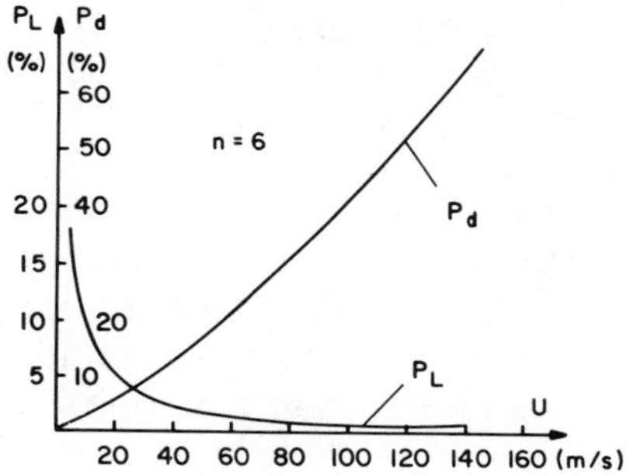

Fig. 11.16. Force pulsation against speed.

= 6 loops per pole, $R_1 = R_t = 3.7 \times 10^{-6}$ Ω. In order to limit the skin effect, the ladder secondary is manufactured of thin transposed conductors embedded in an epoxy resin. Following the procedure described in this section the dependence of levitation and drag forces on speed is obtained as shown in Fig. 11.15. The pulsation rates, p_L and p_d, of levitation and drag forces, respectively, are expressed by

$$p_L = \frac{F_{Lmax} - F_{Lmin}}{F_{Lmax} + F_{Lmin}} \quad \text{and} \quad p_d = \frac{F_{dmax} - F_{dmin}}{F_{dmax} + F_{dmin}} \qquad (41)$$

As shown in Fig. 11.16, the rate of pulsation of the levitation force varies little with speed and generally has low values for this case ($n = 6$). But the drag force pulsation rate, p_d, increases rapidly with speed. The average drag force (Fig. 11.15) decreases also with increasing speed, and thus, the propulsion system of Maglev is not significantly disturbed by these pulsations.

It should be observed that for the same SC coils, as for the case of sheet secondary, the image levitation force is attained at about 60 m/sec. On the other hand, the peak drag force occurs at only 20 m/sec, thus improving the starting performance of Maglev. However, the total costs (manufacture, mounting, and maintenance) are somewhat higher for the ladder secondary, though the quantity of aluminum is about 25% less than that of sheet secondary.

Essentially, because at high speeds (100 to 140 m/sec) the drag force of ladder secondary is about three times smaller than that of sheet secondary, the former should be considered definitely superior.

11.6. EDS Dynamics

Up to this point we have demonstrated the capacity of EDS to levitate vehicles at high speeds--100 to 140 m/sec. Now, we investigate their dynamic behavior and ride comfort. First, the ride-comfort parameters are defined for a high-speed Maglev.

11.6.1. Standard Ride Comfort

The ride-comfort level may be defined in several ways. However, an adequate procedure is based upon the relation between the power spectral densities of guideway irregularities and of vertical acceleration, for a moving Maglev.

The power spectral density $\psi(\Omega')$ of guideway irregularities can be approximated by[8]

$$\psi(\Omega') = \frac{A}{\Omega'^2} \tag{42}$$

where Ω' is the wave number and is given by

$$\Omega' = \frac{2\pi f}{U} \tag{43}$$

In (43), f = frequency of oscillations and U = speed.

A reasonable, practical value for A is $A = 1.5 \times 10^{-6}$ m, corresponding, approximately, to a good, welded, conventional railway. The squared average amplitude of guideway irregularities, z_g, is:

$$z_g = \left[\int_{\Omega'_1}^{\infty} \psi(\Omega') d\Omega' \right]^{1/2} \tag{44}$$

where Ω'_1 is the lowest wave number considered. The frequency domain may be restricted to 0.3 to 50 Hz, according to the normal ride comfort of the passenger.

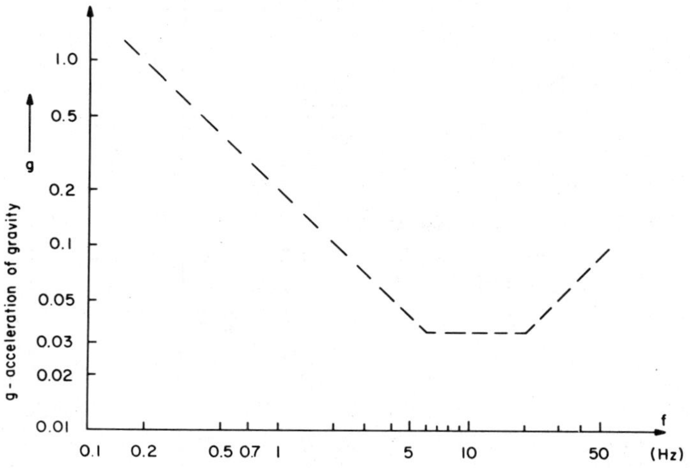

Fig. 11.17. Janeway criterion.

The power spectral density of accelerations, as experienced by the passengers, depends on the frequency and amplitude of guideway irregularities, the damping system, and the linearized natural characteristic of the SC coils. The standard ride comfort may be defined in a number of ways, but it seems adequate to limit the amplitude of vertical accelerations according to their frequency such that the passengers consider the ride as comfortable. Although such an evaluation is subjective, the Janeway criterion (Fig. 11.17) leads to useful practical results:

$$a_{max}(f) = \frac{a_1}{f} \qquad \text{for} \quad 1 \leqq f \leqq 6 \text{ Hz}$$

$$a_{max}(f) = b_1 \qquad \text{for} \quad 6 \leqq f \leqq 20 \text{ Hz}$$

$$a_{max}(f) = c_1 f \qquad \text{for} \quad 20 \leqq f \leqq 60 \text{ Hz} \tag{45}$$

where $a_1 = 0.204$ g; $b_1 = 0.034$ g, $c_1 = 1.7 \times 10^{-3}$ g, and g = gravitational acceleration.

Below 1 Hz a linear dependence[8] may be adopted:

$$a_{max}(f) = a_1 f \qquad \text{for} \quad f \leqq 1 \text{ Hz} \tag{46}$$

Generally the ride comfort is rated as good if the EDS acceleration-frequency curve lies below the Janeway curve (Fig. 11.17). However, if the EDS curve is much below the Janeway curve, the damping system is too strong and thus higher values of A may be accepted; that is, guideways with higher irregularities (and lower cost) may be used.

In order to determine the EDS ride comfort, its dynamic behavior is investigated first.

11.6.2. Dynamic Regime

To study the fundamental aspects, simplified equations are used here. Only the vertical oscillations are studied, where the sheet or ladder secondary has an equivalent resistance, R_e, and an inductance, L_e. Thus the equation of the secondary equivalent circuit is

$$L_e \frac{dI_r}{dt} + R_e I_r = - \frac{d\phi}{dt} \tag{47}$$

where I_r is the secondary equivalent current, and ϕ the magnetic flux crossing the equivalent secondary. The SC coil ampturns, I_0, are considered constant.

As already mentioned the flux, ϕ, may be decomposed into harmonics. Here we retain only the fundamental, ϕ_1:

$$\phi_1 \simeq I_0 M_0 e^{-\gamma_{10} z} e^{-(\pi/L_v)x} \quad ; \quad x = Ut \tag{48}$$

with

$$\gamma_{10} \simeq [(\frac{\pi}{L_v})^2 + (\frac{\pi}{2c})^2]^{1/2}, \quad \text{for the sheet secondary} \tag{49}$$

and

$$\gamma_{10} \simeq \frac{\pi}{L_v}, \quad \text{for the ladder secondary}$$

For small oscillations the airgap, z, is given by

$$z = z_0 + z_1(t) \tag{50}$$

Under these conditions I_r is given by[9]

$$I_r = -\frac{I_0}{L_e} M_0 e^{-j(\pi/L_v)Ut} e^{(-\gamma_{10} z_0)} \frac{1}{\xi}\left[\frac{R_e}{L_e} - j\frac{\pi}{L_v} U \frac{L_e}{R_e}\right.$$

$$\left. + -j\frac{\pi}{L_v} U \frac{L_e}{R_e} \gamma_{10} z_1(t) - \frac{\gamma_{10}}{\xi}\dot{z}_1(t) - \frac{\gamma_{10}}{\xi^2}\ddot{z}_1(t)\right] \quad (51)$$

where

$$\xi = \frac{R_e}{L_e} - j\frac{\pi}{L_v} U \quad (52)$$

If $2\pi f/\xi \ll 1$ the terms of (51) containing derivatives of an order higher than one may be neglected. The above condition is generally fulfilled at medium and high speeds.

The levitation force, F_{1b}, experienced by an SC coil, is given by

$$F_{1b} = -\frac{1}{2}\,\mathrm{Real}\left(I_r I_0 \frac{\partial M_0^*}{\partial z}\right) \quad (53)$$

where $\underline{M}_0$ is

$$\underline{M}_0 = M_0 e^{-\gamma_{10} z} e^{-j(\pi/L_v)x} \quad (54)$$

Finally, F_{1b} becomes

$$F_{1b} \simeq F_{L\infty} \frac{\omega^2\tau^2}{1 + \omega^2\tau^2} [1 - 2\gamma_{10} z_1(t)$$

$$- \frac{1}{\omega^2\tau} \frac{1 - \omega^2\tau^2}{1 + \omega^2\tau^2} \gamma_{1o} \frac{dz_1(t)}{dt}] \tag{55}$$

where $F_{L\infty}$ is the image levitation force, $\omega = (\pi/L_v)U$ and $\tau = L_e/R_e$ is the equivalent time constant of the secondary. This time constant has already been calculated earlier in this chapter for both types of secondary.

11.6.3. Static Stability

It should be noted that the coefficient multiplying $z_1(t)$ in (55) is negative, signifying a restoring force. Thus, the EDS is statically stable. An approximate expression of EDS rigidity, S_e, is

$$S_e = \frac{-\partial F_{Lb}/\partial z_1}{F_{1b}} \simeq 2\gamma_{10} > 0 \tag{56}$$

The higher the L_v, the smaller the γ_{10}, and S_e; that is, the levitation system is softer.

11.6.4. Natural Damping Coefficient

The conventional natural damping coefficient, $\eta_d(\omega\tau)$, is

$$\eta_d(\omega\tau) = F_{L\infty}\gamma_{10}\tau \frac{1 - \omega^2\tau^2}{(1 + \omega^2\tau^2)^2} \qquad (57)$$

For low speeds $\omega\tau < 1$, and thus $\eta_d > 0$. In this case the oscillations are slightly damped. For medium and high speeds $\omega\tau > 1$ and $\eta_d < 0$, that is the oscillations will be slightly amplified in time.

The time constants τ of the two types of secondary for the data given earlier in this chapter are: τ_{1s} = 0.06037 sec, for U = 100 m/sec for sheet secondary; τ_1 = 0.0613 sec, for ladder secondary. With these data, the dependence of the natural damping coefficient, η_d, upon speed, for both types of secondary, is shown in Fig. 11.18. A similar behavior results when the magnetic flux of the SC coils is kept constant.[9] It is evident that above 10 to 15 m/sec a damping system is necessary in order to make the SED dynamically stable.

11.6.5. Damping of Oscillations

Three main damping systems have been proposed:

1. Passive electric damper (PED), Fig. 11.19.

2. Active electric damper (AED), Fig. 11.20.

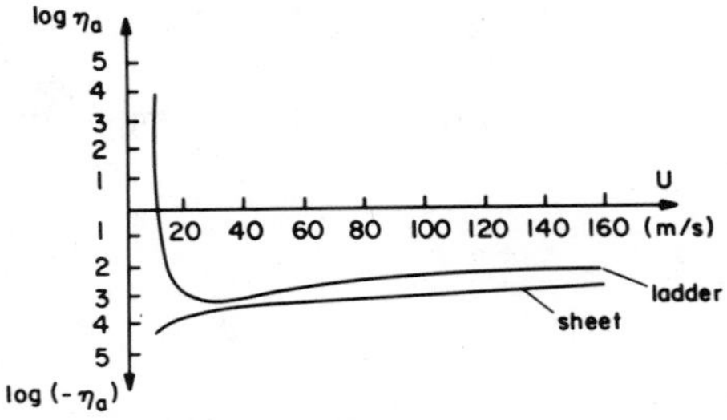

Fig. 11.18. Natural attenuation coefficient.

3. Secondary suspension system (SSS), Fig. 11.21.

It has been proved that the PED[10] consisting of a conductive plate, or short-circuited coils mounted below the SC coil, is able to yield a damping time constant, τ_d, below 1 only if the vertical distance between the two is at least 0.15 m. Thus, the distance between the vehicle and the guideway is reduced to 2 to 5 cm. To avoid the vehicle collision with the guideway, the latter must be very carefully manufactured (at high

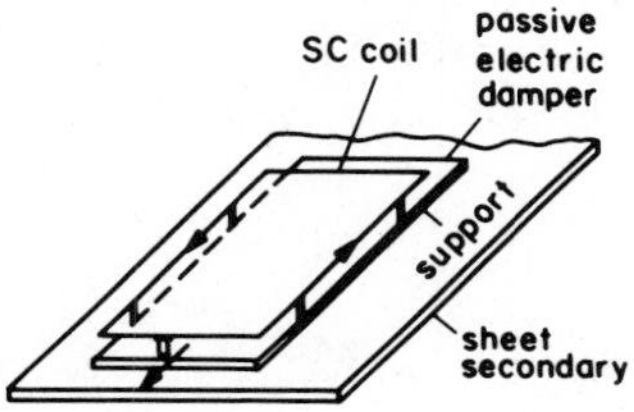

Fig. 11.19. Passive electric damper (PED).

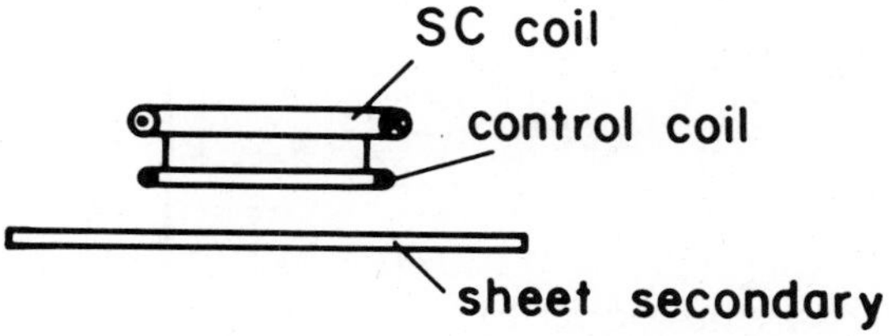

Fig. 11.20. Active electric damper (AED).

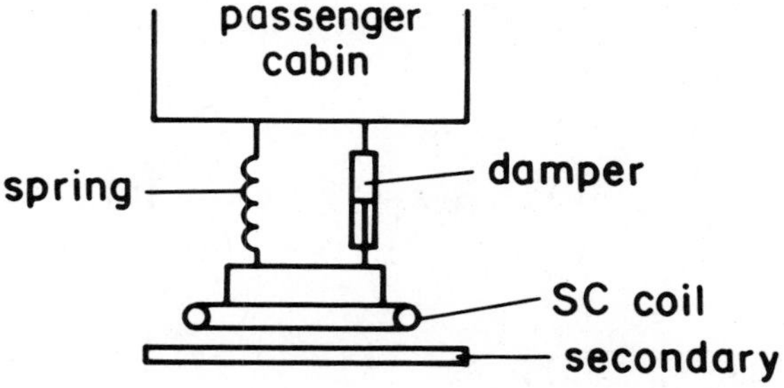

Fig. 11.21. Secondary suspension system (SSS).

costs). Moreover, at low frequencies, 0.3 to 2 Hz, PED is not able to damp the oscillations. For these reasons, this solution is not practical.

The SSS, if active, is potentially capable of fulfilling the Janeway criteria over the entire frequency range.[8] However, this damping system has a hydraulic or pneumatic character and is beyond the scope of this book.

In the following, only the AED is treated. An AED consists of a control coil mounted on the vehicle below the SC coil. The current in the control coil is modified in proportion to the vertical speed and displacement.

11.6.6. AED Design

The equation of SC coil vertical motion about a rated position, z_0, is

$$mZ_1 = F_{1b} - mg - 2I_c(t)I_0 \left(\frac{\partial M'_{Lc}}{\partial z}\right)_{z=z_0+z_c} \frac{\omega^2\tau^2}{1+\omega^2\tau^2} \tag{58}$$

where $I_c(t)$ is the control coil current and M'_{Lc} is the mutual inductance between a control coil and the SC coil image with respect to the secondary. Other image interactions are neglected. The control coil.current is a linear function of the vertical speed, $z_1(t)$, and vertical displacement, $z_1(t)$:

$$I_c(t) = \alpha_v \dot{z}_1(t) - K_z z_1(t) \tag{59}$$

From (55), (58), and (59) we finally obtain

$$m\ddot{z}_1(t) + \frac{\omega^2\tau^2}{1+\omega^2\tau^2}\{[2F_{1\infty}\gamma_{10} - 2K_zI_0\left(\frac{\partial M'_{Lc}}{\partial z}\right)_{z=z_0+z_c}]z_1(t) + [\frac{\gamma_{10}}{\omega^2\tau}F_{1\infty}\frac{1-\omega^2\tau^2}{1+\omega^2\tau^2} + 2\alpha_vI_0\left(\frac{\partial M'_{Lc}}{\partial z}\right)_{z=z_0+z_v}]\dot{z}_1(t)\} = 0 \qquad (60)$$

In (60) it is inferred that at $z = z_0$ the levitation force "cancels" the vehicle weight. The solution to (60) represents, in general, a damped oscillation of time constant, τ_d, and pulsation ω_0. If at $t = 0$ the vertical displacement from z_0 position is z_{10}, then the solution to (60) takes the form

$$z_1(t) = z_{10}\, e^{-t/\tau_d} \cos \omega_0 t \qquad (61)$$

where

$$\tau_d = 2m/\{\frac{\omega^2\tau^2}{1+\omega^2\tau^2}[\frac{\gamma_{10}}{\omega^2\tau}\frac{(1-\omega^2\tau^2)}{(1+\omega^2\tau^2)}F_{L\infty} + 2\alpha_vI_0\left(\frac{\partial M'_{Lc}}{\partial z}\right)_{z=z_0+z_c}]\} \qquad (62)$$

and

$$\omega_0 = \left(\frac{\omega^2\tau^2[2\gamma_{10}F_{L\infty} - 2K_zI_0\left(\frac{\partial M'_{Lc}}{\partial z}\right)_{z=z_0+z_c}]}{m(1+\omega^2\tau^2)}\right)^{1/2} \qquad (63)$$

When $\omega_0^2 < 0$ there will be no oscillations at all, but this case implies very small values of τ_d, or very high values of the control current.

We now return to our numerical example for the sheet secondary with $\omega_0 = 3.6$ rad/sec, $\tau_d = 0.4$ sec, $z_c = 0.1$ m, for U = 100 m/sec.

Finally, the values of α_v and K_z are as follows:

$$\alpha_v = 1.33 \times 10^4 (\text{A} \cdot \text{sec/m}); \quad K_z = 1.16 \cdot 10^5 (\text{A/m}) \tag{64}$$

11.6.7. AED Performance

There are a few criteria to assess the dynamic performance of an AED. Among them, two are considered essential. These are

1. The response to a step change in vertical displacement.
2. The response to guideway irregularities.

Step change in vertical displacement

For a step change, z_i, of $z_1(t)$, the solution to (60) is

$$\dot{z}_1(t) = -z_i \frac{\omega_0^2 + 1/\tau_d^2}{\omega_0} e^{-t/\tau_d} \sin \omega_0 t \tag{65}$$

and thus

$$z_1(t) = z_i(\cos \omega_0 t + \frac{\sin \omega_0 t}{\tau_d \omega_0}) e^{-t/\tau_d} \tag{66}$$

The vertical acceleration $\ddot{z}_1(t)$ can be easily obtained from (65). For $z_i = 0.03$ m, $\omega_0 = 3.6$ rad/sec, $\tau_d = 0.4$ sec, we have calculated and plotted $z_1(t)$ and $\ddot{z}_1(t)$. The final results are given in Figs. 11.22 and 11.23. The effect of an EAD is seen to be significant.

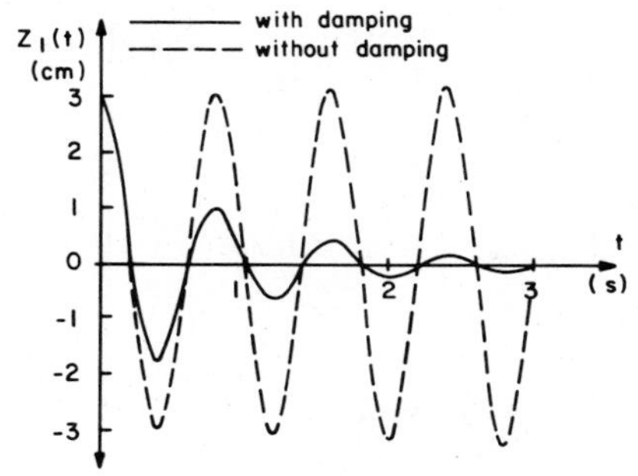

Fig. 11.22. EDS response to step airgap deviation (with AED).

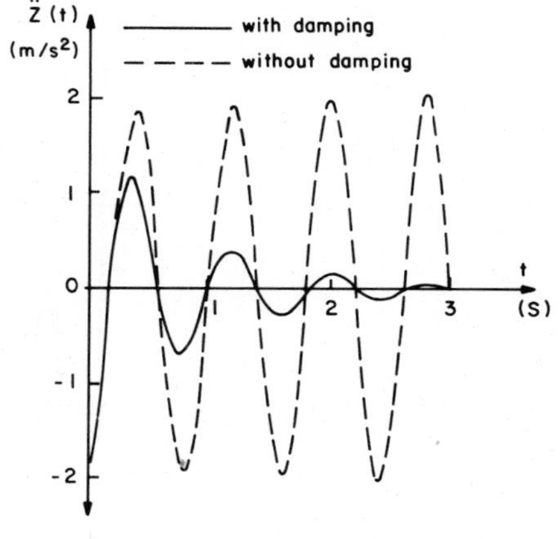

Fig. 11.23. Acceleration response to step airgap deviation (with AED).

EAD response to guideway irregularities

In general, the guideway irregularities are considered as sinusoidal:

$$z_0 = z_g \sin \omega t; \quad \omega = 2\pi f \tag{67}$$

where z_g is obtained from (42) and (44):

$$z_g = \left(\int_{0.2}^{f} \frac{AU}{2\pi f^2} df \right)^{1/2} \tag{68}$$

If this perturbation is introduced in (60) its final solution becomes

$$z_1(t) = z_0 \sin(\omega t + \phi_i) + A_1 e^{-t/\tau_d} \cos(\omega_0 t + \phi_0) \tag{69}$$

$$\phi_i = \arctan \frac{2\omega/\tau_d}{\omega^2 - (\omega_0^2 + 1/\tau_d^2)}$$

$$z_0 = \frac{z_g(\omega_0^2 + 1/\tau_d^2)}{\sin \phi_i(\omega^2/\tan\phi_i + 2\omega/\tau_d - \omega_0^2 - 1/\tau_d^2)} \tag{70}$$

In (70), A_1 and ϕ_0 can be calculated from the initial conditions. Here for simplicity we consider them zero. Hence

$$\tan\phi_0 = \frac{\omega \cos \phi_i + (1/\tau_d) \sin \phi_i}{\omega_0 \sin \phi_i} \tag{71}$$

$$A_1 = - \frac{z_0 \sin \phi_i}{\cos \phi_0} \tag{72}$$

With the above expressions, we can determine the vertical displacement, $z_1(t)$, speed, $\dot{z}_1(t)$, and acceleration, $\ddot{z}_1(t)$, as functions of time, for different speeds and frequencies for a given guideway constant, A.

The peak value, a_m, of the vertical acceleration, $\ddot{z}_1(t)$ can now be calculated for given speed and frequency, where a_m is found by a numerical procedure. Finally, for a given speed, we obtain a function $a_m(f)$. By comparing $a_m(f)$ with the Janeway curve (Fig. 11.24) we determine if the AED is capable of yielding adequate ride comfort.

For $\tau_d = 0.4$ sec, $A = 1.5 \times 10^{-6}$ m, $U = 100$ m/sec, we have calculated $a_m(f)$. The results are presented in Fig. 11.24 together with the Janeway curve. It is obvious that for these data, an AED provides good ride comfort at 100 m/sec. For the same data, by means of (59) the peak values of the control coil current have been

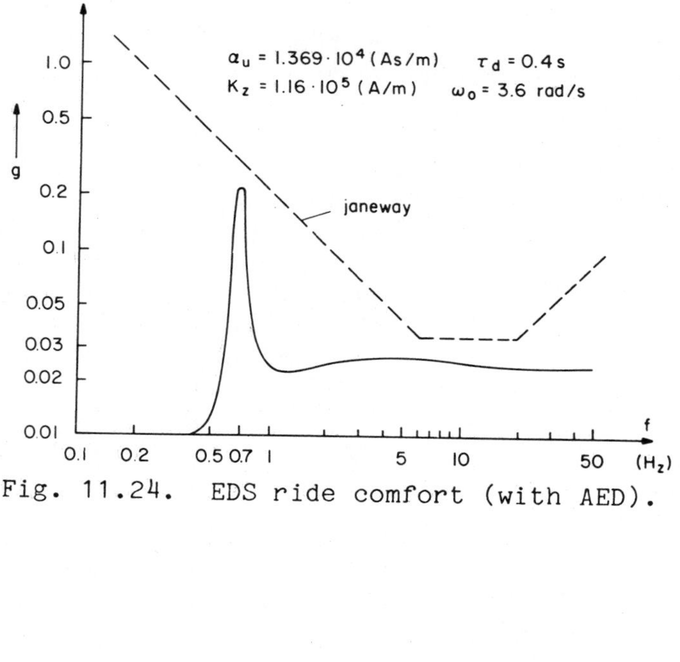

Fig. 11.24. EDS ride comfort (with AED).

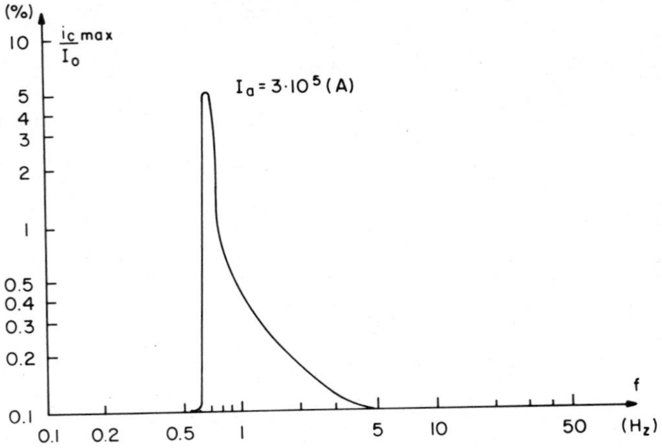

Fig. 11.25. Relative maximum values of control current against frequency of oscillations (in % of SC coil mmf).

determined and plotted for different frequencies (Fig. 11.25). The peak value of $i_c(t)$ is seen to be less than 5.5% of the SC coil ampturns. Keeping in mind that the control coil has the same geometrical dimensions as the SC coil, the control power can be determined. Its maximum, P_c, is about 5 kW/tonne of vehicle. Moreover, taking into account the control coil inductance, the peak value of the voltage (provided by a chopper that feeds the control coil) can be calculated.

However, these details are not considered here. We can now conclude that the AED is capable of providing adequate ride comfort for all frequencies up to high speeds (100 to 140 m/sec), at a moderate control power rate (5 kW/tonne). This does not preclude, however, the possibility of using an SSS instead of an AED. In order to reduce the control power level, it is also possible to mount the control coil on an SSS. The price of this advantage is the cost and weight of the SSS.

11.7. Note on Other Specific Phenomena

To complete the study of the levitation EDS, a few complementary aspects should be evaluated. Some of these are:

1. AC losses in the SC coils in diverse situations imposed by the operating conditions.

2. Passenger cabin screening from the high magnetic field of the SC coils.

3. Liquid helium source and SC coil persistence current mode control.

4. Influence upon the vehicle dynamics and passengers' ride comfort if one (or more) SC coil malfunctions.

5. Levitation and drag forces transients at joint strips in a sheet guideway.[12]

6. Auxiliary power sources on the vehicle.

7. Guideway switching.

Among these aspects the SC coils' losses are of primary importance. It is generally accepted that for an advanced Maglev of 50 tonnes, the loss in the SC coils may be reduced to 1 W/ton. Approximately 1.25 liters of helium would vaporize in order to evacuate 1 Wh of thermal loss from an SC coil dewar. Thus, for a 50-tonne Maglev in 8 hr of working, 500 liters of helium would vaporize, resulting in 3.6 m^3 of gaseous helium at 100 atm. On the other hand, the Maglev may be provided with a helium refrigerator on board. The total power of the refrigerator should be about 12.5 kW only (about 250 W are necessary to evacuate 1 thermal W from the SC coil dewar).

In general, research on the problems listed above is in progress, still requiring much experimental work before a feasible solution is obtained.

11.8. Levitation Equivalent Goodness Factor

The preceding calculations of levitation and drag forces, control power, P_c, and refrigerator power, P_1, for a 50-tonne Maglev, permit us to use (1) in order to obtain the levitation goodness factor, G_e, as a function of speed (Fig. 11.26). Both types of secondary have been considered, using the same initial data as for the previous numerical examples worked out in this chapter. It is evident that the ladder secondary is much better from the energy-conversion point of view. It is, however, more expensive. (This conclusion has been drawn from a fair comparison, that is the SC coils arrangement and ampturns have been considered identical for the two secondaries.)

11.9. Conclusions

Based on "technical field theory" the main performance of EDS levitation systems has been explored for sheet and ladder secondaries. Levitation and drag forces as well as the dynamics, in the presence of an

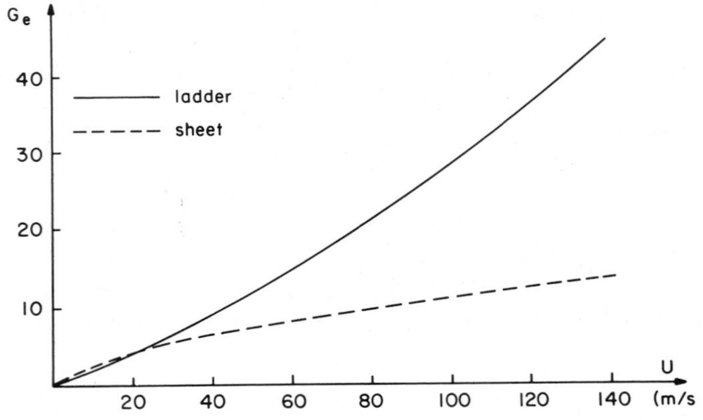

Fig. 11.26. Levitation goodness factor, G_e.

AED, have been evaluated. It has been shown (theoretically) that an AED can provide a good ride comfort at 100 m/sec without an SSS. A levitation goodness factor has been defined as an equivalent energy-conversion criterion. From this point of view the EDS with ladder secondary (or made up of short-circuited coils) is superior to an EDS with sheet secondary at high speeds (100 to 140 m/sec). Finally, the main directions for future research are presented.

References

1. E. Bachelet, The Engineer, Vol. 114, 1912, p. 420.

2. C. P. Parsch, "Erlanger magnetic test facility," Status of Technique, Second State Seminar Federal Report NT 53, NT 248-250, West Germany.

3. S. Foner and B. Schwartz (Editors), Superconducting Machines and Devices, Plenum Press, New York, 1974.

4. B. T. Ooi and E. R. Eastham, "Transverse edge effects on sheet guideways in magnetic levitation," IEEE Trans., Vol. PAS-94, 1975, pp. 72-80.

5. T. Yamada and M. Iwamoto, "Theoretical analysis of lift and drag forces on magnetically suspended high-speed trains," Elec. Eng. Jap., Vol. 92, No. 1, 1973, pp. 53-61.

6. J. R. Reitz, R. H. Bocherts, et al., "Preliminary design studies of magnetic suspensions for high ground transportation," Final Report DOT-FRA-10026, Washington, DC 20590, March 1973, p. 16.

7. S. A. Nasar and I. Boldea, Linear Motion Electric Machines, Wiley-Interscience, 1976, chapter 5.

8. R. H. Borcherts et al., "Preliminary design studies of magnetic suspensions for high speed ground transportation," Report FRA, PB 224893, Washington, DC 29591, June 1973.

9. T. Yamada, M. Iwamoto, and T. Ito, "Magnetic damping force in inductive magnetic levitation system for high speed trains," Elec. Eng. Jap., Vol. 94, No. 1, 1974, pp. 80-84.

10. M. Iwamoto, T. Yamada and E. Ohno, "Magnetic damping force in electrodynamically suspended trains," IEEE Trans., Vol. MAG-10, No. 3, 1974, pp. 458-461.

11. M. Tinkham, "AC losses in superconducting magnet suspensions for high speed transportation," J. of Appl. Phys., No. 5, 1973, pp. 2385-2390.

12. D. L. Atherton, A. R. Eastham and R. E. Tedford, "Joints in strips for electrodynamic magnetic levitation systems," IEEE Trans., Vol. MAG-14, No. 2, 1978, pp. 69-75.

CHAPTER 12

THE ELECTRODYNAMIC WHEEL

12.1. Introduction

A Maglev with superconducting field-winding electrodynamic wheels (EDWs) capable of producing its propulsion, levitation [1,2] and guidance[3] through the currents induced in a conductive guideway is called a Cryobus. The superconducting wheels (EDWs) are rotated either by rotary machines on board or by the synchronous-type interaction between the upper half of wheel poles and an arch-type three-phase stator[3] placed above it, on-board (Fig. 12.1). The main advantage of the Cryobus, in comparison with the electrodynamic levitation (EDL) guidance and the propulsion systems using flat superconducting coils, consists in realizing simultaneously all these functions from standstill to a maximum speed and back down to standstill through regenerative braking.

The propulsion mode corresponds to the case when the peripheral wheel speed is greater than the vehicle speed, whereas the regenerative braking occurs when the wheel speed is less than the vehicle speed. Apparently, by managing all the three functions, the typical magnetic drag force (well known from the EDL systems with

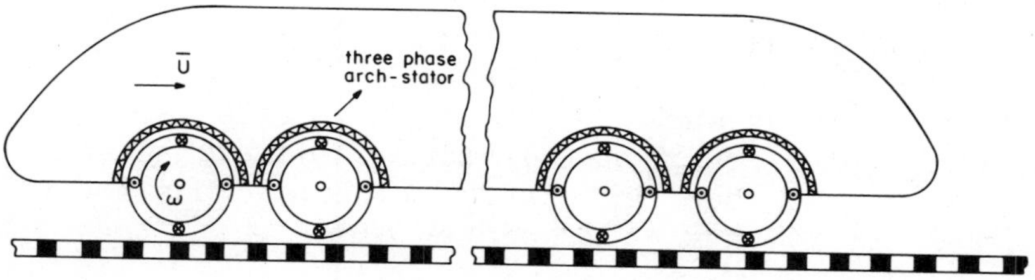

Fig. 12.1a. The cryobus longitudinal view.

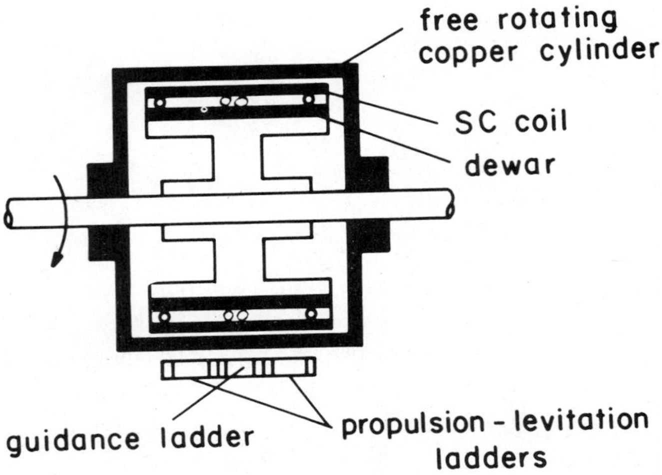

Fig. 12.1b. The cryobus cross section in a superconducting wheel.

flat superconducting coils) disappears in this case. In reality, the magnetic coupling between the superconducting wheel and the flat conductive guideway is weaker than that of flat SC coils, and thus, the overall efficiency of the EDW system is slightly deteriorated.

However, the overall efficiency[2] of EDW, used for propulsion and levitation, could reach values of 60 to 64%, which is greater than the efficiency of a linear synchronous motor with superconducting field winding (LSCM) for propulsion combined with an EDL for levitation. Because of the high complexity of a complete theoretical study of EDW, in this chapter we present only an introductory study of it. First, the construction guidelines are presented.

12.2. Construction Guidelines

A Cryobus has in general eight (or more) superconducting wheels mounted in two axle trucks (Fig. 12.1). A superconducting wheel has four to eight alternating poles along the peripheral direction and two alternating poles along the wheel axis (Fig. 12.1b). Thus the wheel flux density has components along all the three directions of a rectangular reference frame. The wheel may be rotated through a synchronous-type interaction with an arch-type three-phase stator placed along the upper half of the wheel periphery. The arch stator is fed from a rectifier-inverter system. A free wheeling copper cage (or cylindrical sleeve) is used for damping some of the transients due to guideway irregularities.

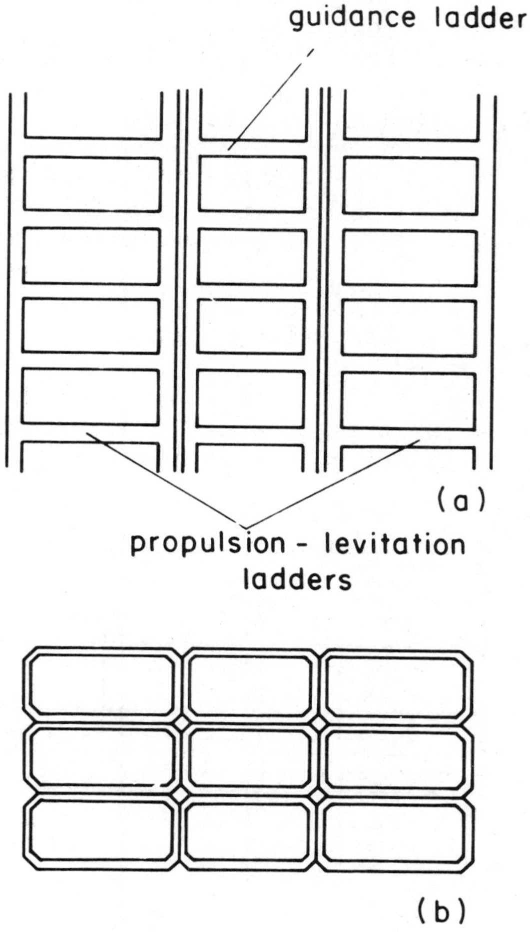

Fig. 12.2. Conductive guideway (one side only). (a) Triple ladder, (b) three rows of short-circuited coils.

A secondary suspension system is necessary to provide levitation-guidance dynamic stability and the required ride comfort for high speeds. The conductive guideway may be manufactured either as a triple ladder (Fig. 12.2a) or from three rows of short-circuited coils (Fig. 12.2b). In the middle ladder (or row of short-circuited coils) induced currents occur only in the case of a lateral displacement from the equilibrium position (caused by random winds, for example). These induced currents produce a centralizing (guidance) force (as in a null-flux-type system). The two lateral ladders (or rows of short-circuited coils) are, mainly, used for integrated propulsion and levitation.

The arch-stator three-phase winding (Fig. 12.1a) has two alternate poles in the axial direction. An example of such a winding (for two poles) is shown in Fig. 12.3.

12.3. Principle of Operation

As already mentioned, the propulsion and levitation of the vehicle is obtained when the peripheral speed of the wheel, $U_s \simeq \omega_1(R + g_0)$ as measured on the conductive guideway surface, gets higher than the vehicle speed,

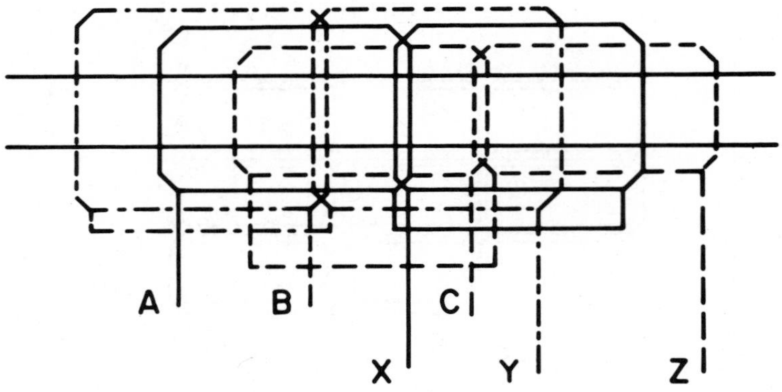

Fig. 12.3. Two-pole three-phase winding of wheel arch stator.

$U(U_s > U)$. It should be noticed that the "speed" of the wheel field in track loops depends on its position with respect to the wheel and no conventional (unique) synchronous speed can be defined here. Because of the relative motion at the speed $U_X = (U_s - U) > 0$, the propulsion and levitation forces are developed for any speed value of U including the standstill situation (Fig. 12.4). On the other hand, when the vehicle is laterally displaced from the position of symmetry, induced currents occur in the guidance loops (Fig. 12.5) producing a centralizing (guidance) force together with additional propulsion and levitation forces. However, in this case the propulsion and levitation forces developed in the lateral loops are slightly reduced. All these aspects should be given consideration in any realistic performance computation approach.

When the superconducting wheels operate as generators, the retrieved energy is taken from the vehicle kinetic energy. The levitation and guidance still hold during entire regenerative braking process down to standstill. In what follows, a performance computation approach is presented, accompanied by a numerical example.

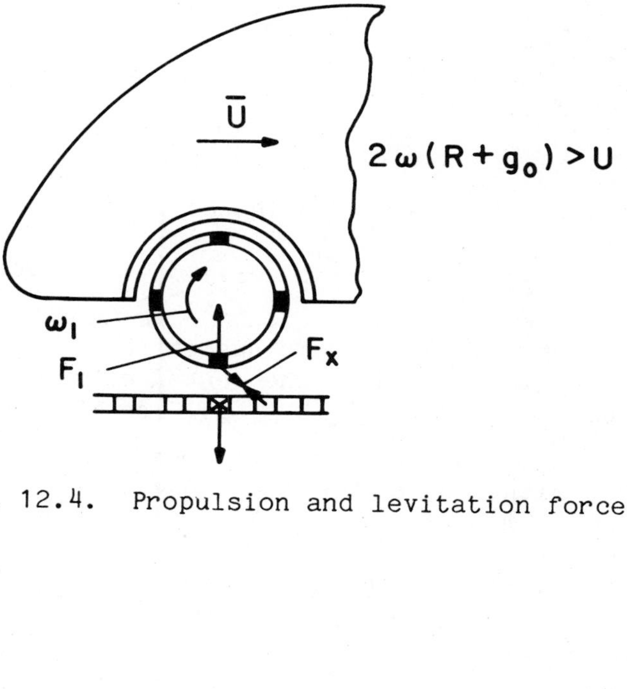

Fig. 12.4. Propulsion and levitation forces.

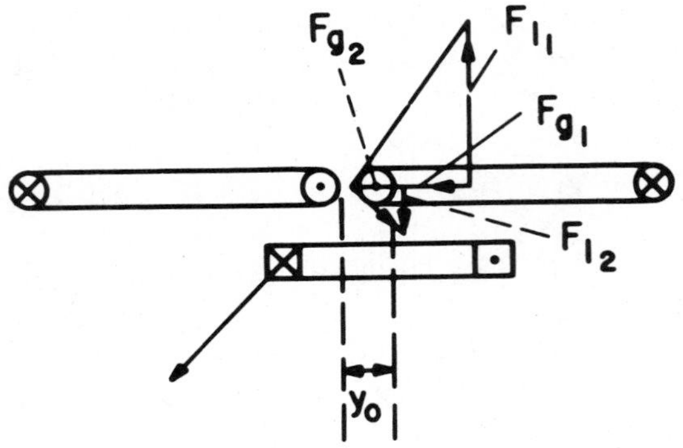

Fig. 12.5. Guidance force.

12.4. Superconducting-Wheel Field Distribution

The vertical component of the wheel field (along Oz), produced at a point $A(x_0 - Ut, y, g_0)$ situated in the guideway plane, is made up of the contributions of all the wheel poles. These contributions are now approximated. Thus, each pole could be replaced by an equivalent rectangular coil whose normal makes an angle α_i with the vertical direction:

$$\alpha_i = \omega_1 t + (i - 1) \frac{2\pi}{N} + \alpha_0 \qquad (1)$$

where $i = 1, 2..., N$; N = the number of poles, and α_0 = the value of α_i for zero time instant.

Let us now denote by (x_{fi}, y, z_{fi}) the coordinates of A with respect to a reference frame attached to a pole, having two axes in the plane of the coil, and having the origin located at the center of the coil. Let $B_{zi}(x_{fi}, y, z_{fi})$ and $B_{xi}(x_{fi}, y, z_{fi})$ be the components of the field produced by the ith poles at the point $A(x - Ut, y, z)$ situated along the guideway plane. The resultant vertical field, B_z, at this point is

$$B_z(x - Ut, y, z)$$

$$= \sum_{i=1}^{N} [B_{zi}(x_{fi}, y, z_{fi}) - B_{zi}(x_{fi}, y + c, z_{fi})]\cos \alpha_i$$

$$+ [B_{xi}(x_{fi}, y, z_{fi}) - B_{xi}(x_{fi}, y + c, z_{fi})]\sin \alpha_i \quad (2)$$

where x_{fi} and z_{fi} are given by (Fig. 12.6)

$$x_{fi} = (R + g_0)\sin \alpha_i + (x - Ut)\cos \alpha_i \quad (3)$$

$$z_{fi} = (R + g_0)\cos \alpha_i - (x - Ut)\sin \alpha_i - R \cos \frac{\pi}{N} \quad (4)$$

The components B_{xi} and B_{zi} are computed using the method of Reference 3. Now, to determine the currents induced

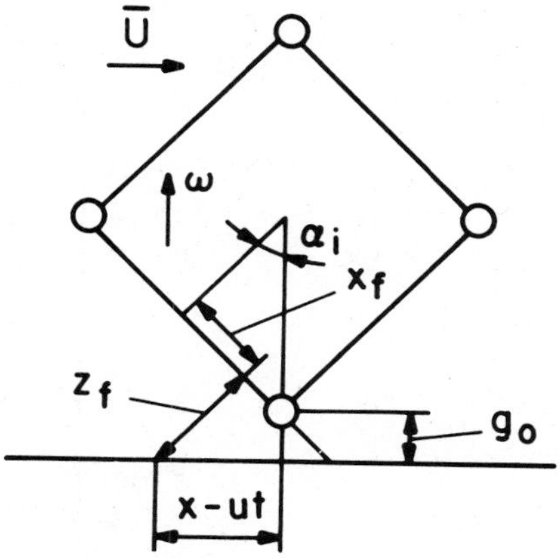

Fig. 12.6. The equivalent coordinates z_{fi} and x_{fi}.

in the guideway, we use the superposition of two particular situations:

1. First, the speed is zero (U = 0) while the wheel is spinning at the angular speed ω_1.

2. Secondly, the wheel does not rotate ($\omega_1 = 0$) while the vehicle moves at the speed U.

12.4.1. Zero Speed (U = 0)

For this case, the magnetic flux penetrating the lateral loop, $\psi_{s,d}$, located at the distance x_0 from the wheel center vertical line (Fig. 12.6), is

$$\psi_{s,d}(x_0,y_0,t)$$

$$= \int_{y_s-d/2}^{y_s+d/2} \int_{x_0-1/2}^{x_0+1/2} B_z(x - Ut,y,z,t)\, dx\, dy \tag{5}$$

where

$$y_s = \left(\frac{c - g - d}{2}\right) + y_0 \;; \quad y_d = y_s + c \tag{6}$$

For a guidance loop, ψ_g, a similar expression holds:

$$\psi_g(x_0, y_0, t)$$

$$= \int_{y_0-g_0/2}^{y_0+g_0/2} \int_{x_0-1/2}^{x_0+1/2} B(x - Ut, y, z, t)\, dx\, dy \tag{7}$$

We have denoted by y_0, the lateral displacement from the position of lateral symmetry, and the term "loop" refers to both ladder and short-circuited coils of the guideway.

Now the voltage induced in a guideway loop, U_{et}, is given by

$$U_{et} = \frac{\partial \psi_{s,d,g}}{\partial t}(x_0, y_0, t)$$

$$= \sum_{\nu=1,2,\ldots}^{\infty} U_{e\nu 1} \cos[\nu(p\omega_1 t - \frac{\pi}{\tau_1} x_0) + \alpha_{\nu 1}] \quad (8)$$

where

$$\tau_1 \simeq 2R \sin(\frac{\pi}{N}) \quad \text{and} \quad p = \frac{N}{2} \quad (9)$$

12.4.2. Wheel at Standstill ($\omega_1 = 0$)

In this case the induced voltage, U_{ed}, is obtained by motion:

$$U_{ed} = -\frac{d}{dt}\{[\psi_{s,d,g}(x_0, y_0, t)]\}$$

$$= \sum_{\nu=1,2}^{\infty} U_{e\nu 2} \cos[\nu(p\omega_2 t - \frac{\pi}{\tau} x_{01}) + \alpha_{\nu 2}] \quad (10)$$

where

$$\tau_2 = \frac{L_a}{2} ; \quad \omega_2 = \frac{\pi}{\tau_2} U$$

The length, L_a, is selected sufficiently large so that the flux attenuates to zero within a distance of $L_a/2$ from the wheel center vertical line. Generally $L_a \simeq 4R$ is an acceptable choice.

12.5. Guideway-Induced Currents

The circuit equation of loop 0, accounting for the interaction of right and left neighboring loops, is (Fig. 12.7)

$$L_0 \frac{d}{dt}[i_0(x_0, t)] + 2R_1[i_0(x_0, t)]$$

$$+ R_t[2i_0(x_0, t)] - i_0(x_0 - 1, t)$$

$$- i_0(x_0 + 1, t) - M_0 \frac{d}{dt}[i(x_0 + 1, t)$$

$$+ i_0(x_0 - 1, t)] = - \frac{d}{dt}[\psi_{s,d,g}] \qquad (11)$$

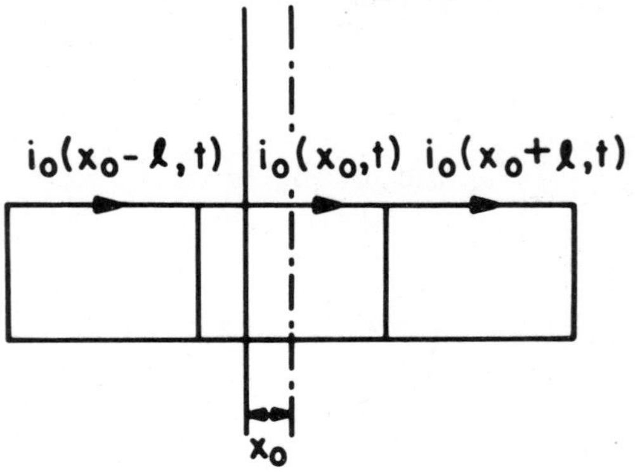

Fig. 12.7. Induced currents.

where L_0 = the loop self inductance and M_0 = the mutual inductance between two neighboring loops.

If the guideway is made of short-circuited coils, the third and fourth terms of do not occur. As the magnetic fluxes crossing neighboring loops do not have equal amplitudes, an exact solution for the time variation of loop currents can only be obtained by a numerical method considering simultaneously all the loops situated in the vicinity of a superconducting wheel. However, an approximate solution may be obtained if the flux amplitude in any three neighboring loops is equal to that of the middle loop:

$$i_{01,2}(x_0, t) =$$

$$U_{e\nu 1,2} \cos[\nu(p\omega_{1,2}t - \frac{\pi}{\tau_{1,2}} x_0) + \alpha_{\nu 1,2} + \phi_{\nu 1,2}]/\{[2R_l$$

$$+ 2R_t(1 - \cos \frac{\pi \nu l}{\tau_{1,2}})]^2$$

$$+ (\gamma\omega_{1,2}p)^2[L_0 - 2M_0 \cos(\frac{\pi \nu l}{\tau_{1,2}})]^2\}^{1/2} \quad (12)$$

For the secondary made of short-circuited coils, the equivalent resistance in (11) is the coil resistance, $R = 2R_l + 2R_t$. Also,

$$\tan \phi_{\nu 1,2} = \frac{p\omega_{1,2}\nu[L_0 - 2M_0 \cos(\pi\nu l/\tau_{1,2})]}{2R_1 + 2R_t[1 - \cos(\pi\nu l/\tau_{1,2})]} \quad (13)$$

The resultant loop current, $i_o(x_o, t)$, is given by

$$i_0(x_0, t) = i_{01}(x_0, t) + i_{02}(x_0, t) \quad (14)$$

12.6. Forces

We now make use of known expressions for forces obtained from the interaction of wheel flux and guideway currents. These expressions are

$$F_x(t) = \sum_{m=-\infty}^{+\infty} i_0(x_0 + ml, t)$$

$$\cdot \frac{\partial}{\partial_{xo}} [\psi_{s,d,g}(x_{01} + ml, t)] \qquad (15)$$

$$F_y(t) = \sum_{m=-\infty}^{+\infty} i_0(x_0 + ml, t)$$

$$\cdot \frac{\partial}{\partial y} [\psi_{y,d,g}(x_{01} + ml, t)] \qquad (16)$$

$$F_z(t) = \sum_{m=-\infty}^{\infty} i_0(x_0 + ml, t)$$

$$\frac{\partial}{\partial z} [\psi_{s,d,g}(x_{01} + ml, t)] \qquad (17)$$

where F_x = propulsion force, F_y = guidance force, F_z = levitation force.

The above expressions for forces hold for levitation-propulsion (lateral) loops as well as for guidance (middle) loops. It should be noticed that due to the discrete character of guideway structure (made of loops) and wheel poles, all forces pulsate in time. Special measures should be taken, for low speeds especially, to attenuate these oscillations (see also Reference 3). Moreover the three forces are strongly interdependent. Therefore, to obtain adequate propulsion force speed profiles, some of the superconducting wheels will have to work in the regenerative braking mode, thus reducing the vehicle propulsion force but maintaining the levitation and guidance forces.

12.6.1. Efficiency

In order to determine the Cryobus efficiency, first the guideway Joule losses, $P_{2s,d,g}$, have to be calculated, as given by

$$P_{2s,d,g} = \sum_{m=-\infty}^{\infty} \frac{1}{T} \int_0^T \{2R_1\, i_0^2(x_{01} + ml, t)$$

$$+ R_t[2i_0(x_{01} + ml, t) - i_0(x_{01} + (m - 1)l, t)$$

$$- i_0(x_{01} + (m + 1)l, t)]^2\}\, dt \qquad (18)$$

Now the mechanical power, P_{1m}, at the wheel shaft is:

$$P_{1m} = (F_{xs} + F_{xg} + F_d)\, U + P_{2s} + P_{2d} + P_{2g} \qquad (19)$$

where the subscripts s and d refer to lateral loops and g to guidance loops. Thus the efficiency, η_e, is

$$\eta_e = \frac{(F_{xs} + F_{xg} + F_{xd})\, U}{P_{1m}} \qquad (20)$$

The mechanical power recovered by regenerative braking, P_{1g}, is

$$P_{1g} = (F_{xs} + F_{xg} + F_{xd})\,U - P_{2s} - P_{2g} - P_{2d} \quad (21)$$

Here, we have not included the efficiency of the arch-stator synchronous driving system of the wheels.

12.7. Numerical Example

Let us consider a Cryobus of 45 tonne (140 passengers) and 12 wheels 1 m in diameter, capable of reaching up to 140 m/sec. Each wheel has four poles and the airgap $g_0 = 0.2$ m. The active length of wheels $2c = 1.3$ m, and the levitation and guidance loops width $d = g = 0.4$ m. There are four loops per pole ($\tau_{1,2}/l = 4$) and the cross section of loop conductors $S = 12\ \text{cm}^2$. Finally, the wheel poles magnetomotive force (mmf) is $I_0 = 6 \times 10^5$ ampturns.

For standstill ($U = 0$), using the above data and $\omega_1 = 2\pi/T = 10$ rad/sec in (2) through (4), the longitudi-

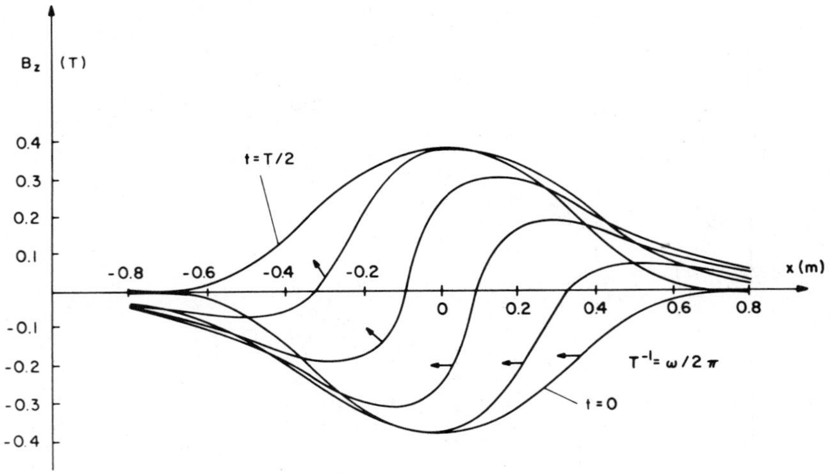

Fig. 12.8. Vertical flux density--longitudinal distribution.

nal distribution of vertical component of flux density, B_z, on the guideway upper surface at y = 0 (Fig. 12.8) is calculated for different instants of time.

As expected a traveling component of flux density distribution is evident in Fig. 12.8. Then, for a few points situated at x_0 horizontal distance from the wheel vertical center line, the time variation of B_z has been explored as shown in Fig. 12.9. An appreciable reduction in the amplitude of the flux density occurs for points far from the wheel vertical centerlines. Also, the time distribution is non-sinusoidal in shape, suggesting the necessity of accounting for higher harmonics. The transverse distribution of B_z is shown in Fig. 12.10.

It is remarkable that B_z varies rapidly along the transverse direction, in the guidance loops, indicating that notable guidance forces may be obtained for small

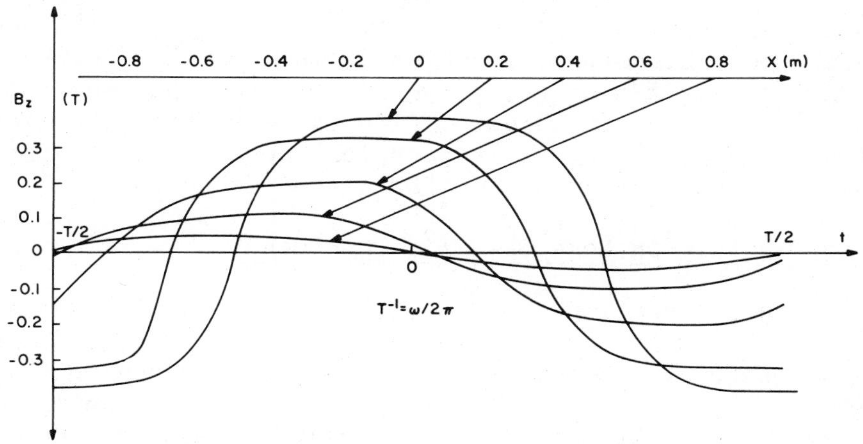

Fig. 12.9. Vertical flux density--time variation.

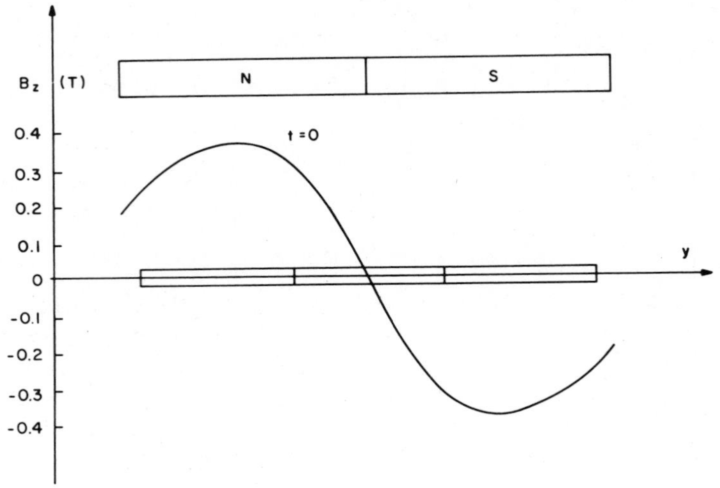

Fig. 12.10. Vertical flux density--transverse distribution.

lateral displacements. Also by increasing the mmfs of the wheel poles to $I_0 = (1 \text{ to } 1.2) \times 10^6$ ampturns, the flux density in the guideway can be increased to 0.7 to 0.8 T. These values are about 2 to 2.5 times greater than those common to flat coils EDL systems. Under these conditions the rather short interaction length of wheels--about 2 m of guideway--is counteracted, and thus the EDW system is capable of fully levitating vehicles of 45 tonnes. Using the expressions developed so far, the complete performance of EDW can be determined.

12.8. A Note on Some Specific Phenomena

The information presented so far in this chapter represents only an introductory study of the Cryobus and, hopefully, a suggestion for future research in this field. Here we indicate some of the specific phenomena thought to be worth research efforts. These phenomena include:

1. Propulsion force-speed profiles obtained by turning some of the vehicle wheels in regenerative braking while maintaining the levitation and guidance forces.

2. Free-wheeling inertial electric damper action during transients.

3. Dynamic stability and ride comfort.

4. Performance of arch-stator synchronous driving of the superconducting wheels.

5. Regenerative braking performance.

6. Vehicle and guideway cost estimation.

7. Technology and reliability of superconducting wheels spinning at super high speeds (100 to 140 m/sec). The potential overall advantages[2] of the Cryobus seem to indicate that a thorough study of this solution is worth the effort.

12.9. Conclusions

In this chapter an electrodynamic integrated system for Maglev propulsion, levitation, and guidance, using superconducting wheels, has been presented. All these three functions are managed from standstill up to very high speeds (100 to 140 m/sec). Also, regenerative braking is obtainable down to standstill.

A general simplified theory based on Fourier series has been developed. The main performance expressions have been derived. By comparison with flat coils (EDL) it was suggested, through a numerical example, that the Cryobus is capable of providing full levitation, guidance, and propulsion of typical 45-tonne vehicles. In-depth future studies are necessary to explore the full capabilities of the system. Some of the essential problems to be tackled in future investigations of EDW are pointed out in Section 12.8.

References

1. L. C., Davis and R. N. Borcherts, "Superconducting paddle wheels, screws and other propulsion units for high speed ground transportation," Scientific Research Staff, Ford Motor Co., Dearborn, Mich., January 22, 1973.

2. R. H. Borcherts and L. C. Davis, "The superconducting paddle wheel as an integrated propulsion levitation machine for high speed ground transportation," Int. Quat. Elec. Mach. Electromechan., Vol. 3, No. 3.4, 1979, pp. 341-356.

3. I. Boldea, Vehicles on Magnetic Cushion (in Romanian), Romanian Academy Publ. House, 1981.

INDEX